# Concepts of Chemical Engineering for Chemists

## 2nd Edition

# Concepts of Chemical Engineering for Chemists

## 2nd Edition

Edited by

## Stefaan Simons

*Brunel University London, UK*
*Email: Stefaan.Simons@brunel.ac.uk*

Print ISBN: 978-1-78262-358-8

A catalogue record for this book is available from the British Library

© The Royal Society of Chemistry 2017

Published by The Royal Society of Chemistry,
Thomas Graham House, Science Park, Milton Road,
Cambridge CB4 0WF, UK

Registered Charity Number 207890

Visit our website at www.rsc.org/books

Printed in the United Kingdom by CPI Group (UK) Ltd, Croydon, CR0 4YY, UK

# Preface

Welcome to the second edition of *Concepts of Chemical Engineering for Chemists*, replacing the version first published in 2007. Much of the original has been retained, whilst two new chapters have been added in order to make the "systems approach" to chemical engineering more complete. The specifics of what has been changed from the first edition are given later in this preface. Before that, it is worth pointing out the premise for the book and to give some background to what makes chemical engineering distinct from chemistry as a discipline.

This book is meant as a handbook and resource guide for chemists (and other scientists) who either find themselves working alongside chemical engineers or who are undertaking chemical engineering-type projects and who wish to communicate with their colleagues and understand chemical engineering principles. The book arose out of a short course, Concepts of Chemical Engineering for Chemists, held annually at UCL from 1999 through to 2009. It was forerunner to the Royal Society of Chemistry's "4 Chemists" series of professional training courses. The course was designed to provide basic information on the main aspects of chemical engineering in a relatively simple, but practical, manner. Hence, while this book tries to emulate this, it also includes worked examples, plus extensive reference lists and bibliographies, in order that the reader can research elsewhere for more detail and for aspects that are not covered in the book.

This book aims to give chemists an insight into the world of chemical engineering, outlining the basic concepts and explaining the terminology of, and systems approach to, process design. It can be said that chemists create new molecules and compounds and chemical engineers manufacture these into useful products on a commercial scale, but, of course, the two disciplines do not work in

Concepts of Chemical Engineering for Chemists: 2nd Edition
Edited by Stefaan Simons
© The Royal Society of Chemistry 2017
Published by the Royal Society of Chemistry, www.rsc.org

isolation; chemistry and chemical engineering are intertwined. One only has to look at the history of chemical engineering and its origins in chemistry (or, more correctly, applied chemistry) to appreciate the close relationship between the two and their shared foundation in molecular behaviour. The reader is referred to Darton *et al.*'s collection of visionary essays on chemical engineering's role in society[1] and the Whitesides report, "Beyond the Molecular Frontier",[2] which reflect on the importance of chemists and chemical engineering working effectively together to tackle the enormous challenges facing the world today. The impacts of the latter are still being felt today, with many chemical engineering courses in the UK now focussed on product, rather than process engineering; that is, from the understanding of the properties and functions of the chemical you wish to produce (and its end uses), how best can you design the process(es) to achieve that product? This is leading to both product and process innovations that are more sustainable and more efficient, all starting from the molecular scale.

To achieve these innovations and work effectively together, chemists and chemical engineers need to know how to communicate. This was the premise for the short course mentioned above and is the basis on which this book is written. Hence, the book does not focus on the derivation of mathematical formulae, but rather on the use of the governing principles in process design. What follows is a brief look at what a typical chemical engineering degree course involves. I will then describe the contents of the book itself.

## What Chemical Engineering Involves

Chemical Engineers are responsible for the design and operation of processes and of products and their application. Since they must consider processes in their entirety, from raw materials to finished products, they use a "systems" approach, enabling the prediction of the behaviour of both the process as a whole and of the individual plant items. Once a plant has been designed, chemical engineers are involved in its construction, commissioning and subsequent operation. Hence, in terms of education, a chemical engineer will undertake courses in:

- Mathematics (the emphasis is on engineering, after all).
- Science (notably, chemistry, biology, material science).
- Process analysis (defining the mass, momentum and energy balances for the entire process).

- Thermodynamics (determining the fundamental parameters on which the process can be analysed).

Once the process has been defined, attention can then be paid to the "unit operations", the reactors, separators, *etc.* that make up the processor ute, or flowsheet. This involves the study of:

- Reaction engineering (the manipulation of molecular behaviour to determine the reaction routes to a specified product).
- Transport processes (the physical manipulations that underlie the process).
- Separation processes (the manner in which products are separated and purified).

Interwoven with all these topics are the crucial areas of safety, risk analysis, plant and equipment design, process control and process economics. A practising chemical engineer will often find that the data he/she requires is either unreliable or incomplete[3] and, hence, he/she must make sound judgements based on mathematics, physics and chemistry to determine appropriate simplifying assumptions, while at the same time satisfying safety, environmental, operational and legal constraints.

Chemical Engineers are profoundly aware of their ethical and social responsibilities, encompassed in the notion of "sustainable development", which is becoming an increasing component of chemical engineering degree programmes. Often a chemical engineering student will supplement his or her degree programme with courses on environmental practice, law, management and entrepreneurship. In addition, a great deal of emphasis is placed on transferable skills training, communication, team working and leadership.

The culmination of a degree programme in chemical engineering is the design project, in which the students work in teams to carry out the preliminary design of a complete process plant. This exercise involves the use of much of the material covered in the topic areas mentioned above, starting from the mass and energy balances and ending with a full economic appraisal, environmental impact and risk assessment (or, increasingly, a sustainability analysis on the socio-economic as well as environmental impacts) and safety analysis. Very often, the teams will include MSc students with first degrees in chemistry, underlining the close relationship between the two disciplines and the need for mutual understanding in the development of effective and appropriate plant designs.

Sustainable development is defined as "development that meets the needs of the present without compromising the ability of future generation to meet their own needs". As stated in ref. 4, technological change must be at the heart of sustainable development. In order to achieve a sustainable future, the basic principles that should guide technological (and societal) development are that consumption of resources should be minimised, while that of non-renewable materials should cease, with preference given to renewable materials and energy sources. It is chemical engineers, together with chemists and other scientific disciplines, who will lead this technological revolution.[5] Although beyond the scope of this book, these issues must be at the forefront of all process and product developments and the concepts covered here are the bases on which to found this progress.

## What this Book Contains

The book begins with the cornerstone of any process design, the development of mass and energy balances, based on the simple conservation principle. From such balances, a chemical engineer will then go on to add more detail in order to come up with a fully optimised process flowsheet, providing the most efficient, safe and economic route to the production of the specified chemicals within the constraints of thermodynamics, material properties and environmental regulations. All chemical processes are dynamic in nature, involving the flow of material (fluid flow or momentum transfer) and the transfer of heat and mass across physical and chemical interfaces. Chapter 1, revised from the first edition, describes some of the basic principles for developing mass and energy balances.

The central part of any chemical process is the reactor, in which chemicals are brought together to produce the precursors to the eventual products. A chemical engineer must be able to predict and manipulate chemical reaction kinetics to be able to design such reactors. Hence, the concepts of chemical reaction engineering in relation to reactor design are considered in Chapter 2. The equally important principles behind momentum, heat and mass transfer are then covered in the following three chapters. Moving often huge quantities of material around a plant, through pipelines and in and out of vessels, requires a knowledge of fluid mechanics and its use in the design of the appropriate pumps and measurement equipment (Chapter 3). Maintaining rates of reactions and product quality requires the transfer of heat to and from chemicals, often through physical boundaries (*e.g.* pipe and vessel walls), and heat exchangers are a common means in which this transfer is achieved. The formulation of heat-transfer rate equations is described in Chapter 4 and their use demonstrated in the design of shell and tube heat exchangers. Mass transfer typically occurs in the separation of chemical components to remove impurities from the desired product. Distillation is a common separation technique employed in chemical plants and is described in Chapter 5.

Often, chemical plant design is informed by laboratory and pilot-scale experimentation. While the initial chapters in this book will inform the reader of the most important design parameters that need to be measured and determined from such experiments, how to then ensure that these parameters perform in the same way in a large-scale plant is the subject of Chapter 6. Although it is desirable to conduct the experimental work in the system for which the results are required, this is not always easy. The system of interest may be hazardous or expensive to build and run, while the fluids involved may be corrosive or toxic. In this case, scale models are used, which overcome the above problems and allow extensive experimentation. In the majority of cases the model will be smaller in size than the actual, desired plant, but sometimes, due to the nature of the materials to be handled, the fluids involved may also be different. Scale-up is only possible if the model and plant are physically similar and, hence, the procedure is based on dimensionless groups. How to develop and use these groups is described in Chapter 6.

Many chemicals, at some stage in a process (*i.e.* whether as raw materials, intermediates, products and by-products) are in powder form. Often, the handling of powders is mistakenly assumed to be relatively straightforward, leading to disastrous consequences (*e.g.* clogging of storage vessels, dust explosions, *etc.*). The characterisation and handling of powders is described in Chapter 7, with particular emphasis given to fluidisation as a common process operation. Particle science and technology is a rapidly maturing field but, surprisingly, there is still much reliance on empiricism in chemical engineering practice.

Throughout the design of a chemical plant, issues relating to safety, economics and environmental impact must be considered. By doing so, the risks associated with the plant can be minimised before actual construction. The same principle applies, whatever the scale of the process. The field of process control (Chapter 8) considers all these issues and is, indeed, informed by the type of hazard analyses described in Chapter 10. The objectives of an effective control system are the safe and economic operation of a process plant within the constraints of environmental regulations, stakeholder requirements and what is physically possible. Processes require control in the first place because they are dynamic systems, so the concepts covered in the earlier chapters of this book are central to process control (*i.e.* control models are based on mass, energy and momentum balances derived with respect to time). Chapter 8 focuses on the key aspects of control systems.

The economic assessment of a proposed plant, known as project appraisal, is necessary at the design stage in order to determine its viability, the capital requirements and the expected return on investment. Such an analysis can kill a project at any stage in the design. Chapter 9 discusses how the planning for profitable operation is undertaken.

Safety is the number one concern for chemical engineers and the reader should not confuse the fact that it is the focus of the penultimate chapter in this book with its order of importance. However, in order to analyse the hazards and risks associated with a particular process, one must understand the concepts on which a process flowsheet is developed, and these are covered in the preceding chapters. Even the best run processing facilities are inherently dangerous places. Often operating at elevated temperatures and pressure, and with chemicals that may be toxic or corrosive, loss of containment can lead to lost production, loss of the facility, damage to the environment and even loss of life. Chapter 10, which is new to this edition, introduces the concepts of hazards and risks, and the importance of considering safety during the design and operation of a processing facility. It replaces a chapter that was focussed on the specific hazard analysis technique employed by a particular chemical company. Whilst the principles could be adapted for any process, I felt that, for this edition, a chapter describing a broader approach may be more useful.

Chapter 11 is the second new chapter in this edition and is entirely new to the book. Having gained an understanding from the previous chapters of the various unit operations that make up a process flowsheet, this chapter attempts to pull everything together in describing how units that interact with each other can be optimised in relation to certain process variables, such as temperature, or performance measures, such as costs. The holistic approach to process design and optimisation that exploits the interactions between the different units to employ resources effectively and to minimise costs is known as *process integration*. Often, statistical tools are employed.

This chapter introduces the basic concepts and demonstrates their use with the example of the optimisation of a process designed to produce a certain polymer.

The chapter authors of this book have tried to keep mathematical derivations to a minimum and have assumed that their readers have background knowledge of thermodynamics, since the latter is ubiquitous to chemistry, as well as chemical engineering. The authors either were or are full-time members of academic staff at universities from around the world, with extensive experience in the teaching and research of the subjects of their respective chapters. My thanks go to them for all their hard work in preparing the contents of this book.

Stefaan Simons
Brunel University London

# References

1. *Chemical Engineering: Visions of the World*, ed. R. Darton, D. Wood and R. Prince, Elsevier Science, Amsterdam, The Netherlands, 2003.
2. Committee on Challenges for the Chemical Sciences in the 21st Century, *Beyond the Molecular Frontier: Challenges for Chemistry and Chemical Engineering*, National Research Council, National Academy of Sciences, Washington D.C., USA, 2003.
3. *An Introduction to Industrial Chemistry*, ed. C. A. Heaton, Leonard Hill, Glasgow, 1984.
4. *Sustainable Development for Engineers*, ed. K. Mulder, Greenleaf Publishing, Sheffiled, UK, 2006.
5. *Sustainable Development in Practice: Case Studies for Engineers and Scientists*, ed. A. Azapagic, S. Perdan and R. Clift, Wiley, Chichester, 2004.

# Contents

Concepts of Chemical Engineering for Chemists: 2nd Edition
Edited by Stefaan Simons
© The Royal Society of Chemistry 2017
Published by the Royal Society of Chemistry, www.rsc.org

# 4   An Introduction to Heat Transfer                              99
*Tim Elson and Paola Lettieri*

# 5   An Introduction to Mass-transfer Operations                  141
*Eva Sorensen*

# 6   Scale-up in Chemical Engineering                             175
*Tim Elson*

# 7 An Introduction to Particle Systems 207
*Paola Lettieri*

# 8 An Introduction to Process Control 253
*Eva Sorensen*

# 9 Economic Appraisal of Large Projects 277
*Ken Sutherland*

# 10   Process and Personal Safety                    314

*D. C. Shallcross*

# 11   Engineering Statistics, Process Integration, and Experimental Design                    352

*Joseph Menicucci, Elizabeth Aisenbrey and Daniella Cirillo*

# Subject Index                                        387

# 1 Process Analysis – The Importance of Mass and Energy Balances

Eric S. Fraga

## 1.1 Introduction

Process engineering includes the generation, study and analysis of process designs. All processes must obey some fundamental laws of conservation. We can group these into conservation of matter and conservation of energy.[†] Mass and energy balances are fundamental operations in the analysis of any process. This chapter describes some of their basic principles.

### 1.1.1 Nomenclature and Units of Measurement

In carrying out any analysis, it is important to ensure that all units of measurement used are consistent. For example, mass may be given in kg (kilogrammes), in lb (pounds) or in many other units. If two quantities are given in different units, one quantity must be converted to the same units as the other quantity. Any book on chemical

---

[†]These two laws are separate in non-nuclear processes. For nuclear processes, we of course have the well-known equation, $E = mc^2$, which relates to mass and energy. For this chapter, we will consider only non-nuclear processes, but the same fundamental principles apply to all processes.

---

Concepts of Chemical Engineering for Chemists: 2nd Edition
Edited by Stefaan Simons
© The Royal Society of Chemistry 2017
Published by the Royal Society of Chemistry, www.rsc.org

**Table 1.1**  Some of the quantities encountered in process analysis with typical notation and units of measurement.

| Quantity | Notation | Dimension | Units |
|----------|----------|-----------|-------|
| Time | $t$ | T | s |
| Mass | $m$ | M | kg |
| Mass flow | $\dot{m}$ | $MT^{-1}$ | $kg\ s^{-1}$ |
| Mole | $n$ | M | mol |
| Molar flow | $\dot{n}$ | $MT^{-1}$ | $mol\ s^{-1}$ |
| Pressure | $P$ | $MT^{-2}\ L^{-2}$ | bar |
| Energy | $H, Q, W$ | $ML^2\ T^{-2}$ | Joule |

engineering (or physics and chemistry) will have conversion tables for standard units.

There are 7 fundamental quantities which are typically used to describe chemical processes: mass, length, volume, force, pressure, energy and power, although some of these can be described in terms of others in the list. For example, volume is length raised to the power 3; power is energy per unit time, pressure is force per area or force per length squared, and so on.

Chemical engineering uses some standard notation for many of the quantities we will encounter in process analysis. These are summarized in Table 1.1, where the dimensional terms are T for time, M for mass and L for length.

In describing processes, the variables that describe the condition of a process fall into two categories:

**extensive** variables, which depend on (are proportional to) the size of the system, such as mass and volume, and

**intensive** variables, which do not depend on the size of the system, such as temperature, pressure, density and specific volume, and mass and mole fractions of individual system components.

The number of intensive variables that can be specified independently for a system at equilibrium is known as the **degree of freedom** of the system.

## 1.2   Mass Balances

Chemical processes may be classified as batch, continuous or semibatch and as either steady-state or transient. Although the procedure required for performing mass, or material, balances depends on the type of process, most of the concepts translate directly to all types.

The general rule for the mass balance in a *system box* (a box drawn around the complete process or the part of the process of interest) is

$$\text{input} + \text{generation} - \text{output} - \text{consumption} = \text{accumulation} \qquad (1.1)$$

where

**input**: the material entering through the system box. This will include feed and makeup streams.

**generation**: material produced within the system, such as the reaction products in a reactor.

**output**: the material which leaves through the system boundaries. These will typically be the product streams of the process.

**consumption**: material consumed within the system, such as the reactants in a reactor.

**accumulation**: the amount of material that builds up within the system.

In a steady-state continuous process, the accumulation should always be zero, which leads to a more simple mass balance equation:

$$\text{input} + \text{generation} = \text{output} + \text{consumption} \qquad (1.2)$$

In the case of systems with no reaction, where mass is neither generated nor consumed, the result is even simpler:

$$\text{input} = \text{output} \qquad (1.3)$$

## 1.2.1 Process Analysis Procedure

The analysis of the mass balance of a process typically follows a number of steps:

1. Draw and label a diagram for the process, clearly indicating the information given by the problem definition and the values that have been requested.
2. Choose a basis of calculation, if required. If no extensive variables (*e.g.* amount or flow rate of a stream) have been defined, a basis of calculation is required and this must be an extensive variable.
3. Write down appropriate equations until the number of equations equals the number of unknown variables and such that all

the desired unknown variables are referred to in the equations. Possible sources of equations include the following.

(a) Mass balances. For a system with $n$ species, $n$ mass balance equations may be written down. These mass balance equations may be drawn from a total mass balance and from individual species mass balances. The only exception is where the system box contains only a splitter, where all the streams have the same composition and differ only in extensive variables. In this case, only one mass balance equation can be included.

(b) Process specifications and conditions, such as, for example, the separation achieved by a distillation unit or the conversion in a reactor.

(c) Definitions, such as the relationship between density, mass and volume or the relationship between mole fraction and total mass.

4. Identify the order in which the equations should be solved.
5. Solve the equations for the unknown values.

These steps are illustrated by the following example.

### Example 1. Mass balance on a continuous distillation process

Suppose that 1000 kmol h$^{-1}$ of a feed stream, consisting of 30.0% by mole $n$-pentane and the remainder $n$-hexane, is to be separated into 95.0 molar% pentane and 95.0 molar% hexane streams using a distillation column. Determine the flow rates of the output streams through the use of mass balances, assuming steady state operation. We will assume three digits of signficance for this example.

### Solution

The first step is to draw and label a flowsheet diagram, indicating the process steps and all the streams. Figure 1.1 shows the layout of the distillation unit labelled with both the known variables and the variables we wish to determine. The system box for this example is the whole process, *i.e.* the distillation unit. The streams we wish to consider are those that intersect this system box and consist of the feed stream and the two output streams. All other streams can be ignored in solving this example.

Each of the three streams has been numbered and subscripts are used on every variable associated with streams. $\dot{n}$ is a molar flowrate and $x$ is a mole fraction. We know one flow rate, that of the feed, and wish to determine the flow rates of the two product or output streams, $\dot{n}_2$ and $\dot{n}_3$. The p subscript refers to pentane and h to hexane.

System box

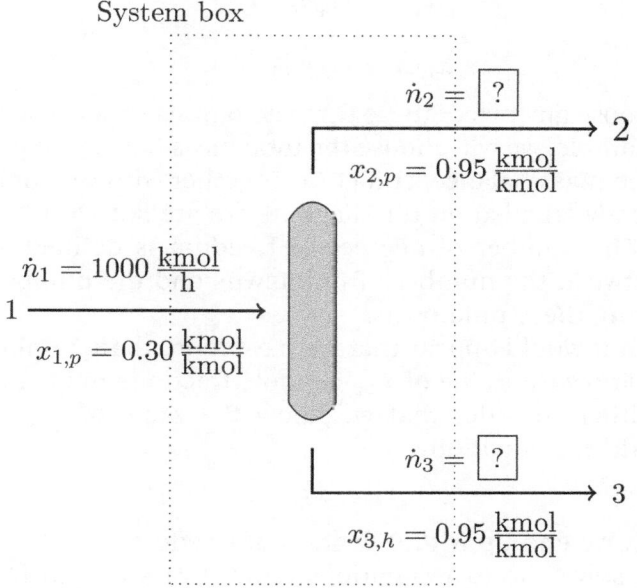

**Figure 1.1**   Distillation process diagram for Example 1, showing the system box and the streams that intersect this system box.

There are two unknowns and, as there are two species involved, we can write down two mass balance equations. We should be able to solve this problem.

As this is the simplest case described above (steady state, continuous and no reaction), we can use the simplest mass balance equation:

$$input = output$$

As indicated above, we have two unknowns. Therefore, we need to generate two independent equations that will allow us to solve for these unknowns. For mass balance problems, the general rule is that we can define $n_c$ equations if there are $n_c$ components in the streams involved in the mass balance problem defined by the system box chosen. The one exception is that if the system box includes only a splitter, there is only one independent equation that can be defined. This is because the splitter does not change the compositions of the streams involved, only the amounts or flows.

For this example, we have two components, so we can define two mass balance equations. The choice of equations is a total mass balance and two individual component balances:

$$\dot{n}_1 = \dot{n}_2 + \dot{n}_3 \tag{1.4}$$

$$x_{1,p}\dot{n}_1 = x_{2,p}\dot{n}_2 + x_{3,p}\dot{n}_3 \tag{1.5}$$

$$x_{1,h}\dot{n}_1 = x_{2,h}\dot{n}_2 + x_{3,h}\dot{n}_3 \tag{1.6}$$

We can choose any two of these three equations to solve our problem. For this example, we will choose the total mass balance, eqn (1.4), and the pentane mass balance, eqn (1.5). Together with the mole fraction values already labelled on the diagram, we are left with 0 degrees of freedom. The number of degrees of freedom is defined as the difference between the number of unknowns and the number of equations relating these unknowns.

Although it would appear that we have more than 2 unknowns, we implicitly know the value of $x_{3,p}$ as mole fractions must add up to 1. We can either consider that we know the value of $x_{3,p}$ or we can explictly add the equation

$$x_{3,p} + x_{3,h} = 1$$

Either way, we end up with 0 degrees of freedom.

We can solve the two equations, eqn (1.4) and eqn (1.5), by rearranging the first equation to have $\dot{n}_2$ alone on one side and then replace $\dot{n}_2$ in the second equation with the expression on the other side. The second equation can be solved for $\dot{n}_3$. This value can then be used to solve the first equation for $\dot{n}_2$.

$$
\begin{aligned}
(1.4) \quad &\Rightarrow \quad \dot{n}_2 = \dot{n}_1 - \dot{n}_3 \\
(1.5) \quad &\Rightarrow \quad x_{1,p}\dot{n}_1 = x_{2,p}(\dot{n}_1 - \dot{n}_3) + x_{3,p}\dot{n}_3 \\
&\Rightarrow \quad (x_{3,p} - x_{2,p})\dot{n}_3 = x_{1,p}\dot{n}_1 - x_{2,p}\dot{n}_1 \\
&\Rightarrow \quad \dot{n}_3 = \frac{x_{1,p} - x_{2,p}}{x_{3,p} - x_{2,p}}\dot{n}_1 \\
&\qquad\quad = \frac{x_{1,p} - x_{2,p}}{(1 - x_{3,h}) - x_{2,p}}\dot{n}_1 \hspace{3cm} (1.7)\\
&\qquad\quad = \frac{0.3 - 0.95}{(1 - 0.95) - 0.95} \times 1000 \text{ kmol h}^{-1} \\
&\qquad\quad = 722 \text{ kmol h}^{-1} \\
(1.7) \quad &\Rightarrow \quad \dot{n}_2 = 1000 \text{ kmol h}^{-1} - 722 \text{ kmol h}^{-1} = 278 \text{ kmol h}^{-1}
\end{aligned}
$$

The example has been solved. We can now use the unused mass balance equation, in this case, the hexane component mass balance, eqn (1.6), to provide a check for consistency:

$$
\begin{aligned}
&\quad\ \ x_{1,h}\dot{n}_1 = x_{2,h}\dot{n}_2 + x_{3,h}\dot{n}_3 \\
\therefore\ &\quad (1 - x_{1,p})\dot{n}_1 = (1 - x_{2,p})\dot{n}_2 + x_{3,h}\dot{n}_3 \\
\therefore\ &\quad 0.7 \times 1000 \text{ kmol h}^{-1} = 0.05 \times 278 \text{ kmol h}^{-1} + 0.95 \times 722 \text{ kmol h}^{-1} \\
\checkmark\ &\quad 700 \text{ kmol h}^{-1} = 700 \text{ kmol h}^{-1}
\end{aligned}
$$

(assuming three significant digits in the calculations) so the results are at least consistent, which gives some confidence in their correctness.

## 1.2.2 Processes with Reactions

For processes involving reactions, the mass balance equation, eqn (1.3) used in the first example, is not sufficient. Eqn (1.1) or (1.2) must be used. The presence of reactions means that the *generation* and *consumption* terms in these equations are non-zero. The first key difference between a simple separation process and a process involving reactions is the need to define these extra terms in terms of the amounts of the components in the system. Extra equations are required to satisfy the extra degrees of freedom introduced by having these extra terms present. There is another key difference: each reaction has associated with it a degree of freedom, one which essentially describes the extent to which the reaction takes place. Therefore, the analysis of a process involving a reaction will require extra equations to satisfy degrees of freedom introduced by the reactions taking place.

Two concepts are often used to describe the behaviour of a process involving reactions: *conversion* and *selectivity*. Conversion is defined with respect to a particular reactant and describes the extent of the reaction that takes place relative to the amount that could take place. If we consider the *limiting reactant*, the reactant that would be consumed first based on the stoichiometry of the reaction, the definition of conversion is straightforward:

$$\text{conversion} \equiv \frac{\text{amount of reactant consumed}}{\text{amount of reactant fed}}$$

Selectivity is a concept that applies to processes with multiple simultaneous reactions. It is used to quantify the relative rates of the individual reactions. However, any discussion about multiple reactions and the analysis of these is beyond the scope of this chapter. Refer to the further reading material identified at the end of this chapter for more information. In any case, the specification of conversion and/or selectivity will often provide the extra information required to satisfy those extra degrees of freedom introduced by having reactions present in the process.

**Example 2. Mass balance on a process with reaction**
Suppose that an initially empty tank is filled with 1000 mol of ethane $(C_2H_6)$ and the remainder is air. A spark is used to ignite this mixture

and the following combustion reaction takes place:

$$2C_2H_6 + 7O_2 \rightarrow 4CO_2 + 6H_2O \tag{1.8}$$

Assume that the amount of air provides twice the stoichiometric requirement of oxygen for this reaction and that air is composed of 79% nitrogen and the remainder is oxygen. Suppose that the reaction reaches a 90% conversion. What is the composition of the mixture in the tank at the end?

**Solution**
As in the first example, the first step is to draw and label a diagram, as shown in Figure 1.2. This example is a batch problem, so arrows indicate material flow in the sense of loading and discharging the reactor. The subscripts for mass amounts are the stream index (1 for the original contents of the tank and 2 for the final contents after combustion) and the species involved. The key is E for ethane, O for oxygen, N for nitrogen, C for carbon dioxide and W for water.

As in the first example, the boxed question marks highlight the variables which need to be determined. As there are five highlighted variables, we will require at least five equations. As there are five species involved, five mass balance equations can be defined. In this case, we should use eqn (1.1), but with *accumulation* set to zero to indicate that the tank is empty to start with and also empty at the end. This means that the mass balance equation we should use is identical in form to eqn (1.2):

$$input + generation = output + consumption$$

key: $\boxed{2E + 7O \longrightarrow 4C + 6W}$

$1 \xrightarrow{\quad\quad\quad\quad} \quad\quad\quad \xrightarrow{\quad\quad\quad} 2$
$n_{1,E} = 1000\,mol$

$n_{2,E} = \boxed{?}$
$n_{2,O} = \boxed{?}$
$n_{2,C} = \boxed{?}$
$n_{2,W} = \boxed{?}$
$n_{2,N} = \boxed{?}$

**Figure 1.2**  Diagram for the reaction mass balance problem for the combustion of ethane.

However, as there is one reaction taking place, we will need another equation to satisfy our degrees of freedom. We will use the specification of conversion to provide this extra equation.

The result is 6 equations:

$$
\left.
\begin{array}{ll}
n_{1,E} + n_{g,E} = n_{2,E} + n_{c,E} & \text{(ethane balance)} \\
n_{1,O} + n_{g,O} = n_{2,O} + n_{c,O} & \text{(oxygen balance)} \\
n_{1,N} + n_{g,N} = n_{2,N} + n_{c,N} & \text{(nitrogen balance)} \\
n_{1,C} + n_{g,C} = n_{2,C} + n_{c,C} & \text{(carbon dioxide balance)} \\
n_{1,W} + n_{g,W} = n_{2,W} + n_{c,W} & \text{(water balance)} \\
\text{conversion} = \dfrac{n_{1,E} - n_{2,E}}{n_{1,E}} & \text{(specification)}
\end{array}
\right\} \quad (1.9)
$$

where the subscript $g$ is used to indicate an amount generated and the subscript $c$ indicates an amount consumed. At this point, all the variables except for $n_{1,E}$ are unknown. We have 19 unknown variables. As we have just defined 6 equations, we have 13 degrees of freedom remaining.

To solve this problem, therefore, it would seem that we need to define at least 13 more equations. We can write down new equations relating the unknown and known variables by making use of the stoichiometric coefficients given by eqn (1.8). It is helpful to write all of these in terms of one of the consumption or generation terms. In this case, given that the key process specification is the conversion of ethane, it helps to write the equations in terms of the amount of ethane consumed:

$$
\left.
\begin{array}{ll}
n_{g,E} = 0 & \text{(no ethane is generated)} \\
n_{g,O} = 0 & \text{(no oxygen is generated)} \\
n_{c,O} = \dfrac{7}{2} n_{c,E} & \\
n_{g,N} = 0 & \text{(no nitrogen is generated)} \\
n_{c,N} = 0 & \text{(no nitrogen is consumed)} \\
n_{g,C} = \dfrac{4}{2} n_{c,E} & \\
n_{c,C} = 0 & \text{(no carbon dioxide is consumed)} \\
n_{g,W} = \dfrac{6}{2} n_{c,E} & \\
n_{c,W} = 0 & \text{(no water is consumed)}
\end{array}
\right\} \quad (1.10)
$$

This set of 9 equations reduces the degrees of freedom to 4 as no new variables have been introduced.

Further equations can be defined based on the specifications of the feed:

$$
\left.
\begin{array}{ll}
n_{1,O} = 2 \times \dfrac{7}{2} n_{1,E} & \text{(twice as much as required)} \\
n_{1,N} = 0.79 \times \dfrac{n_{1,O}}{0.21} & \text{(nitrogen is remainder of air)} \\
n_{1,C} = 0 & \text{(no carbon dioxide in the feed)} \\
n_{1,W} = 0 & \text{(no water in the feed)}
\end{array}
\right\} \quad (1.11)
$$

This set of 4 equations also introduces no new unknown variables. The result is that we have 19 equations, comprising the three sets of equations above, (1.9)–(1.11). These can now be solved as follows. Given $n_{1,E} = 1000$ mol, the initial amount of ethane, we evaluate the equations in set (1.11):

$$n_{1,O} = 2 \times \frac{7}{2} n_{1,E} = 7 \times 1000 \text{ mol} = 7000 \text{ mol}$$

$$n_{1,N} = 0.79 \times \frac{n_{1,O}}{0.21} = 0.79 \frac{7000 \text{ mol}}{0.21} \approx 26333 \text{ mol}$$

$$n_{c,E} = 0.90 \times n_{1,E} = 0.90 \times 1000 \text{ mol} = 900 \text{ mol}$$

Now determine the amounts generated and consumed for each species using the set of equations, eqn (1.10):

$$n_{c,O} = \frac{7}{2} n_{c,E} = \frac{7}{2} \times 900 \text{ mol} = 3150 \text{ mol}$$

$$n_{g,C} = \frac{4}{2} n_{c,E} = \frac{4}{2} \times 900 \text{ mol} = 1800 \text{ mol}$$

$$n_{g,W} = \frac{6}{2} n_{c,E} = \frac{6}{2} \times 900 \text{ mol} = 2700 \text{ mol}$$

Finally, we use the mass balance equations, eqn (1.9), to determine the amount of each species in the output:

$$n_{2,E} = n_{1,E} + n_{g,E} - n_{c,E} = 1000 \text{ mol} + 0 - 900 \text{ mol} = 100 \text{ mol}$$
$$n_{2,O} = n_{1,O} + n_{g,O} - n_{c,O} = 7000 \text{ mol} + 0 - 3150 \text{ mol} = 3850 \text{ mol}$$
$$n_{2,N} = n_{1,N} + n_{g,N} - n_{c,N} = 26333 \text{ mol} + 0 - 0 = 26333 \text{ mol}$$
$$n_{2,C} = n_{1,C} + n_{g,C} - n_{c,C} = 0 + 1800 \text{ mol} - 0 = 1800 \text{ mol}$$
$$n_{2,W} = n_{1,W} + n_{g,W} - n_{c,W} = 0 + 2700 \text{ mol} - 0 = 2700 \text{ mol}$$

## 1.2.3  More Complex Process Configurations

The two examples above consist of a single processing unit or step. More realistic problems will consist of multiple steps. When there are multiple steps connected to each other, we have not only input and output streams, but also internal streams. In such a case, we have alternative system boxes and we often need to consider more than one such box to solve a problem. The key concept is that the input and output streams in the mass balance equations are only those streams that enter or leave the particular system box. Streams internal to the system box are not involved at all.

## Example 3. Process with recycle

Consider the production of cyclohexane from 1-hexene, both $C_6H_{12}$, but with different configurations of the atoms:

This reaction does not achieve 100% conversion, so the *effluent* (the output) from the reactor needs to be processed to separate the reactant from the product. This is undertaken using a distillation unit. Any unused reactant is then recycled back to the reactor.

The feed to the continuous process, which should operate at steady state, is 10 mol s$^{-1}$ pure 1-hexene. The product is to be cyclohexane at 95 molar% purity. The reactor effluent will be fed to a distillation unit that has two outputs: a distillate stream and a bottoms stream. The distillate product is to contain 90 mol% 1-hexene and the bottoms product should meet the process requirements. The distillate is combined with the feed to the process before the mixture is sent to the reactor. This mixture will be 95 molar% 1-hexene.

We want to find the composition of all streams in the process and the conversion that is achieved in the reactor.

## Solution

We start by drawing and labelling the process diagram, including all the information we have about the process. This is shown in Figure 1.3. There are 5 streams involved at this level of view.

If we take the whole process as the system box, we have one input stream, the feed, and one output stream, the bottoms product of the distillation unit. There are two unknowns: the amount of the output and the extent of the reaction overall. We can write two equations, one for each species using eqn (1.2):

$$x_{1,h}\dot{n}_1 = (1 - x_{5,c})\dot{n}_5 + \dot{n}_{c,h} \quad \text{(1-hexane mass balance)} \quad (1.12)$$

$$(1 - x_{1,h})\dot{n}_1 + \dot{n}_{g,c} = x_{5,c}\dot{n}_5 \quad \text{(cyclohexane mass balance)} \quad (1.13)$$

where $\dot{n}_{c,h}$ is the rate at which 1-hexene is consumed and $\dot{n}_{g,c}$ the rate at which cyclohexane is generated. These two equations have three unknowns, so we need at least another equation to be able to solve them. This equation comes from the stoichiometry, which tells us

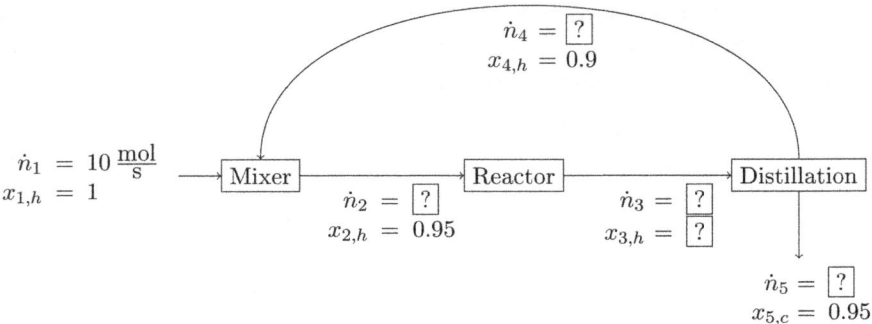

**Figure 1.3** Process diagram for Example 3. h as a subscript refers to 1-hexene and c as a subscript refers to cyclohexane.

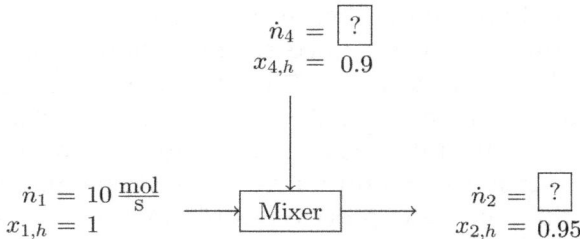

**Figure 1.4** System box containing only the mixer for the process in Example 3.

that for each mole of 1-hexene consumed, 1 mole of cyclohexane is generated:

$$\dot{n}_{c,h} = \dot{n}_{g,c} \tag{1.14}$$

Solving this set of three equations, given the input flow rate and the mole fractions of 1-hexene in the feed stream and cyclohexane in the output stream, we find that $\dot{n}_5 = 10$ mol s$^{-1}$. The rate at which 1-hexene is consumed and cyclohexane is generated are both 9.5 mol s$^{-1}$.

To determine the flows and compositions of the other streams in the process, we have to consider different system boxes. For instance, we can use a system box that contains only the mixer, as shown in Figure 1.4. Given the specification on the inlet to the reactor; that the composition is 95% 1-hexene, we have two unknowns and we can write two mass balance equations for this system box:

$$\dot{n}_1 + \dot{n}_4 = \dot{n}_2 \quad \text{(total mass balance)} \tag{1.15}$$

$$x_{1,h}\dot{n}_1 + x_{4,h}\dot{n}_4 = x_{2,h}\dot{n}_2 \quad \text{(1-hexene mass balance)} \tag{1.16}$$

We can solve these two equations to get $\dot{n}_2 = 20$ mol s$^{-1}$ and $\dot{n}_4 = 10$ mol s$^{-1}$.

$$\dot{n}_4 = 10\,\frac{\text{mol}}{\text{s}}$$
$$x_{4,h} = 0.9$$

$$\dot{n}_3 = \boxed{?}$$
$$x_{3,h} = \boxed{?}$$

Distillation

$$\dot{n}_5 = 10\,\frac{\text{mol}}{\text{s}}$$
$$x_{5,c} = 0.95$$

**Figure 1.5** System box containing only the distillation unit for the process in Example 3.

With stream 4 known, we can now determine stream 3 by considering a system box around the distillation unit, as shown in Figure 1.5. We again have two unknowns, $\dot{n}_3$ and $x_{3,h}$, and we can write two mass balance equations:

$$\dot{n}_3 = \dot{n}_4 + \dot{n}_5 \quad \text{(total mass balance)} \tag{1.17}$$

$$x_{3,h}\dot{n}_3 = x_{4,h}\dot{n}_4 + (1 - x_{5,c})\dot{n}_5 \quad \text{(1-hexene mass balance)} \tag{1.18}$$

which yields $\dot{n}_3 = 20$ mol s$^{-1}$ and $x_{3,h} = 0.475$.

We now have determined all the stream compositions and can therefore calculate the single pass conversion of the reactor:

$$\text{conversion} = \frac{\text{reactant consumed}}{\text{reactant fed}} \tag{1.19}$$

$$= \frac{x_{2,h}\dot{n}_2 - x_{3,h}\dot{n}_3}{x_{2,h}\dot{n}_2} \tag{1.20}$$

$$= \frac{0.95 \times 20 \text{ mol S}^{-1} - 0.475 \times 20 \text{ mol S}^{-1}}{0.95 \times 20 \text{ mol S}^{-1}} \tag{1.21}$$

$$= 0.5 \tag{1.22}$$

which is 50%.

## 1.3   Energy Balances

Energy balances can be treated in much the same way as material balances. The only fundamental difference is that there are three types of energy (for non-nuclear processes):

**Kinetic:** Energy due to the translational motion of the system as a whole, relative to some frame of reference (the earth's surface, for instance) or to the rotation of the system about some axis.

**Potential:** Energy due to the position of the system in a potential field. In chemical engineering, the potential field will typically be gravitational.

**Internal:** All energy possessed by the system, other than kinetic or potential. For example, the energy due to the motion of molecules relative to the centre of mass of the system and to the motion and interactions of the atomic and subatomic constituents of the molecules.

Energy may be transferred between a system and its surroundings in two ways:

1. As *heat*, or energy that flows as a result of a temperature difference between a system and its surroundings. The direction of flow is always from the higher temperature to the lower. Heat is defined as *positive* when it is transferred to the system from its surroundings.
2. As *work*, or energy that flows in response to any driving force other than a temperature difference. For example, if a gas in a cylinder expands and moves a piston against a restraining force, the gas does work on the piston. Energy is transferred as work from the gas to its surroundings, including the piston. Positive work means work done by the system on its surroundings, although this convention is sometimes not followed and one should be careful to note the convention used by other people.

In this chapter, we will deal solely with heat.

Energy and work have units of force times distance, such as a Joule (J), which is a Newton-metre (N m). Energy is sometimes measured as the amount of heat that must be transferred to a specified mass of water to raise the temperature of the water a specified temperature interval at a specified pressure (*e.g.* 1 kcal corresponds to raising 1 kilogramme of water from 15 °C to 16 °C).

As in material balances, energy must be conserved.[‡] We have two energy balance equations: one for *closed* systems and one for *open* systems. A closed system is one in which no mass comes in or goes out of the process during its operation. An example of this would be a batch process with an initial state and a final state. An open system has mass coming in and/or going out during the operation. A continuous process is an example of an open system. In either case, the energy balance equations are based on the first law of thermodynamics.

The full energy balance equation for a *closed* system is

$$\Delta U + \Delta E_k + \Delta E_p = Q - W \tag{1.23}$$

where $\Delta U$ is the difference in internal energy of all the streams at the end of the process in relation to those streams at the start of the process, $\Delta E_k$ the change in kinetic energy of the system, $\Delta E_p$ the change in potential energy of the system, $Q$ the amount of heat put into the system and $W$ the amount of work done by the system.

For continuous processes, we wish to consider rates of energy, so the equivalent equation is

$$\Delta \dot{H} + \Delta \dot{E}_k + \Delta \dot{E}_p = \dot{Q} - \dot{W}$$

where $H$ is the *enthalpy* of the inlet and outlet streams and the changes in kinetic and potential energies refer to the streams coming out of the process and the streams coming into the process. Specific enthalpy (enthalpy per unit mass), denoted by $\hat{H}$, is defined as the combination of internal energy and *flow work*, the work that is required to move material into and out of a process:

$$\hat{H} = \hat{U} - P\hat{V}$$

where $F$ is the system pressure and $\hat{V}$ the specific volume of the material. The actual rate of enthalpy for a stream, $\dot{H}$, will be calculated by $\dot{n}\hat{H}$.[§] In working with changes of enthalpy, we will use a reference state, a combination of temperature and pressure for which the enthalpy is assumed to be zero. Enthalpy data in property tables is typically given with reference to a specified reference state.

---

[‡]Noting again that we are refering only to non-nuclear processes.
[§]This calculation assumes that the specific enthalpy is given in units of energy per mole. However, many tables will have units of energy per unit mass, so a unit conversion using molecular weights may be required.

In this chapter we will introduce some of the basic properties required to perform energy balances on a process. Specifically, we will describe how to estimate changes in the rate of enthalpy due to changes in temperature and phase for streams. We will consider open systems and will ignore changes in kinetic and potential energies. For systems in which kinetic and potential energies are assumed to not change or where the changes in these energies are assumed to be negligible, the energy balance equation can be reduced to

$$\Delta \dot{H} = \dot{Q} - \dot{W}$$

for an open system.

In working with materials, there are two types of heat:

**sensible** heat, which signifies heat that must be transferred to raise or lower the temperature of a substance, assuming no change in phase (solid, liquid, gas), and

**latent** heat, which is the heat necessary to change from one phase to another.

The calculation of sensible heat is based on the *heat capacity* (at constant pressure) of the substance, $C_p(T)$, which is in units of heat (energy) per unit mass. Heat capacity information is typically in the form of coefficients for a polynomial expression:

$$C_p(T) = a + bT + cT^2 + dT^3$$

The values of the coefficients are given in physical property tables found in most chemical engineering reference books. To determine the change in enthalpy in heating a substance from one temperature, $T_1$, to another temperature, $T_2$, without a phase change, we integrate the polynomial over the temperature range:

$$\int_{T_1}^{T_2} C_p(T) \mathrm{d}T = \left[ aT + b\frac{T^2}{2} + c\frac{T^3}{3} + d\frac{T^4}{4} \right]_{T_1}^{T_2} \tag{1.24}$$

We evaluate the right-hand side of this equation as the difference of the polynomial evaluated at $T_2$ and at $T_1$. The relation between the heat capacity (integrated over a temperature interval) and the change in enthalpy is exact for ideal gases, exact for nonideal gases only if the pressure is indeed constant, and a close approximation for solids and liquids.

For latent heat, we look up the corresponding entry in the tables for either the *latent heat of vapourization* (or simply the heat of vapourization) or the *heat of fusion*, depending on the type of phase change encountered: liquid to vapour and solid to liquid, respectively. These quantities are in units of energy per unit mass and are given for a specific reference state, often the 1 atm boiling point or melting point of the substance.

The calculation of the change in enthalpy from one temperature to another for a given substance will often be a multi-step process. The main principle is to identify a path of pressure and temperature changes that goes from the initial state to the final state, passing through states at which we have reference data (*e.g.* the heat of vapourization) available. Any sections of the path that do not involve phase changes will simply require the calculation of sensible heat using the heat capacity equation given above. Phase changes will then require the use of the appropriate latent heat quantity.

For process analysis including energy balances, the same procedure defined for mass balances is followed. The only change is that the source of equations, described in step 3 of the procedure, now includes the energy balance equation, as well as the definition of the different energy terms. Again, this procedure will be illustrated by an example.

### Example 4. Energy balance on distillation column

We again consider the distillation unit introduced in Example 1, updated with temperature information for each of the streams, and now including detail about the inner workings on the unit. Specifically, the process now includes the total condenser, which cools the vapour coming out of the top of the column into liquid, and the reboiler, which boils up the liquid from the bottom of the column into vapour (see Figure 1.6). These two processing steps were previously considered to be internal to the system box used in solving the mass balance problem.

The temperatures noted on the diagram have been estimated using a physical property estimation system. There are a number of computer-based tools for estimating physical properties and most simulation software systems will include appropriate methods. Figure 1.6 shows these temperatures, as well as the results obtained earlier. As more streams have been included in this diagram, we have new unknowns. Specifically, we now have the vapour stream, $V$, from the top of the column to the condenser and the liquid reflux stream, $L$, from the condenser back into the column. Both of these streams have the same composition as the distillate product, stream 2, and their flow rates will be denoted by $\dot{n}_V$ and $\dot{n}_L$, respectively. The

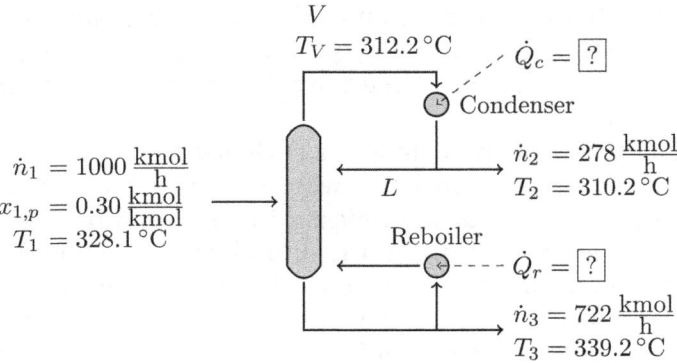

**Figure 1.6** Distillation unit including a total condenser and a partial reboiler with streams annotated with temperatures.

relationship between the liquid reflux stream back into the column and the actual distillate product stream ($\dot{n}_2$) is given by the *reflux ratio*:

$$R = \frac{\dot{n}_L}{\dot{n}_2} \tag{1.25}$$

For this particular configuration, the reflux ratio required to achieve the separation is $R = 1.6$.

Neglecting the effect of pressure on enthalpy, we wish to estimate the rate at which heat must be supplied to the reboiler.

**Solution**
Figure 1.6 is the flowsheet diagram for this problem, with all the streams labelled, both with known quantities and with an indication of what we need to estimate to solve this problem. The specific answer to the question posed is the value of $\dot{Q}_r$, the rate of heat supplied to the reboiler. To determine this amount, we will need to determine the change in enthalpy of the output streams relative to the feed stream and the amount of cooling done in the condenser, $\dot{Q}_c$. We will consider two system boxes: one around the whole unit, essentially the same system box as was used in solving Example 1, and one around the condenser section of the unit, shown in Figure 1.7. The latter system box will be useful for determining $\dot{Q}_c$ and will be the first we consider.

For this system box, there are three streams: an input vapour stream, which originates in the top of the column and two output streams, the distillate product of the unit and the liquid reflux going back into the column. The system box includes a splitter and a condenser. There are three unknowns indicated in the figure: the amount

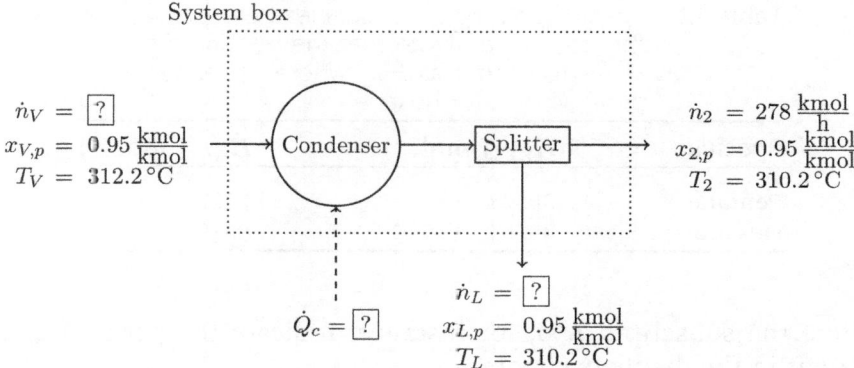

**Figure 1.7** System box containing the condenser of the distillation unit with a splitter. The input stream is vapour from the top of the distillation column. The output streams are the distillate product (indicated by subscript 2) and the liquid reflux. There is also a heat input, $\dot{Q}_c$, for the condenser.

of cooling required, $\dot{Q}_c$ and the flow rates of the vapour, $\dot{n}_V$, and liquid streams, $\dot{n}_L$. Material and energy balances must be addressed simultaneously. As the only step that changes stream amounts is a splitter, we will only be able to write down one mass balance equation for this system box. Together with the appropriate energy balance equation, we have

$$\dot{n}_V = \dot{n}_2 - \dot{n}_L \tag{1.26}$$

$$\Delta \dot{H} = \dot{Q}_c \tag{1.27}$$

As there are three unknowns in the diagram, we will need at least one more equation. We will use eqn (1.25), which relates the flows of the liquid reflux and the distillate product stream. We also need to calculate the change in enthalpy, which we do by estimating the sensible and latent heats for each of the streams involved, relative to a reference temperature. For this problem, we will use $0\,^\circ\text{C} = 273.15$ K as the reference temperature.

The estimate of the specific enthalpy of each stream requires data, available from a number of sources as discussed above, for the heat capacity equation and the latent heat of vapourisation. For each stream, we will have a contribution for each compound present. For instance, the enthalpy of the distillate stream will be composed of the enthalpy contribution of pentane and of hexane:

$$\dot{H}_2 = x_{2,\text{p}} \dot{n}_2 \hat{H}_{2,\text{p}} + x_{2,\text{h}} \dot{n}_2 \hat{H}_{2,\text{h}} \tag{1.28}$$

**Table 1.2**  Specific enthalpies for species before and after the condenser. Note that the specific enthalpy $\hat{H}_{L,i}$ is the same as $\hat{H}_{2,i}$, where $i$ is either p for pentane or h for hexane.

| Species | $\hat{H}_{V,i}$ (J mol$^{-1}$) | $\hat{H}_{2,i}$ (J mol$^{-1}$) |
|---|---|---|
| Pentane | 38521 | 7377 |
| Hexane | 38601 | 8013 |

where the subscripts 2, p, for instance, indicate the pentane component in the distillate stream.

The results of these calculations, using eqn (1.24) and data from Felder & Rousseau (2000) (see further reading section below), are shown in Table 1.2. The table presents the specific enthalpies for the two species before and after the condenser. The specific enthalpies for the species in the liquid reflux stream will be the same as in the distillate output stream. The difference between the enthalpies in the vapour stream and the distillate stream is primarily due to the heats of vapourisation, but there is a small contribution from the drop in temperature across the condenser.

Using the specific enthalpies in Table 1.2 to get the actual stream enthalpies allows us to then solve for the remaining 3 unknowns, with the following results: $\dot{n}_V = 722$ kmol h$^{-1}$, $\dot{n}_L = 444$ kmol h$^{-1}$ and $\dot{Q}_c = -2.25 \times 10^7$ kJ h$^{-1}$. The negative sign on $\dot{Q}_c$ indicates that the condenser is removing heat from the process, as expected.

To determine the amount of heating required in the reboiler, we now look at the first system box, the box that includes the whole distillation unit: the column, the condenser and the reboiler. This box also has three streams, the feed stream and the two output streams, and two heat inputs, $\dot{Q}_c$ and $\dot{Q}_r$. All mass flow rates are known and we also know the condenser requirements, so we have one unknown, the reboiler heat duty. We have only one equation, the energy balance around the column

$$\Delta \dot{H} = \dot{Q}_c + \dot{Q}_r \tag{1.29}$$

We again need to calculate the enthalpies of the stream, noting that we have already calculated the enthalpy of the distillate product stream. If we calculate the enthalpies of the feed and bottom product streams and solve eqn (1.29), we get $\dot{Q}_r = 2.20 \times 10^7$ kJ h$^{-1}$, a positive value indicating that heat is being added to the process, again as expected.

## 1.4  Summary

This chapter has introduced the concepts of mass and energy balances. These are essential steps in the analysis of any process. Simple examples have been used to illustrate the different steps, including not only mass and energy balances, but also simultaneously solving mass and energy balances together.

The key to doing process analysis is the identification of the extra equations that it may be necessary to solve for the unknown variables. These equations will come from a number of sources, including the balance equations themselves (eqn (1.1) and (1.23)), process specifications (such as the purity of output streams and the reflux ratio), physical relations (such as the definition of enthalpy for liquid and vapour streams) and other constraints imposed by the problem. Once a full set of equations has been developed, the equations can be solved, usually with little difficulty, and the desired results obtained.

## Further Reading

J. Coulson, J. F. Richardson, J. R. Backhurst and J. H. Harker, *Coulson & Richardson's Chemical Engineering Volume 1: Fluid Flow, Heat Transfer and Mass Transfer*, Butterworth-Heinemann, 5th edn, 1997.

R. M. Felder and R. W. Rousseau, *Elementary Principles of Chemical Processes*, John Wiley & Sons, New York, 3rd edn, 2000.

*An Introduction to Industrial Chemistry*, ed. C. A. Heaton, Leonard Hill, Glasgow, 1984.

R. H. Perry, D. W. Green and J. O. Maloney, *Perry's Chemical Engineers' Handbook*, McGraw-Hill, 7th edn, 1997.

# 2 Introduction to Chemical Reaction Engineering

George Manos

## 2.1 Introduction

The major objective of chemical reaction engineering is the analysis and design of equipment for carrying out desirable chemical reactions, *i.e.* chemical reactors.

Apart from choosing operating conditions and type of reactor, including provisions for heat exchange, the design of a reactor involves calculating its size to accomplish the desired extent of a reaction. This requires knowledge of the rate of the chemical reaction, *i.e.* how fast the chemical reaction occurs. In other words, it requires knowledge of the kinetics of the chemical reaction. Very often, physical processes, such as mass and energy transfer, play an important role and influence the overall rate of a chemical reaction. The principles governing the physical processes are as important as those governing the chemical kinetics.

The essential feature in chemical reactor analysis and design is the formulation of conservation equations for mass and energy for the chosen type of reactor. These equations can be either algebraic or differential and their solution gives the extent of the reaction.

This chapter gives an introduction to the subject of chemical reaction engineering. The first part introduces basic definitions and concepts of chemical reaction engineering and chemical kinetics and the importance of mass and heat transfer to the overall chemical

Concepts of Chemical Engineering for Chemists: 2nd Edition
Edited by Stefaan Simons
© The Royal Society of Chemistry 2017
Published by the Royal Society of Chemistry, www.rsc.org

reaction rate. In the second part, the basic concepts of chemical reactor design are covered, including steady-state models and their use in the development of reactor design equations. Worked examples are given in the text to demonstrate these concepts.

## 2.1.1 Classification of Reactors

There are different ways to group chemical reactors. First of all they can be classified according to their operation, either *batch* or *continuous*, with *semibatch* or *semicontinuous* being a category between them. A batch reactor has no stream continuously flowing into or out of it and is charged with the reactants in a pure form or in a solution, heated to the reaction temperature and left for the reaction to occur for a specific time. At the end of the reaction the reactor is emptied, cleaned, maintained (if necessary) and then filled with reactants again. Batch reactors operate in such cycles and to calculate production rates the overall time, including filling and emptying of the reactor, must be considered. A semibatch reactor has either an inlet or an outlet stream only. An example is a hydrogenation reactor using a liquid reactant to produce a liquid product. Both liquid reactant and liquid product stay during the whole reaction time in the reactor and only hydrogen flows into it. In continuous reactors an inlet and an outlet stream flow continuously. Continuous reactors normally operate at steady state.

Another classification of chemical reactors is according to the phases being present, either *single phase* or *multiphase* reactors. Examples of multiphase reactors are gas–liquid, liquid–liquid, gas–solid or liquid–solid catalytic reactors. In the last category, all reactants and products are in the same phase, but the reaction is catalysed by a solid catalyst. Another group is gas–liquid–solid reactors, where one reactant is in the gas phase, another in the liquid phase and the reaction is catalysed by a solid catalyst. In multiphase reactors, in order for the reaction to occur, components have to diffuse from one phase to another. These mass transfer processes influence and determine, in combination with the chemical kinetics, the overall reaction rate, *i.e.* how fast the chemical reaction takes place. This interaction between mass transfer and chemical kinetics is very important in chemical reaction engineering. Since chemical reactions either produce or consume heat, heat removal is also very important. Heat transfer processes determine the reaction temperature and, hence, influence the reaction rate.

Finally, chemical reactors can be classified according to the mode of heat removal. We can have either *isothermal* or *non-isothermal* reactors, a sub-category of which is *adiabatic reactors*. Often it is desirable to use the heat released by an exothermic reaction for an endothermic reaction, in order to achieve higher heat integration.

## 2.2 Chemical Reaction Kinetics

### 2.2.1 Definitions

A chemical reaction can generally be written as

$$\sum_{i=1}^{N} \nu_i A_i = 0$$

where $A_i$ $(i = 1, 2, \ldots, N)$ are the chemical species participating in the chemical reaction and $\nu_i$ are their stoichiometric coefficients. Stoichiometric coefficients of products are positive and of reactants, negative.

For example, the simple full oxidation of methane

$$CH_4 + 2O_2 \rightarrow CO_2 + 2H_2O \tag{2.1}$$

can be written as

$$CO_2 + 2H_2O - CH_4 - 2O_2 = 0 \tag{2.2}$$

In this example, the chemical species and respective stoichiometric coefficients are $A_1 = CO_2$, $A_2 = H_2O$, $A_3 = CH_4$, $A_4 = O_2$ and $\nu_1 = 1$, $\nu_2 = 2$, $\nu_3 = -1$, $\nu_4 = -2$.

#### 2.2.1.1 Extent of Reaction

The extent of a reaction $(\xi)$ is defined as the change in the number of moles of any reaction compound (reactant or product) due to the chemical reaction divided by its stoichiometric coefficient.

Let us consider the full oxidation of methane again. As the various components react according to the reaction stoichiometry, at any time the change in the number of moles of the reaction components has the following relationship:

$$\Delta n_{H_2O} = 2\Delta n_{CO_2} = -\Delta n_{O_2} = -2\Delta n_{CH_4} \tag{2.3}$$

Note: The change in the number of moles $(\Delta n_i)$ during a reaction is negative for a reactant and positive for a product.

Therefore, the extent of reaction will have the same value, whatever compound we choose for its calculation

$$\xi = \frac{\Delta n_{CO_2}}{1} = \frac{\Delta n_{H_2O}}{2} = \frac{\Delta n_{CH_4}}{-1} = \frac{\Delta n_{O_2}}{-2} \tag{2.4}$$

In a complex reaction network of $R$ reactions occurring simultaneously, the extent of the reaction $\xi_j$ is

$$\xi_j = \frac{\Delta n_{ij}}{v_{ij}} \qquad (2.5)$$

where $\Delta n_{ij}$ is the change in the number of moles of the reaction component $i$ due to the reaction $j$ and $v_{ij}$ is the stoichiometric coefficient of the reaction component $i$ in the reaction $j$.

The total change in the number of moles of $i$ is thus

$$\Delta n_i = \sum_{j=1}^{R} \Delta n_{ij} = \sum_{j=1}^{R} v_{ij} \xi_j \qquad (2.6)$$

### 2.2.1.2 Conversion of a Reactant

The conversion of a reactant A $(X_A)$ is the fraction of reactant A converted/transformed to products.

$$X_A = \frac{n_{A0} - n_A}{n_{A0}} = \frac{|\Delta n_A|}{n_{A0}} = \frac{\left| \sum_{j=1}^{R} v_{Aj} \xi_j \right|}{n_{A0}} \qquad (2.7)$$

where $n_{A0}$ is the initial number of moles of A.

The conversion is defined with reference to a reactant and could be different for different reactants. As a matter of fact, in a general reaction the conversions of reactants A and B will only be equal when the initial (or feed) ratio of moles of A to B is equal to the stoichiometry of the reaction $a/b$, where $a$ and $b$ are the stoichiometric coefficients of A and B, respectively.

For example, if during a reaction at some time from the initial 50 mol of reactant A only 20 mol have been left unreacted, *i.e.* 30 mol have reacted to products, the conversion at that time would be

$$X_A = \frac{50 - 20}{50} = 0.60 = 60\%$$

The conversion can be defined for a continuous reactor as well. For example, 50 mol min$^{-1}$ of reactant A enter a continuous reactor. At the outlet of the reactor only 20 mol min$^{-1}$ of A exit the reactor. The conversion is then

$$X_A = \frac{50\,\text{mol min}^{-1} - 20\,\text{mol min}^{-1}}{50\,\text{mol min}^{-1}} = 0.60 = 60\%$$

### 2.2.1.3  Yield

The yield of a specific product K ($Y_K$) is the amount of K produced expressed as a fraction of the maximum amount of K (according to the stoichiometry) that could be produced. This is the amount of reactant A converted to the specific product:

$$Y_K = \frac{n_K - n_{K0}}{n_{A0}} \frac{|v_A|}{|v_K|} \tag{2.8}$$

For example, let us consider the following reaction scheme for decomposition/isomerisation reactions

$$A \rightarrow 2K, \quad A \rightarrow B$$

From 100 initial moles of A, 30 are converted to produce 60 mol of K and 50 are converted to B. The total number of moles of A converted (to either K or B) is 80. The (total) conversion of A is 80%, but only 30% is converted to K. Hence, the yield of K is

$$Y_K = \frac{60}{100} \frac{1}{2} = 0.30 = 30\% \tag{2.9}$$

### 2.2.1.4  Selectivity

We now introduce the term selectivity for reaction networks. Selectivity can be defined in two ways.

First, the selectivity to a product P is the fraction of the converted reactant A (not the initial A) that is converted to the specific product K.

$$S_P = \frac{n_P - n_{P0}}{n_{A0} - n_A} \frac{|v_A|}{|v_P|} = \frac{Y_P}{X_A} \tag{2.10}$$

In the previous example the selectivity to P is

$$S_P = \frac{60}{80} \frac{1}{2} = 0.375 = 37.5\%$$

*i.e.* 37.5% of the converted A, not the initial A, is reacted to P.

The yield to a product is equal to the conversion multiplied by the selectivity

$$Y_K = S_K \times X_A \tag{2.11}$$

Secondly, selectivity can be defined with reference to another product. Overall or integrated selectivity is the ratio of the amount of one product (B) produced to the amount of another (C). For example, for reactions in series or in parallel

$$A \rightarrow B \rightarrow C \quad \text{or} \quad A \rightarrow B, A \rightarrow C$$

$$S_{B/C} = \frac{n_B - n_{B0}}{n_C - n_{C0}} \frac{|v_C|}{|v_B|} = \frac{Y_B}{Y_C} \tag{2.12}$$

Point selectivity can be defined as the ratio of the reaction rates of B and C at a specific point in a reactor at a specific time. We introduce the term 'reaction rate' in the next section.

$$S_{B/C} = \frac{r_B}{r_C} \tag{2.13}$$

### 2.2.1.5   Reaction Rate

The reaction rate ($r$) is the change of the extent of the reaction per unit time per unit volume. Consider a point somewhere in a reactor (continuous or batch) and some volume dV around it. Imagine that you are in this volume and can measure how many moles of a reaction species $i$ ($dn_{ij}$) have been produced[†] during some time dt *via* reaction $j$.

The change of the extent of this reaction is

$$d\xi_j = \frac{dn_{ij}}{v_{ij}} \tag{2.14}$$

and the rate of reaction $j$ is

$$r_j = \frac{d\xi_j}{dVdt} = \frac{1}{v_{ij}} \frac{dn_{ij}}{dVdt} \tag{2.15}$$

This is the general definition of reaction rate. Only in special cases is this equal to d$C$/dt ($C$ = concentration).

In a reaction network of $R$ reactions, the total change in the number of moles of a reaction component $i$ is due to many reactions and given by

$$dn_i = \sum_{j=1}^{R} dn_{ij} = \left( \sum_{j=1}^{R} r_j v_{ij} \right) dVdt \tag{2.16}$$

The above definition of rate refers to a chemical reaction. If we would like to use a similar term for a component, its reaction rate, or better, the rate of change in the number of moles of $i$ is

$$R_i = \frac{dn_i}{dVdt} = \sum_{j=1}^{R} v_{ij} r_j \tag{2.17}$$

---

[†]Reactant consumption is considered as a negative production.

We use capital $R$ for a component in order to distinguish it from small case $r$ which symbolises the rate of a chemical reaction.

## 2.2.2  Chemical Reaction Thermodynamics

Consider the reversible reaction

$$a\text{A} + b\text{B} \Leftrightarrow c\text{C} + d\text{D}$$

At equilibrium, the rate of reaction from left to right is equal to the rate of the reaction from right to left, so that the net overall reaction rate is zero, *i.e.* nothing changes. The same amount of A and B react to C and D as C and D react to give A and B.

The equilibrium conversion is the conversion of a reversible reaction at its equilibrium. This is the maximum conversion that can be achieved. Reversible reactions cannot progress further than the equilibrium.

The equilibrium constant ($K$) of a gaseous reaction will be defined here for the case of gaseous reaction components behaving like ideal gases

$$K = K_\text{P} = \prod_{i=1}^{N} P_i^{v_i} \tag{2.18}$$

where $P_i$ is the partial pressure of $i$. $K$ and $K_\text{P}$ have no units. The reason for this is that the more rigorous definition for ideal gases is

$$K_\text{P} = \prod_{i=1}^{N} \left( \frac{P_i}{P_0} \right)^{v_i} \tag{2.19}$$

where $P_0$ is a standard pressure. Normally $P_0 = 1$ bar so that arithmetically the first equation is valid. It is widely used because of its simplicity.

For the gaseous reversible reaction of the example above, eqn (2.19) can be rewritten as

$$K_\text{P} = \frac{P_\text{C}^c P_\text{D}^d}{P_\text{A}^a P_\text{B}^b} \tag{2.20}$$

$K$ or $K_\text{P}$ can be calculated from the Gibbs free energy of the reaction, $\Delta G_R^\circ$

$$\Delta G_{RT}^\circ = -RT \ln K \tag{2.21}$$

where $R$ is the universal gas constant.

### 2.2.2.1 Temperature Dependence of $K_P$ and Equilibrium Conversion

The temperature dependence of $K_P$ is given by the equation

$$\frac{d \ln K_P}{dT} = \frac{\Delta H^{\circ}_{RT}}{RT^2} \tag{2.22}$$

or the integrated form

$$\ln\left(\frac{K_{P2}}{K_{P1}}\right) = \frac{\Delta H^{\circ}_{R}}{R}\left(\frac{1}{T_1} - \frac{1}{T_2}\right) \tag{2.23}$$

with the assumption that the reaction enthalpy $\Delta H^{\circ}_R$ is independent from the temperature; *i.e.* it is almost constant between $T_1$ and $T_2$.

For exothermic reactions (*e.g.* oxidation, hydrogenation, polymerisation reactions, *etc.*) the equilibrium constant decreases with temperature. Consequently, the equilibrium (maximum) conversion decreases with temperature.

Note: Chemical reactions – even exothermic reactions – become faster at higher temperatures. Only their equilibrium conversion decreases with temperature.

For endothermic reactions, *e.g.* dehydrogenation, cracking reactions, *etc.*, the equilibrium constant and the equilibrium conversion increase with temperature.

### 2.2.2.2 Pressure Dependence of Equilibrium Conversion

For ideal gases, $K_P$ is independent of pressure. However, the equilibrium conversion depends upon pressure:

$$K_P = \prod_{i=1}^{N} P_i^{v_i} = \left(\prod_{i=1}^{N} y_i^{v_i}\right) P_{\text{tot}}^{\left(\sum_{i=1}^{N} v_i\right)} \tag{2.24}$$

since $P_i = y_i P_{\text{tot}}$.

By introducing $K_y = \prod_{i=1}^{N} y_i^{v_i}$ we can write

$$K_P = K_y P_{\text{tot}}^{\left(\sum_{i=1}^{N} v_i\right)} \tag{2.25}$$

The higher $K_y$ becomes, the higher is the equilibrium conversion (more products than reactants). To examine the influence of pressure, we have to consider three cases:

(a) More reactants than products $\sum_{i=1}^{N} v_i < 0$ (*e.g.* polymerisation, hydrogenation, hydration, alkylation, *etc.*). The equilibrium conversion increases with pressure; the higher the pressure, the more products at equilibrium.

(b) The same amount of products as reactants $\sum\limits_{i=1}^{N} v_i = 0$ (*e.g.* isomerisation). No pressure effect.

(c) More products than reactants $\sum\limits_{i=1}^{N} v_i > 0$ (*e.g.* cracking/decomposition, dehydrogenation, dehydration, dealkylation, *etc.*). The equilibrium conversion decreases with pressure.

### 2.2.2.3  Reaction Networks

In the case of a network of simultaneous reactions we use the following methodology:

(i) We calculate the equilibrium constant of each reaction according to the equation

$$\ln K_j = -\frac{\Delta G_R}{RT} \quad \text{or} \quad K_j = \exp\left\{-\frac{\Delta G_R}{RT}\right\} \tag{2.26}$$

(ii) We calculate $K_{yj}$ for each reaction from the equation

$$K_j = K_{Pj} = K_{yj} P_{\text{tot}}^{\left(\sum\limits_{i=1}^{N} v_i\right)} \tag{2.27}$$

where

$$K_{yj} = \prod\limits_{i=1}^{N} y_i^{v_{ij}} \tag{2.28}$$

(iii) It is advisable to express the mole-fractions $y_i$ as functions of the equilibrium extents of the different reactions ($\xi_{j,\text{eq}}$); in other words, use the extents of the reactions at equilibrium as independent variables.

(iv) Solve the system of the $R$ algebraic eqn (2.28) to calculate all ($\xi_{j,\text{eq}}$) and then the mole-fractions of the reaction components at equilibrium.

## 2.2.3  Kinetics

### 2.2.3.1  Fundamentals – Effect of Concentration

The simplest relation between reaction rate and concentrations of the reactants is a power law.

For the reaction

$$a\text{A} + b\text{B} \rightarrow c\text{C} + d\text{D}$$

the rate of reaction can be written as

$$r = kC_A^m C_B^n \tag{2.29}$$

where $k$ is the kinetic constant dependent upon reaction conditions (temperature and pressure), $m$ is the order of the reaction with respect to A, $n$ is the order of the reaction with respect to B and $(m + n)$ is the total reaction order.

Generally, $m \neq a$, $n \neq b$. If $m = a$, $n = b$, then the reaction is called elementary. The reaction mechanism is the network of elementary steps (reactions) that add up to the reaction. The reaction mechanism describes how, and in what steps, the overall reaction occurs. An elementary reaction cannot be analysed in simpler steps; it occurs exactly as the reaction states.

From the reaction mechanism, we can derive a reaction rate equation (kinetics) for the overall reaction. Below, we will discuss two such methods: rate limiting step and quasi-stationary state (pseudo-steady-state).

### 2.2.3.2 Rate Limiting Step

According to this concept, the overall reaction rate is equal to the reaction rate of the slowest step in the reaction mechanism. If one of the elementary steps occurs at a much lower rate than the others, then that slowest step determines the overall reaction rate. The other steps are assumed to be so fast that they are already at equilibrium.

We will apply this concept on the decomposition of $N_2O_5$ to explain the experimentally found first order of the reaction:

$$2N_2O_5 \rightarrow 4NO_2 + O_2$$

Reaction mechanism:

(1) $N_2O_5 \underset{k'_1}{\overset{k_1}{\rightleftharpoons}} NO_2 + NO_3 (2 \times /\text{occuring twice})$

Two molecules of $N_2O_5$ have to be decomposed for the right amount of product to be produced (right stoichiometry). The formalism we use differs from a reaction equation with all the stoichiometric coefficients 2 (or $-2$), in that the equilibrium and/or kinetic equation for this step would then be different.

(2) $NO_2 + NO_3 \xrightarrow{k_2} NO + O_2 + NO_2 (\text{slowest step})$

(3) $NO + NO_3 \xrightarrow{k_3} 2NO_2$

The second reaction step is much slower than the others, so that the first reaction is in equilibrium.

The reaction of $NO_2$ and $NO_3$ to NO, $O_2$, $NO_2$ (second reaction) is very slow. What happens then is that a lot of $NO_2$ and $NO_3$ are produced from the fast first reaction but cannot react further. This will force these products to react back to $N_2O_5$ until an equilibrium is established. Now when a little of $NO_2$ and $NO_3$ react in reaction 2, immediately some more $NO_2$ and $NO_3$ will be produced by the fast first reaction, so that equilibrium will be established again.

Using the equilibrium constant for the first reaction step, we can write

$$K_1 = \frac{C_{NO_2} C_{NO_3}}{C_{N_2O_5}}$$

Note: At equilibrium, as the reaction rate from left to right is equal to the reaction rate from right to left. This means $K_1 = (k_1/k_1')$.

The overall reaction rate is equal to the reaction rate of the slowest step (2):

$$r = r_2 = k_2 C_{NO_2} C_{NO_3}$$

How fast the products are produced depends on how fast the second reaction occurs. It does not help if the first reaction is very fast. For the products of the overall reaction to be produced, we have to wait for the second reaction. On the other hand, when the second reaction occurs, NO and $NO_3$ react very fast in the third reaction. The third reaction does not build up a bottleneck.

Since from the equilibrium equation $C_{NO_2} C_{NO_3} = K_1 C_{N_2O_5}$, the reaction rate can be written as

$$r = k_2 K_1 C_{N_2O_5} \quad \text{or}$$

$$r = k C_{N_2O_5} \quad \text{where} \quad k = k_2 K_1 = k_2 \frac{k_1}{k_1'}$$

Thus, the kinetics are first order; the reaction rate is proportional to the concentration of the reactant. We have seen that even complicated reaction mechanisms can lead to simple effective kinetics.

### 2.2.3.3  Quasi-stationary State

The assumption in this method is that the concentration of intermediates is constant, after a short initial period. This means that the rate of formation of intermediates is equal to the rate of their disappearance/consumption.

We consider the same reaction mechanism as before. In this mechanism $NO_3$ and $NO$ are intermediate components and therefore

$$\frac{dC_{NO_3}}{dt} = 0 \quad \text{and} \quad \frac{dC_{NO}}{dt} = 0$$

For the rate of formation of the intermediates we can use the individual kinetics of the elementary reactions. $NO_3$ is formed and disappears in the first reversible reaction step and disappears in the second and third steps. Therefore,

$$\frac{dC_{NO_3}}{dt} = k_1 C_{N_2O_5} - k_1' C_{NO_2} C_{NO_3} - k_2 C_{NO_2} C_{NO_3}$$

$$- k_3 C_{NO} C_{NO_3} = 0$$

Applying the same procedure for the NO that is formed in the second reaction and that disappears in the third one, we get

$$\frac{dC_{NO}}{dt} = k_2 C_{NO_2} C_{NO_3} - k_3 C_{NO} C C_{NO_3} = 0$$

Solving this system of two equations for $C_{NO}$ and $C_{NO_3}$ we have

$$C_{NO} = \frac{k_2}{k_3} C_{NO_2} \quad \text{and} \quad C_{NO_3} = \frac{k_1 C_{N_2O_5}}{(2k_2 + k_1') C_{NO_2}}$$

Finally,

$$r = \frac{dC_{O_2}}{dt} = k_2 C_{NO_2} C_{NO_3} = \left( \frac{k_1 k_2}{2k_2 + k_1'} \right) C_{N_2O_5}$$

or

$$r = k_* C_{N_2O_5} \quad \text{with} \quad k_* = \frac{k_1 k_2}{2k_2 + k_1'}$$

This derived kinetic expression also explains the experimental observation of first-order kinetics in respect to $N_2O_5$, but the first-order kinetic constant is different from the one derived by the rate-limiting step method.

## 2.2.3.4 Temperature Dependence of Kinetic Constant

The dependence of $k$ on temperature for an elementary reaction follows the Arrhenius equation

$$k = A \exp\left\{ -\frac{E_A}{RT} \right\} \tag{2.30}$$

**Figure 2.1**   Activation energy of a chemical reaction.

where *A* is known as the pre-exponential factor and $E_A$ is the activation energy. The meaning of activation energy is illustrated in Figure 2.1.

A reaction $A + B \rightarrow C$ occurs *via* an intermediate activated complex $(AB)^*$.

$$A + B \rightarrow (AB)^* \rightarrow C$$

The activation energy actually represents the energy barrier the reactants have to overcome to form this activated intermediate.

The activation energy can be estimated from kinetic experimental data at different temperatures. The Arrhenius equation can also be written as

$$\ln k = \ln A - \frac{E_A}{R}\frac{1}{T} \tag{2.31}$$

If $\ln k$ is plotted against $1/T$, the graph is a straight line with slope equal to $(-E_A/R)$. By calculating the slope of the line, the activation energy $E_A$ can be estimated.

## 2.2.4   Importance of Mass and Heat Transfer Processes

We will now examine how important the mass and heat transfer phenomena are.[‡] We will do so with the example of a solid catalysed reaction. In a catalytic reaction the reactants have to first form intermediate species with the active sites of the catalyst. The products are then formed much faster than in the absence of catalyst. What actually happens is that the activation energy of the route through the

---

[‡]The reader is referred to Chapters 4 and 5 for more detail on these subjects.

intermediate catalytic complex is much lower than the respective activation energy through the intermediate route in the absence of catalyst.

Let us examine the various steps of a catalytic reaction on a solid catalyst particle:

(i) Mass transfer of reactants from the bulk fluid phase through the boundary layer to the external surface of the catalyst particle.

A boundary layer is formed between the two phases (fluid and solid). This is a stagnant film that represents a layer of less movement of the fluid and hence builds up a zone with resistance to mass transfer. The mass transfer coefficient and generally the mass transfer rate depend on the fluid dynamics of the system. Higher fluid velocities significantly reduce the thickness of the film.

Normally mass transfer between two phases is modelled using a mass transfer coefficient ($k_m$). The flux (mol s$^{-1}$ m$_{ext.surf}^{-2}$) of a component $i$ is given by the formula

$$\hat{N}_i = k_m(C_{if} - C_{is}) \tag{2.32}$$

where $C_{if}$ is the concentration of the component i in the fluid phase and $C_{is}$ is the concentration on the external surface of the catalytic particle.

The above rate is expressed per unit of external surface. To express the rate per gramme of catalyst the flux has to be multiplied by the catalyst specific area ($m_{ext.surf}^2 g_{cat}^{-1}$).

(ii) Diffusion of the reactants into the pores of the particle where the majority of the active catalytic sites reside.
(iii) Adsorption of the reactants on the active sites.
(iv) Surface reaction to produce the products adsorbed on the active sites.
(v) Desorption of the products from the active sites.
(vi) Pore diffusion of the products to the external surface of the catalyst particle.
(vii) Mass transfer of the products from the external surface of the catalytic particle through the stagnant film to the bulk fluid phase.

The overall reaction rate, *i.e.* how fast the products are formed, depends on how fast is each one of the above phenomena.

Let us first examine the case where the catalytic reaction is very slow compared to the mass transfer (both external – from the fluid through the film to the solid particle – and internal – pore diffusion). In this case, the concentration in the particle is the same as in the bulk fluid phase. As the molecules move into the particle and react, the relatively fast mass transfer can replace the reacted molecules with new ones, keeping the level of the concentration constant.

The picture is completely different when the reaction is very fast. Now the mass transfer cannot cope with the rate of reacting molecules. Mass transfer cannot supply the new molecules to the inner part of the catalytic particle as fast, with the result that the concentration in the particle is much lower than in the bulk fluid phase. If, from the mass transfer phenomena, pore diffusion is the slowest step, additionally a falling concentration profile is formed in the particle. In both cases, the reaction occurs with a concentration much lower than the one we believe occurs, namely the one in the bulk fluid phase.

The picture is similar for heat transfer. As we know, reaction rates depend strongly upon temperature. Since the chemical reaction occurs in the catalytic particle, the transfer rate of the reaction heat will determine the temperature level in the particle. A fast heat transfer, compared with the rate of heat generation by the reaction, will lead to the particle temperature being equal to that in the bulk fluid phase, the one at which we believe the reaction takes place. However, if the heat transfer is very slow, the two temperatures will be different.

A major difference to mass transfer is that now the real reaction rate can be higher than the apparent one. The concentration in the particle is always lower or equal to the concentration in the bulk fluid phase and therefore, from this perspective, the real rate is always lower or, at best, equal to the apparent rate. However, in strong exothermic reactions heat transfer cannot cope with the high rates of generation of reaction heat and the temperature in the catalyst particle can be much higher than in the fluid phase, resulting in a much higher reaction rate than the apparent one.

Regarding mass transfer, the slowest step is normally pore diffusion rather than external mass transfer, whilst for heat transfer the slowest step is the interphase heat transfer between the particle and the fluid phase rather than the internal heat transfer in the solid particle.

## 2.2.5 Kinetics of a Catalytic Reaction

We conclude the first part of this chapter with a kinetic example. We will derive, using the rate limiting step approach, the Langmuir–Hinselwood kinetics of a simple catalytic (isomerisation) reaction: A ⇔ B.

This reaction occurs *via* the following steps: A is adsorbed onto an active catalytic site, the adsorbed species AS undergoes a surface reaction to form BS and finally B is desorbed to free the active site that can now catalyse another cycle. We assume that the surface reaction is the slowest and therefore the limiting step. The kinetic expression we derive is valid only under this assumption and if the assumption changes, another expression needs to be derived. Since the surface reaction is the slowest step, we assume that all other reaction steps are so fast that they are at equilibrium.

Reaction mechanism:

- Adsorption of reactant A:

$$A(g) + S(s) \rightleftharpoons AS(s)$$

$K_A$ is the adsorption equilibrium constant of A.
- Surface reaction:

$$AS(s) \rightleftharpoons BS(s)$$

Slow surface reaction (rate-limiting step). $k_1$, $k_2$ are the kinetic constants of the forward and backward reactions.
- Desorption of product B:

$$BS(s) \rightleftharpoons B(g) + S(s)$$

$K_B$ is the adsorption equilibrium constant of B.

In the above mechanism, (g) shows that the respective species is in the gas phase and (s) in the solid phase (catalyst). We symbolise with $C$ the concentrations in the gas phase and $\theta$ the concentration in the solid phase, usually in units of mol $g_{cat}^{-1}$.

As the surface reaction is the rate limiting step, the overall reaction rate is equal to the rate of this step:

$$r = k_1\,\theta_{AS} - k_2\theta_{BS}$$

This equation is not in the right kinetic form because it contains concentrations in the solid $\theta_i$, which cannot be measured. We have to find a way to replace them with known quantities. The final kinetic equation should contain only concentrations of A or B in the gas phase and kinetic constants.

The other steps are at equilibrium and thus we can write

$$K_A = \frac{\theta_{AS}}{C_A \theta_S} \Rightarrow \theta_{AS} = K_A C_A \theta_S$$

$$K_B = \frac{\theta_{AS}}{C_A \theta_S} \Rightarrow \theta_{BS} = K_B C_B \theta_S$$

We wrote the same equation for B because the adsorption equilibrium constant is not the equilibrium constant of the final reaction step but of the reverse reaction.

The reaction rate equation can then be written as

$$r = (k_1 K_A C_A - k_2 K_B C_B)\theta_S$$

$\theta_S$ is the concentration of free active sites, that is, active sites that are not occupied by A or B. Therefore, it is not allowed to be in the kinetic expression. It is not a constant and we must replace it with a function of $\theta_{tot}$, the total number of active sites, that is characteristic of the catalyst.

For $\theta_{tot}$ we can write

$$\theta_{AS} + \theta_{BS} + \theta_S = \theta_{tot} \Rightarrow K_A C_A \theta_S + K_B C_B \theta_S + 1 = \theta_{tot}$$

$$\Downarrow$$

$$\theta_S = \frac{1}{K_A C_A + K_B C_B + 1} \theta_{tot}$$

$$\text{and} \quad r = \frac{(k_1 K_A C_A - k_2 K_B C_B)\theta_{tot}}{1 + K_A C_A + K_B C_B}$$

We can simplify the above equation by lumping $\theta_{tot}$ into the kinetic constants

$$k_1{}^* = k_1 \theta_{tot} \quad \text{and} \quad k_2{}^* = k_2 \theta_{tot}$$

The final Langmuir–Hinselwood kinetic expression is therefore

$$r = \frac{k_1{}^* K_A C_A - k_2{}^* K_B C_B}{1 + K_A C_A + K_B C_B}$$

## 2.3   Concepts of Chemical Reactor Design

In order to determine design characteristics of chemical reactors, we have to formulate mole balance equations.

## 2.3.1 Mole Balances for Chemical Reactors

### 2.3.1.1 General Mole Balance Equation

The mole balance equation is applied for individual reaction mixture components. The first decision we have to make is the choice of the system, *i.e.* reactor space, in which we apply the mole balance. This system could be the whole reactor or a differential part of it.

The general mole balance equation for a reaction component *i* at a time *t* in words is

(Change in number of moles of *i* within the system between time *t* and *t* + d*t*)

= (Number of moles of *i* flowing into the system during the time interval d*t*)

− (Number of moles of *i* flowing out of the system during the time interval d*t*)

+ (Number of moles of *i* formed by chemical reactions inside the system during the time interval d*t*)

If *i* is a reactant, its consumption can be considered as a negative formation. In terms of the right hand side, we should include all incoming and outgoing streams where the component *i* is present, as well as all reactions where *i* participates, being either a reactant or product.

The above general mole balance equation can be expressed in terms of rates as

(Rate of accumulation of moles of *i* within the system)

= (Rate of molar flow of *i* into the system)

− (Rate of molar flow of *i* out of the system)

+ (Rate of formation of *i* by chemical reactions inside the system)

In short

(Accumulation) = (In) − (Out) + (Generation)

From the above mole balance equation we can develop the design equation for various reactor types. By solving the design equation we can then determine the time required for a batch reactor system or a reactor volume for a continuous flow system to reach a specific conversion of the reactant to products.

Before we develop the design equation, let us have a closer look at the terms of the balance equation.

(i) The rate of accumulation of component $i$ in the system can be expressed as

$$\frac{dN_i}{dt} = \frac{d(C_i V_R)}{dt} \qquad (2.33)$$

if the mole balance is applied to the whole reactor;
or it can be expressed as

$$\frac{dN_i}{dt} = \frac{d(C_i dV_R)}{dt} \qquad (2.34)$$

if the mole balance is applied to a differential part of the reactor where $C_i$ is the concentration of component $i$ and $V_R$ is the reactor volume.

(ii) The molar flow rate of component $i$ can be expressed as

$$F_i = C_i \dot{V} \qquad (2.35)$$

where $\dot{V}$ is the volumetric flow rate of the specific stream, inlet or outlet.

(iii) Since the reaction rate is defined as the number of moles reacting per unit time as well as per unit volume, to find the rate of formation of component $i$ we should multiply the reaction rate by the volume of the system $dV_R$. Furthermore, we should take into account all reactions where $i$ participates. Each reaction rate term should be multiplied by the stoichiometric coefficient of $i$ in each reaction $j$

$$R_i = \sum_j v_{ij} r_j \qquad (2.36)$$

In the following sections we will examine the following three reactor types and develop the mole balance/design equations for each:

- Perfectly mixed batch reactor
- Continuous stirred tank reactor (CSTR)
- Tubular plug flow reactor (PFR).

### 2.3.1.2   Batch Reactor

In a batch reactor there are no inlet or outlet streams: In = Out = 0. The total feed is charged into the reactor at the beginning and no withdrawal is made until the desired conversion level has been reached. Hence a reaction process occurring in a batch reactor is an unsteady one. All variables change with time. In addition, we assume that it is a perfectly mixed batch reactor, so that the concentrations of the reaction components, reactants or products are the same over the whole reactor volume. This assumption allows us to consider applying the mole balance equation across the whole reactor. With the term reactor we mean the space where the reaction(s) take place. For liquid phase reactions the reactor volume is smaller than the size of the physical reactor. It is the volume of the liquid phase, where the reaction(s) take(s) place.

Without inlet or outlet terms the mole balance becomes

$$\frac{dN_i}{dt} = \frac{d(C_i V_R)}{dt} = \left( \sum_j v_{ij} r_j \right) V_R \tag{2.37}$$

Only in the special case of constant reactor volume ($V_R = $ const) can eqn (2.37) be simplified into the more familiar form in terms of the concentration derivative

$$\frac{dC_i}{dt} = \sum_j v_{ij} r_j \tag{2.38}$$

### 2.3.1.3 Continuous Stirred Tank Reactor

A CSTR is a continuous flow reactor. It is a tank reactor, *i.e.* its height is comparable to its diameter, with powerful stirring. We assume that the stirring is so thorough that the reaction mixture is perfectly mixed. This means that each of the concentrations of the reaction mixture components has the same value everywhere inside the reactor. This assumption allows us again to formulate the mole balance equations in the whole reactor volume. In addition, the thorough stirring ensures that the temperature is the same everywhere in the reactor.

We are going to formulate the mole balance equation in a CSTR operating at steady state, where nothing changes with time. This means that there is no accumulation inside the reactor and that the inlet as well as outlet streams are at steady state. Following the general equation the mole balance for a component $i$ in a CSTR is

$$0 = F_{i,\text{in}} - F_{i,\text{out}} + \left( \sum_j v_{ij} r_j \right) V_R \tag{2.39}$$

or in concentration terms

$$0 = C_{i,\text{in}} \dot{V}_{\text{in}} - C_{i,\text{out}} \dot{V}_{\text{out}} + \left( \sum_j v_{ij} r_j \right) V_R \tag{2.40}$$

where the subscripts 'in' and 'out' indicate, respectively, the inlet and the outlet characteristics (molar flow rates $F_i$ or volumetric flow rate $\dot{V}$).

### 2.3.1.4  Plug Flow Reactor

A PFR is a continuous flow reactor. It is an ideal tubular type reactor. The assumption we make is that the reaction mixture stream has the same velocity across the reactor cross-sectional area. In other words, the velocity profile across the reactor is a flat one. In a PFR there is no axial mixing along the reactor. The condition of plug flow is met in highly turbulent flows, as is usually the case in chemical reactors.

Since the concentration changes along the PFR, we cannot apply the mole balance equation to the whole reactor. We would not know what concentrations to assign to the reaction rate term. We are going instead to choose a differential slice of the reactor, between lengths $z$ and $(z + dz)$. Instead of length we are going to use volume positions, namely $V$ and $(V + dV)$. For this system the molar flow rates at the positions $V$ and $(V + dV)$ are the inlet and outlet terms, accordingly, in the mole balance equation. For the steady-state condition (*i.e.* no accumulation) the mole balance becomes

$$0 = (F_i)_V - (F_i)_{V+dV} + \left(\sum_j v_{ij} r_j\right) dV \tag{2.41}$$

which can be written as

$$dF_i = \left(\sum_j v_{ij} r_j\right) dV \quad \text{or} \quad \frac{dF_i}{dV} = \sum_j v_{ij} r_j \tag{2.42}$$

Introducing concentrations, the above equation becomes

$$\frac{d(C_i \dot{V})}{dV} = \sum_j v_{ij} r_j \tag{2.43}$$

## 2.3.2  Reactor Design Equation

Integrating the mole balance equations gives us the design equations for the different reactor types:

- Batch reactor:

$$t = \int_{N_{i0}}^{N_i} \frac{dN_i}{\left(\sum_j v_{ij} r_j\right) V_R} \tag{2.44}$$

- CSTR:

$$V_R = \frac{F_{i,\text{in}} - F_{i,\text{out}}}{-\sum_j \nu_{ij} r_j} \tag{2.45}$$

- PFR:

$$V_R = \int_{F_{i,\text{in}}}^{F_{i,\text{out}}} \frac{dF_i}{\sum_j \nu_{ij} r_j} \tag{2.46}$$

In the next paragraph we are going to rewrite the above equations using conversion and introduce the term 'residence time'.

### 2.3.2.1 Design Equations Using Conversion

Recall from Section 2.2.1.2 the definition of conversion of a reactant A:

- Batch reactor:

$$X_A = \frac{N_{A0} - N_A}{N_{A0}} \quad \text{and} \quad N_A = N_{A0}(1 - X_A) \tag{2.47}$$

- Continuous flow reactor:

$$X_A = \frac{F_{A,\text{in}} - F_{A,\text{out}}}{F_{A,\text{in}}} \quad \text{and} \quad F_{A,\text{out}} = F_{A,\text{in}}(1 - X_A) \tag{2.48}$$

By substituting for $N_i$ and $F_i$ in the design equations, these are transformed into the following:

- Batch reactor:

$$t = C_{A0} \int_0^{X_A} \frac{dX_A}{-\sum_j \nu_{Aj} r_j} \tag{2.49}$$

- CSTR:

$$V_R = \frac{F_{A,\text{in}} X_A}{-\sum_j \nu_{Aj} r_j} \tag{2.50}$$

- PFR:

$$V_R = F_{A,\text{in}} \int_0^{X_A} \frac{dX_A}{-\sum_j \nu_{Aj} r_j} \tag{2.51}$$

The above equations are valid for the case of individual reactors. For the case of a series of continuous reactors, where the conversion at any point is defined with reference to the feed of the first reactor $(F_{A,0})$, the equations change to

- CSTR:

$$V_R = \frac{F_{A,0}(X_{A,\text{out}} - X_{A,\text{in}})}{-\sum_j \nu_{Aj} r_j} \tag{2.52}$$

- PFR:

$$V_R = F_{A,0} \int_{X_{A,\text{in}}}^{X_{A,\text{out}}} \frac{dX_A}{-\sum_j \nu_{Aj} r_j} \tag{2.53}$$

These equations are particularly useful for graphical reactor design, when the plots of $1/(-r_j)$ *versus* $X_A$ are given.

### 2.3.2.2 Design of a Batch Reactor

As seen above, for a batch reactor the design equation estimates only the time required for a reaction to proceed to a desired conversion level. If we want to design a batch reactor, *i.e.* estimate its volume, we have to consider other factors such as production rate, as demonstrated in the following example.

*Example*:
Chemical B is to be manufactured in a batch reactor at an average rate of 5 kmol h$^{-1}$ through the liquid phase isomerisation reaction A $\rightarrow$ B. At the end of each run, it takes 5 h to empty the reactor, carry out any repairs and maintenance work and refill the reactor. For a working year of 7000 h, determine the volume of the reactor and the number of batches.

*Data*:
Reaction rate constant of the first-order reaction: $k = 0.1$ h$^{-1}$; concentration of A in solution: $C_{A0} = 1$ kmol m$^{-3}$; final conversion of A: $X_A = 90\%$.

*Solution*:
Applying the design equation for a batch reactor with $r_A = kC_A = kC_{A0}(1 - X_A)$, $\nu_A = -1$, we estimate the reaction time needed to

convert 90% of the reactant:

$$t = \int_0^{0.9} \frac{C_{A0} dX_A}{k C_{A0}(1 - X_A)} = -\frac{\ln 0.1}{k} = 23 \text{ h}$$

What we have found so far is the reaction time, the time needed for 90% of A to be converted to B. We have to add to this the time needed for emptying the reactor, repairing/maintaining and refilling it, what we will call operation time (5 h), in order to find the total batch cycle time: 28 h.

Having calculated the total batch cycle, 28 h, we can estimate how much B we should produce with one batch during these 28 h. This is equal to the average production rate multiplied by the batch time: $N_B = 5$ kmol h$^{-1} \times 28$ h $= 140$ kmol of B. We know on the other hand that this amount is equal to the amount of A converted, $1:1$ stoichiometry: $N_B = N_{A0}(1 - X_A) = (1 - X_A)C_{A0}V_R$. Hence:

$$V_R = \frac{N_B}{(1 - X_A)C_{A0}} = 156 \text{ m}^3$$

Obviously, the number of batch cycles per working year is 7000 h/ 28 h $= 250$ batches per year.

### 2.3.2.3  *Residence Time of Continuous Flow Reactors*

The definition of the residence time of a continuous flow reactor is simply the ratio of its volume to its volumetric flow rate: $\tau = (V_R/\dot{V})$. It is the time that every reaction component element, *i.e.* every molecule, stays in the reactor.

In order to be able to express the equations in terms of residence time rather than reactor volume, the assumption is made that the volumetric flow rates of the inlet and outlet streams are the same. Dividing then both sides of the design eqn (2.45), (2.46) and (2.50), (2.51), by the total volumetric flow rate, we get

- CSTR:

$$\tau = \frac{C_{i,\text{in}} - C_{i,\text{out}}}{-\sum_j v_{ij} r_j} \quad \text{or} \quad \tau = \frac{C_{A,\text{in}} X_A}{-\sum_j v_{Aj} r_j} \tag{2.54}$$

- PFR:

$$\tau = \int_{C_{i,\text{in}}}^{C_{i,\text{out}}} \frac{dC_i}{\sum_j v_{ij} r_j} \quad \text{or} \quad \tau = C_{A,\text{in}} \int_0^{X_A} \frac{dX_A}{-\sum_j v_{Aj} r_j} \tag{2.55}$$

It is obvious that, for continuous flow reactors, designing the reactor means estimating its residence time rather than volume. Continuous flow reactors of the same type with different volumes but the same residence time will give the same conversion.

### 2.3.3 Comparison between Continuous Stirred Tank Reactor and Plug Flow Reactor

In order to compare the performance of the two types of continuous flow reactors, let us consider an example of a first-order reaction of a reactant A into products $(v_A = -1)$ with a reaction rate constant at the reaction temperature of $k = 0.5$ min$^{-1}$. In order to estimate the residence time of a CSTR needed for a conversion, for example, equal to $X_A = 80\%$, we only have to apply eqn (2.54), where $r = kC_{A,out}$ and $C_{A,out} = (1 - X_A)C_{A,in}$.

Thus,

$$\tau_{CSTR} = \frac{X_A}{k(1 - X_A)} = 8 \text{ min}$$

Let us now see how much conversion can be achieved with a PFR of the same residence time.

From the mole balance for A in the PFR we have:

$$\frac{dC_A}{d\tau} = -kC_A \quad \text{or} \quad \frac{dX_A}{d\tau} = k(1 - X_A)$$

whose solution is $X_A = 1 - \exp(-k\tau_{CSTR}) = 98\%$.

Obviously, in a PFR much more reactant is converted than in a CSTR of the same size. What is the reason for this? As the reaction fluid enters a CSTR, the reactant concentration jumps from its inlet value to the outlet value. In a CSTR all the reaction takes place at the same low concentration value across the whole reactor. Hence, the reaction rate has overall in the reactor a relatively low value achieving a relatively low conversion. In a PFR the reactant concentration decreases gradually (Figure 2.2). At and near the inlet it is relatively high, resulting in a high value of reaction rate at that part of the reactor. With the reaction being faster, the overall conversion is higher. This happens where we have positive kinetics, *i.e.* kinetics where the reaction rate depends positively on the reactant concentration; the reaction rate increases as the concentration increases, as occurs with first-order kinetics and, generally, positive order kinetics.

The above example brings us to another point that seems a little odd. In a CSTR, how does the reactant concentration jump from its

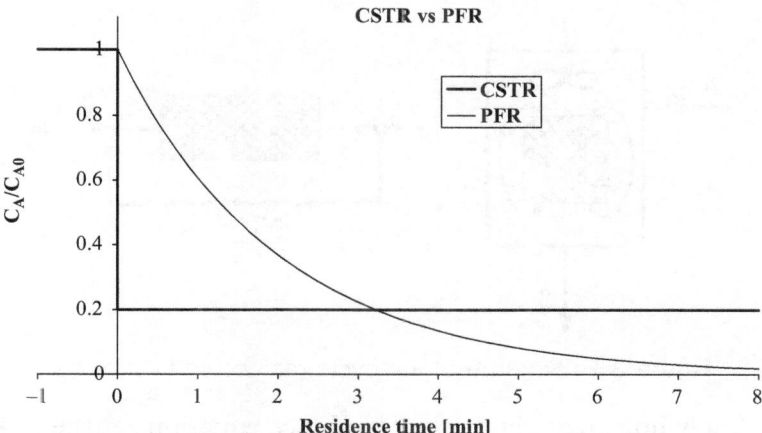

**Figure 2.2** Comparison of reactant exit concentration of a CSTR and PFR of the same residence time.

inlet value to its outlet value? In order to explain this apparent discontinuity, we have to consider what happens inside a CSTR rather than considering it as a 'black box'. We know that a CSTR is a thoroughly stirred reactor. This means that inside this reactor a high flow rate stream, many times more than the incoming flow rate, recirculates continuously. With such a high internal recirculation stream, the incoming stream looks like a 'drop in the ocean'. A drop in the ocean does not affect the concentration of the ocean water. A red drop is not going to change the shade of the pink water inside the CSTR. The internal concentration of a CSTR where the reaction mixture is vigorously recirculated is not going to take notice of the inlet concentration. Actually, we can use a model to represent a CSTR consisting of a tubular reactor, PFR, with a high recycle stream. Such a system is going to be considered in the next section.

### 2.3.4 Recycle Reactor

A recycle reactor, *i.e.* a tubular reactor with a recycle stream (Figure 2.3) could be a real system, not just a model for a CSTR.

The tubular reactor will be modelled like a PFR and for simplicity we are going to consider no change in volumetric flow rate due to reaction; in other words, the inlet volumetric flow rate and the outlet volumetric flow rate are considered the same, $\dot{V}$. Again we are going to consider first-order kinetics. We are going to represent the recycle volumetric flow by $(R\dot{V})$, where $R =$ (recycle volumetric flow/outlet volumetric flow) and is called "recycle ratio".

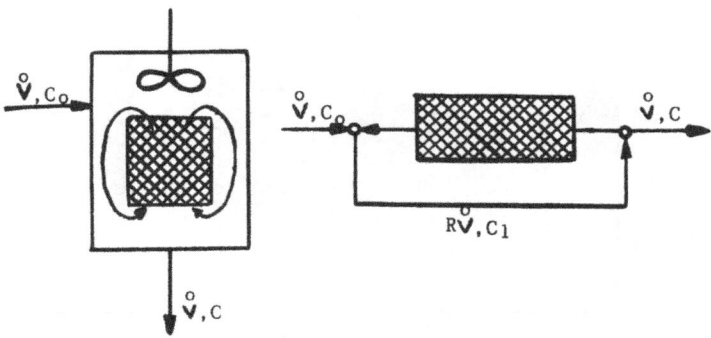

**Figure 2.3** Scheme of a recycle reactor and its model.

We know how to apply the mole balance equation for the PFR, but we should be careful. The inlet concentration for the PFR is not the same as the inlet concentration for the whole system, the recycle reactor, since the inlet stream is mixed with the recycle stream before the combined stream enters the PFR. It would probably be better to write down the balance equation around the mixing point:

$$\dot{V}C_{A0} + R\dot{V}C_A = (R+1)\dot{V}C_{A1} \quad \text{or} \quad C_{A1} = \frac{C_{A0} + RC_A}{R+1} \tag{2.56}$$

The mole balance for PFR, taking into account that the volumetric flow rate through the PFR is $(R+1)\dot{V}$, can be written as

$$(R+1)\dot{V}\frac{dC_A}{dV_R} = -kC_A \quad \text{or} \quad (R+1)\frac{dC_A}{d\tau} = -kC_A \tag{2.57}$$

where $\tau = (V_R/\dot{V})$ is the residence time of the whole recycle reactor system, not the residence time of the PFR alone, on its own.
     Hence

$$\frac{dC_A}{C_A} = -\frac{k}{R+1}d\tau \quad \text{or} \quad \frac{C_A}{C_{A1}} = \exp\left(-\frac{k\tau}{R+1}\right) \tag{2.58}$$

Going back to the mixing eqn (2.57), we can replace $C_{A1}$ from the above equation

$$C_A\exp\left(\frac{k\tau}{R+1}\right) = \frac{C_{A0} + RC_A}{(R+1)} \tag{2.58*}$$

Solving for $C_A$, we get the following equation:

$$C_A = \frac{C_{A0}}{(R+1)\left[\exp\left(\dfrac{k\tau}{R+1}\right) - R\right]} \quad \text{or}$$

$$\frac{C_A}{C_{A0}} = \frac{1}{(R+1)\left[\exp\left(\dfrac{k\tau}{R+1}\right) - R\right]} \tag{2.59}$$

When the recycle ratio $R$ becomes very large (theoretically infinite) the above expression becomes

$$\frac{C_A}{C_{A,0}} = \frac{1}{1 + k\tau}$$

which is the expression for a CSTR.

Obviously the other extreme case is when $R = 0$, when the above formula becomes identical with that of the PFR: $(C_A/C_{A0}) = \exp(-k\tau)$.

## 2.3.5 CSTRs in Series

We consider now a system of $n$ CSTRs in series (Figure 2.4). The outcoming stream from a CSTR is the incoming stream of the next one. For simplicity, we are going to assume the same volumetric flow rate and we are going to examine the reactor system for a reaction of first-order kinetics.

Let us write down the mole balance for reactant A for each reactor $i$:

$$0 = \dot{V}C_{i-1} - \dot{V}C_i - kC_iV_i \quad \text{or} \quad \frac{C_i}{C_{i-1}} = \frac{1}{1 + k\tau_i} \tag{2.60}$$

where $\tau_i = (V_i/\dot{V})$ is the residence time of the $i$-th reactor.

Multiplying each of these equations for $i = 1, 2, \ldots, n$ results in a correlation between the outlet concentration of the system, *i.e.* outlet concentration of the $n$-th reactor, and inlet concentration of the system $C_0$, as well as an expression for the overall conversion:

$$\frac{C_n}{C_0} = \frac{1}{(1 + k\tau_1)(1 + k\tau_2)(1 + k\tau_3) \ldots (1 + k\tau_n)} \tag{2.61}$$

$$X_A = 1 - \frac{1}{(1 + k\tau_1)(1 + k\tau_2)(1 + k\tau_3) \ldots (1 + k\tau_n)} \tag{2.62}$$

**Figure 2.4** CSTRs in series.

In the case of equal volumes $V_1 = V_2 = V_3 = \cdots = V_n = V$ and, hence, equal residence times $\tau_1 = \tau_2 = \tau_3 = \cdots = \tau_n = \tau = (V/\dot{V})$, the above equations are simplified to

$$\frac{C_n}{C_0} = \frac{1}{(1 + k\tau)^n} \tag{2.63}$$

and

$$X_A = 1 - \frac{1}{(1 + k\tau)^n} \tag{2.64}$$

Let us now apply the above equation in order to calculate the conversion of a varying number of CSTRs in series with total volume $V_{tot} = 0.6$ m$^3$ and volumetric flow rate $\dot{V} = 0.1$ m$^3$ h$^{-1}$ for a first-order reaction with reaction rate constant $k = 0.25$ h$^{-1}$. The results are presented in Table 2.1.

It is obvious from the table that the conversion of the series of CSTRs increases with the number of CSTRs (total volume is constant),

**Table 2.1**  Conversion of a system of CSTRs in series (total volume $= 0.6$ m$^3$, volumetric flow rate $= 0.1$ m$^3$ h$^{-1}$) for a first-order irreversible reaction with kinetic constant equal to 0.25 h$^{-1}$.

| Number of CSTRs | Volume of one reactor (m$^3$) | Residence time (h) | Conversion |
|---|---|---|---|
| 1 | 0.600 | 6.000 | 0.600 |
| 2 | 0.300 | 3.000 | 0.673 |
| 3 | 0.200 | 2.000 | 0.704 |
| 4 | 0.150 | 1.500 | 0.720 |
| 5 | 0.120 | 1.200 | 0.731 |
| 6 | 0.100 | 1.000 | 0.738 |
| 7 | 0.086 | 0.857 | 0.743 |
| 8 | 0.075 | 0.750 | 0.747 |
| 9 | 0.067 | 0.667 | 0.750 |
| 10 | 0.060 | 0.600 | 0.753 |
| 11 | 0.055 | 0.545 | 0.755 |
| 12 | 0.050 | 0.500 | 0.757 |
| 13 | 0.046 | 0.462 | 0.758 |
| 14 | 0.043 | 0.429 | 0.759 |
| 15 | 0.040 | 0.400 | 0.761 |
| 16 | 0.038 | 0.375 | 0.762 |
| 17 | 0.035 | 0.353 | 0.762 |
| 18 | 0.033 | 0.333 | 0.763 |
| 19 | 0.032 | 0.316 | 0.764 |
| 20 | 0.030 | 0.300 | 0.765 |
| 100 | 0.006 | 0.060 | 0.774 |
| PFR | 0.6 | 6.000 | 0.777 |

approaching an upper limit that is the value of the conversion of a PFR with the same volume as the total volume of the CSTR system. The PFR itself can be considered as an infinite number of differential homogeneous slices, each one of which behaves like a CSTR. The increase in conversion with the number of CSTRs is at first very rapid, levelling off later and hence, in practice, a number of CSTRs between 10 and 20 has an overall conversion similar to the equivalent total volume PFR. How quickly the conversion of the CSTRs in series approaches that of the PFR depends upon the residence time as well as how fast the reaction is.

What we are going to consider now is to solve the same problem for arbitrary kinetics:

$$r_i = f(C_i) \tag{2.65}$$

Let us write the mole balance equation for the reactant in the $i$-th CSTR:

$$0 = \dot{V}C_{i-1} - \dot{V}C_i - r_i V_i$$

Solving for $r_i$ we get

$$r_i = \left(\frac{\dot{V}}{V_i}\right)C_{i-1} - \left(\frac{\dot{V}}{V_i}\right)C_i \tag{2.66}$$

$C_i$ is the solution of the system of eqn (2.65) and (2.66). Graphically this means that $C_i$ is the intercept of the kinetics curve with the straight line of gradient $(\dot{V}/V_i)$ and the $X$-intercept is the entrance concentration of the $i$-th CSTR, *i.e.* the exit concentration of the $(i-1)$th CSTR. The estimation of the concentrations in the CSTRs one after the other is shown in Figure 2.5.

## 2.3.6 Multiple Reactions

So far we have considered reactor systems where only one reaction is occurring. However, this is seldom the case. Take as an example the partial oxidation reaction over a selective catalyst. Often this reaction is accompanied by full oxidation reactions of the reactant besides the desired partially oxidised product. Full oxidation of reactant decreases the amount of reactant converted to the desired product, while full oxidation of the desired product converts high value products to low value products or waste. Both side reactions result in the decrease of the overall yield to the desired product.

We are going to consider two basic types of multiple reactions

- Series: A → B → C
- Parallel: $\begin{array}{l} A \to B \\ A \to C \end{array}$

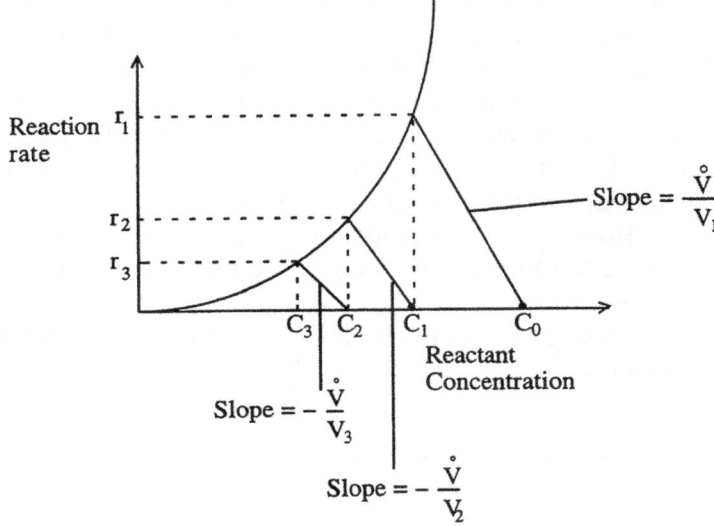

**Figure 2.5**  Solution of the problem of CSTRs in series for arbitrary kinetics.

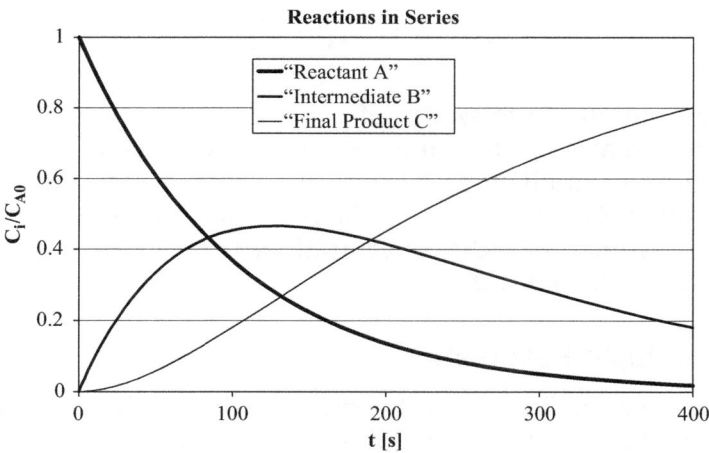

**Figure 2.6**  Concentration profiles of reaction components in a batch reactor for a system of reactions in series.

For a series reaction network the most important variable is either time in batch systems or residence time in continuous flow systems. For the reaction system A → B → C the concentration profiles with respect to time in a batch reactor (or residence time in a PFR) are given in Figure 2.6.

Obviously the reactant concentration decreases with time, exponentially if the reactions are first order, as in the example used in the plot. The concentration of intermediate B will increase in the beginning, as a lot of B is formed from reactant A, whose concentration is still high, but only a little is consumed to C as there is still little B to power this second reaction. As more and more B is formed however, while at the same time less and less A is left, the reaction rate of the second reaction will become greater than that of the first reaction, resulting in a maximum of the intermediate concentration.

The relative position of the maximum, as well as its value, depends on the ratio of the kinetic constants, *i.e.* which of the two reactions is faster and by how much.

In a parallel reaction network of first-order reactions, the selectivity does not depend upon reaction time or residence time, since both products are formed by the same reactant and with the same concentration. The concentration of one of the two products will be higher, but their ratio will be the same during reaction in a batch reactor or at any position in a PFR. The most important parameters for a parallel reaction system are the reaction conditions, such as concentrations and temperature, as well as reactor type. An example is given in the following section.

## 2.3.7 PFR with Continuous Uniform Feed of Reactant Along the Whole Reactor

Consider a product C being produced in a PFR by a second-order reaction:

$$A + B \rightarrow C, \quad r_C = k_1 C_A C_B \tag{2.67}$$

However, an undesired product $A_2$ is also formed by the second-order dimerisation reaction:

$$2A \rightarrow A_2, \quad r_{A_2} = k_2 C_A^2 \tag{2.68}$$

For the maximum selectivity of C compared to $A_2$ ($S_{C/A_2}$), we need to keep the concentration of A as low as possible. To minimise the concentration of A along the reactor, a different feeding arrangement is realised. Only B is fed into the inlet of the PFR, while A is added uniformly along the whole reactor length, see Figure 2.7.

In the following paragraph we are going to formulate the mole balance equations for A and B. In this reactor arrangement we have two feeds; direct feed into the reactor inlet of a solution of B with volumetric flow rate $\dot{V}_B$ and concentration $C_{B0}$, and sidewise feed of a solution of A with concentration $C_{A0}$ and a uniform rate of feed of $f_A$, expressed as the volumetric flow rate of solution A per unit volume of

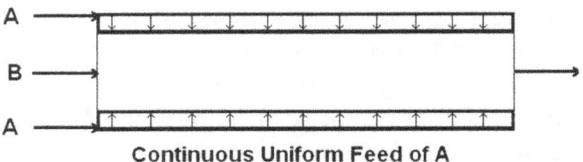

**Continuous Uniform Feed of A**

**Figure 2.7**  Tubular reactor with continuous feed of one of the reactants along the whole reactor.

reactor. The total volumetric flow rate increases along the reactor, since more solution A is added along the whole reactor, and is given by the following formula:

$$\dot{V} = \dot{V}_B + \int_0^V f_A dV = \dot{V}_B + f_A V \tag{2.69}$$

or

$$\frac{d\dot{V}}{dV} = f_A \tag{2.70}$$

The mole balance for A applied to a differential slice of the reactor is as follows. Remember that there are two inlet terms for A, one term expressing A carried forward with the main flow inside the reactor and a second term through the wall: $f_A dV C_{A0}$.

$$0 = (\dot{V}C_A)_V + f_A C_{A0} dV - (\dot{V}C_A)_{V+dV} - (k_1 C_A C_B + 2k_2 C_A^2) dV \tag{2.71}$$

In differential form this equation is

$$\frac{d(\dot{V}C_A)}{dV} = f_A C_{A0} - (k_1 C_A C_B + 2k_2 C_A^2)$$

$$\dot{V}\frac{dC_A}{dV} + f_A C_A = f_A C_{A0} - (k_1 C_A C_B + 2k_2 C_A^2) \tag{2.72}$$

$$(\dot{V}_B + f_A \dot{V})\frac{dC_A}{dV} = -(k_1 C_A C_B + 2k_2 C_A^2) + f_A(C_{A0} - C_A)$$

Mole balance for B:

$$0 = (\dot{V}C_B)_V - (\dot{V}C_B)_{V+dV} - k_1 C_A C_B dV$$

$$\dot{V}\frac{dC_B}{dV} + f_A C_A = -k_1 C_A C_B \tag{2.73}$$

$$(\dot{V}_B + f_A \dot{V})\frac{dC_B}{dV} = -k_1 C_A C_B - f_A C_A$$

In order to design this reactor, eqn (2.72) and (2.73) must be solved simultaneously. However, their solution is outside the scope of these notes.

## Recommended Reading

1. S. Fogler, *Elements of Chemical Reaction Engineering*, Prentice Hall, Englewood Cliffs, 3rd edn, 1998.
2. O. Levenspiel, *Chemical Reaction Engineering*, Wiley, New York, 3rd edn, 1998.
3. J. M. Smith, *Chemical Engineering Kinetics*, McGraw Hill, New York, 3rd edn, 1981.
4. C. Hill, *An Introduction to Chemical Engineering Kinetics and Reactor Design*, Wiley, New York, 1977.
5. G. F. Froment and K. B. Bischoff, *Chemical Reactor Analysis and Design*, Wiley, New York, 2nd edn, 1990.

# 3 Concepts of Fluid Flow

Tim Elson

## 3.1 Introduction

Fluid flow, also known as fluid mechanics, momentum transfer and momentum transport, is a wide-ranging subject, fundamental to many aspects of chemical engineering. It is impossible to cover the whole field here, so we will concentrate upon a few aspects of the subject.

The objective of this chapter is to introduce you to some of the fluid flow problems encountered in the process design of a typical pipe-flow system, illustrated in Figure 3.1, where a fluid is pumped from one vessel into another through a section of pipework.

In particular we will be looking at the following:

- modes of fluid flow;
- pipe flow;
- pressure drops in pipes and pipe fittings;
- flow meters;
- pumps.

Some other types of flow situation are also mentioned. Sample calculations are given in Section 3.10, illustrating the use of some of the principles presented here.

There are many standard texts of fluid flow, *e.g.* Coulson & Richardson,[1] Kay and Neddermann[2] and Massey.[3] Perry[4,5] is also a useful reference source of methods and data. Schaschke[6] presents a

Concepts of Chemical Engineering for Chemists: 2nd Edition
Edited by Stefaan Simons
© The Royal Society of Chemistry 2017
Published by the Royal Society of Chemistry, www.rsc.org

**Figure 3.1** Pipe-flow system.

large number of useful worked examples in fluid mechanics. In many recent texts, fluid mechanics or momentum transfer has been treated in parallel with the two other transport or transfer processes, heat and mass transfer. The classic text here is Bird, Stewart and Lightfoot.[7]

## 3.2  Dimensionless Groups

Chemical engineering in general, and fluid flow in particular, utilises many dimensionless groups, which are discussed in more detail in Chapter 6 "Scale-up in Chemical Engineering". Since we will use a piping system as an example in this chapter, we will now consider the pertinent dimensionless groups for pipe flow.

### 3.2.1  Example: Pipe Flow

Neglecting end effects, *i.e.* $l \gg d$, the pressure drop per unit length of pipe due to friction $(p_1 - p_2)/l = \Delta p_f/l$, will be a function of the pipe diameter $d$, the surface roughness $e$, average flow velocity $\bar{u}$ across the cross-section, and fluid density $\rho$ and viscosity $\mu$ (Figure 3.2).
Thus

$$\frac{\Delta p_f}{l} = f(d, e, \bar{u}, \rho, \mu) \tag{3.1}$$

These six variables (considering $\Delta p_f/l$ as a single variable), $m = 6$, may be expressed in terms of $n = 3$ fundamental dimensions, *i.e.* mass, length and time. Using Buckingham's $\Pi$ theorem, these six

**Figure 3.2** Flow of a fluid in a pipe.

variables may be rearranged into $m - n = 6 - 3 = 3$ new dimensionless variables:

$$\frac{\Delta p_f}{l} \frac{d}{\rho \bar{u}^2} = f\left(\frac{\rho \bar{u} d}{\mu}, \frac{e}{d}\right) \tag{3.2}$$

The procedure for obtaining these dimensionless groups is illustrated in Chapter 6 and in many chemical engineering textbooks.[1,3]

We will return to these dimensionless groups later in Section 3.6, where we will see experimental results of pressure-drop measurements in pipes shown in dimensionless form (Figure 3.7), and go on to use them to calculate pressure drops in the pipework, Section 3.10.

## 3.3 Viscosity

### 3.3.1 Newton's Law of Viscosity

Let us consider a fluid of viscosity $\mu$ held between two large flat parallel plates at distance $Y$ apart, as illustrated in Figure 3.3.

The upper plate is kept in motion in the $x$-direction at a velocity $U$ by applying a force $F$ to the plate in the $x$-direction. The lower plate is kept stationary. The force per unit area required to maintain the upper plate (of area $A$) in motion is given by

$$\frac{F}{A} = \mu \frac{U}{Y} \tag{3.3}$$

or, in differential form

$$\tau = -\mu \frac{\mathrm{d}u}{\mathrm{d}y} \tag{3.4}$$

where $\tau$ is the shear stress, the shear force per unit area, and $\mathrm{d}u/\mathrm{d}y$ is the velocity gradient, also called the *rate of shear* or *shear rate*. Eqn (3.4) is known as *Newton's Law of Viscosity*.

Viscosity, $\mu$, is a property of the fluid and is a measure of its ability to transfer shear stresses. Note that $\mu \, [=] \, \mathrm{M} \, \mathrm{L}^{-1} \, \mathrm{T}^{-1}$, Pa s, N s m$^{-2}$.

**Figure 3.3** Flow between parallel plates.

### 3.3.2 Dynamic and Kinematic Viscosity

The viscosity $\mu$ referred to above is more correctly called the *dynamic viscosity*, and should not be confused with $v$, the *kinematic viscosity*. The kinematic viscosity is the ratio of dynamic viscosity to fluid density, $v = \mu/\rho$ [=] $L^2 T^{-1}$, $m^2 s^{-1}$.

### 3.3.3 Typical Values of Viscosity

#### 3.3.3.1 Liquids

The viscosity of different liquids varies widely.[8] Their viscosity decreases with increasing temperature. Generally, liquid viscosity is very sensitive to temperature. For non-polar and non-associating liquids, viscosity follows Arrhenius type behaviour, $\mu = \mu_0 \exp(-E_\mu/RT)$.

Water at 20 °C has a dynamic viscosity, $\mu$, of 0.001 Pa s = 1 mPa s = 1 cP [centiPoise). The Poise (P) is the c.g.s. unit of dynamic viscosity. The kinematic viscosity, $v$, of water at 20 °C is $10^{-6}$ $m^2$ $s^{-1}$ = 1 cS (centiStoke). The Stoke (S) is the c.g.s. unit of kinematic viscosity. Viscosity data are often quoted in units of cP, P, cS and S in the literature.

Typical values of viscosities of liquids and gases are given in Table 3.1. It can be seen that viscosities of liquids vary widely over many orders of magnitude.

#### 3.3.3.2 Non-Newtonian Liquids

Newtonian liquids are those that obey Newton's law of viscosity, *i.e.* viscosity does not vary with rate of shear $\mu \neq f(du/dy)$. Many liquids, especially those containing significant quantities of particles or high molecular weight polymers, have viscosities that vary with rate of shear, $\mu = f(du/dy)$ (at constant $T$ and $p$), and so do not obey Newton's law. These liquids are called *non-Newtonian*. Common examples are

**Table 3.1** Typical values of viscosity of some gases and liquids at 20 °C, 1 atm.

| Substance | Dynamic viscosity (Pa s) | Kinematic viscosity ($m^2$ $s^{-1}$) |
|---|---|---|
| Gases | | |
| Hydrogen | 0.0000088 | 0.00010 |
| Air | 0.000018 | 0.000015 |
| | | |
| Liquids | | |
| Benzene | 0.00064 | 0.00000073 |
| Water | 0.001 | 0.000001 |
| Mercury | 0.0016 | 0.00000012 |
| Ethylene glycol | 0.021 | 0.0000019 |
| Olive oil | 0.1 | 0.0001 |
| Castor oil | 0.6 | 0.0006 |
| Glycerine | 1.4 | 0.0011 |
| Ball-point pen ink | ~11 | ~0.01 |
| Corn syrup | ~100 | ~0.07 |
| Bitumen | ~1000 | ~1 |

toothpaste, drilling muds, non-drip paints, wallpaper paste, egg white, polymer solutions and melts, silicone rubbers, cement slurries and printing inks. Some non-Newtonian fluids, especially polymer solutions and melts, and silicone rubbers, have elastic as well as viscous properties and these are called *viscoelastic*. The behaviour of non-Newtonian fluids is beyond the scope of this chapter, but there are many textbooks on the subject, for example, Chhabra and Richardson.[9]

### 3.3.3.3 Gases

At moderate pressures, gas viscosity increases with increasing temperature, $\mu \propto T^{0.7-1.0}$, and is relatively independent of pressure.

Air at STP, $\mu \approx 1.7 \times 10^{-5}$ Pa s, $\nu \approx 1.4 \times 10^{-5}$ $m^2$ $s^{-1}$.

## 3.4 Laminar and Turbulent Flow

There are two distinct modes of flow, laminar and turbulent. Fluid inertia tends to allow fluctuations to grow and give rise to turbulent eddies. Viscosity on the other hand, tends to damp out these fluctuations. A ratio of forces, inertial to viscous, is used to characterise the nature of the flow and is called the *Reynolds Number*, Re. For pipe flow this takes the form:

$$\mathrm{Re} = \frac{\rho \bar{u} d}{\mu} = \frac{\bar{u} d}{\nu} \tag{3.5}$$

The Reynolds Number being a ratio of like quantities, forces in this case, is dimensionless. At low Re, flows tend to be laminar, and at high Re, flows tend to be turbulent. As we shall see later, the limiting values of Re for laminar and turbulent flow and the transition between these modes of flow is a function of the flow geometry.

Table 3.2 compares the characteristics of laminar and turbulent flow with particular reference to flow in a pipe. Many of the characteristics described also apply to other flow situations.

### 3.4.1  Boundary Layers

Although the bulk of the flow may be turbulent, the presence of an interface or solid boundary causes boundary layers to form. The presence of a solid wall dampens turbulent fluctuations in the fluid layers next to the wall and close to the wall the flow becomes laminar in nature. At the wall, fluid is brought to a halt. This is known as the non-slip boundary condition.

#### 3.4.1.1  Universal Velocity Profile

The Universal Velocity Profile is often used to describe the velocity near a wall. This profile is divided into three sections or layers and is shown in Figure 3.4. Here, dimensionless velocity, $u^+ = \bar{u}\sqrt{\rho/\tau_w}$, is plotted against dimensionless distance from wall, $y^+ = y\sqrt{\tau_w \rho/\mu}$, where $\tau_w$ is the shear stress at the wall.

- *Laminar sub-layer*: $0 < y^+ < 5$, Newton's Law of Viscosity (3.4), describes the flow and gives $u^+ = y^+$; here laminar flow and transfer by molecular diffusion dominate.
- *Buffer zone*: $5 \leq y^+ \leq 30$, $u^+ = 5.0 \ln y^+ - 3.05$; here both laminar and turbulent flow considerations are important; and
- *Turbulent core*: $30 \leq y^+$, $u^+ = 2.5 \ln y^+ + 5.5$; here turbulent flow and transfer by eddy motion dominate.

Alternatively, a single laminar boundary layer, with $u^+ = y^+$ reaching up to $y^+ \approx 11.6$, is often used, beyond which there is the turbulent core.

**Table 3.2** Comparison of laminar and turbulent flow in a pipe.

| Laminar flow | Turbulent flow |
|---|---|
| • Laminar flows predominantly occur in higher viscosity fluids and flows in small diameter vessels | • Turbulent flow is the most common type of flow found industrially |
| • $Re_{pipe} \lesssim 2000$ | • $Re_{pipe} \gtrsim 5000$ |
| • Many flow problems can be solved analytically | • Flows are much less well defined than in laminar flow, and turbulence is an empirically based subject |
| • Laminar flow is characterised by streamlines – stratified layers of fluid – with transport by molecular diffusion (due to viscosity) between them | • Superimposed upon an average flow in the x-direction is a random three-dimensional flow caused by turbulent eddies. Eddies dominate the transport processes |
| • A coloured dye trace injected into pipe shows a clearly defined thread extending over the whole length of the pipe | • The dye trace thread is broken down and becomes mixed with the fluid, eventually colouring the fluid over the whole width of the pipe |
| • Viscous forces dominate | • Inertial forces dominate |
| • Frictional losses $\propto \mu\, u$ | • Frictional losses $\propto \mu\, u^2$ |
| • Laminar velocity profile | • Turbulent velocity profile |

$$\frac{u(r)}{u_{max}} = 1 - \left(\frac{r}{r_0}\right)^2$$

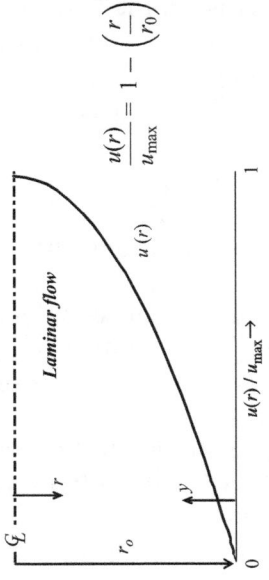

$$\frac{\tilde{u}(r)}{\tilde{u}_{max}} = \left(\frac{y}{r_0}\right)^{1/n} = \left(1 - \frac{r}{r_0}\right)^{1/n}$$

where $\tilde{u}$ is the mean velocity at a point $(r)$, $n = f(Re)$, experimentally $6 < n < 10$. At $Re = 10^5$, $n = 7$ "prandtl 1/7th power law"

• Mean velocity in turbulent flow:

$$\bar{u} = \frac{Q_v}{A} \approx 0.82\, \tilde{u}_{max}$$

• Mean velocity in laminar flow:

$$\bar{u} = \frac{Q_v}{A} = 0.5\, u_{max}$$

**Figure 3.4** Boundary layers and the universal velocity profile.

**Table 3.3** The conservation equations.

| Equation | Conserved property | Conservation law |
|---|---|---|
| Continuity | Mass | Conservation of mass |
| Momentum | Momentum | Newton's 2nd law |
| Energy | Energy | 1st law of thermodynamics |

## 3.5 Balance or Conservation Equations

### 3.5.1 General Form of the Conservation Equations

Three balance equations are used in solving fluid flow problems. Each is a conservation equation, conserving mass, momentum and energy respectively. They are listed in Table 3.3, with the respective conserved property and conservation law.

The general form of the conservation equations is:

$$
\begin{matrix}
\text{Rate of} \\
\text{input} \\
\text{of property}
\end{matrix}
\;-\;
\begin{matrix}
\text{Rate of} \\
\text{output} \\
\text{of property}
\end{matrix}
\;+\;
\begin{matrix}
\text{Rate of} \\
\text{generation} \\
\text{of property}
\end{matrix}
\;=\;
\begin{matrix}
\text{Rate of} \\
\text{accumulation} \\
\text{of property}
\end{matrix}
\quad (3.6)
$$

Note that the conservation equations are rate equations. In the examples below we consider only the steady-state forms of the conservation equations. Thus the accumulation terms will be zero. In most cases we will just present the final forms of the equations.

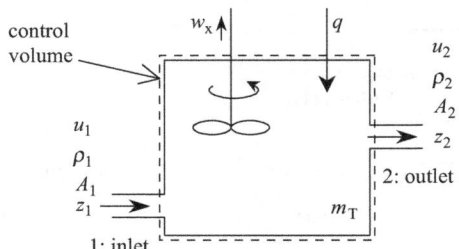

**Figure 3.5** Generalised control volume.

## 3.5.2 Control Volumes

These conservation equations are applied to control volumes. Figure 3.5 shows a generalised control volume, with one inlet (subscript 1) and one outlet (subscript 2), which we will use.

Here $A$ represents the cross-sectional area of the inlet and outlet, $z$ their elevations, $u$ the fluid's mean velocity and $\rho$ its density, $m_T$ is the total mass within the control volume, $q$ rate of heat input to the control volume per unit mass and $w_x$ shaft work output from the control volume per unit mass. Note this is the *Engineers' convention* for shaft work. In the *Chemists' convention*, positive shaft work is an input to the control volume, *i.e.* the direction of positive shaft work is opposite.

## 3.5.3 Continuity Equation (Mass Balance)

Considering total mass and steady state:

| Rate of input of mass | | Rate of output of mass | | Rate of generation of mass | | Rate of accumulation of mass | |
|---|---|---|---|---|---|---|---|
| Rate of input of mass | $-$ | Rate of output of mass | $+$ | Rate of generation of mass | $=$ | Rate of accumulation of mass | (3.7) |
| $G_1$ | $-$ | $G_2$ | $+$ | 0 | $=$ | 0 | |

or

$$G_1 = G_2 \tag{3.8}$$

*i.e.*

$$\rho_1 u_1 A_1 = \rho_2 u_2 A_2 \tag{3.9}$$

If the fluid is incompressible, *e.g.* a liquid, $\rho_1 = \rho_2 = \rho$, then

$$u_1 A_1 = u_2 A_2, \quad \text{or} \quad Q_{v1} = Q_{v2} \tag{3.10}$$

Thus, at steady state the overall mass flow rate in equals the overall mass flow rate out and if the fluid is incompressible, the volumetric flow rate in equals the volumetric flow rate out.

If one is considering a chemical reaction and performing a component mass balance, the generation term of the components taking part in the reaction will not be zero, but related to the rate of reaction. However, if one considers total mass, the generation term will be zero unless the reaction is nuclear.

[See *Derivation of head loss in a sudden expansion* later in this section and the *Venturi Meter* in Section 3.7 for examples of the use of the Continuity Equation.]

### 3.5.4 Steady-state Momentum (Force) Balance Equation

| Rate of input of momentum | − | Rate of output of momentum | + | Rate of generation of momentum | = | Rate of accumulation of momentum |
|:---:|:---:|:---:|:---:|:---:|:---:|:---:|
| $Gu_1$ | − | $Gu_2$ | + | $\Sigma \boldsymbol{F}$ | = | 0 |

$$(3.11)$$

As momentum is a vector, the steady-state momentum equation above is a vector equation and, in general, needs to be applied in each of the three directions $(x, y, z)$ in which momentum (*i.e.* velocity) has a component. The vector parameters in the momentum equations are shown here emboldened.

As the rate of momentum transfer is equal to a force, momentum balances are equivalent to force balances.

The generation term is equal to $\Sigma \boldsymbol{F}$, the sum of forces acting on the control volume, which is comprised of the three following terms:

$p_1 \boldsymbol{A}_1 - p_2 \boldsymbol{A}_2$: the net force exerted on the fluid within the control volume by the pressure of the surrounding fluid at the inlet (point 1) and at the outlet (point 2); $\boldsymbol{A}$ is the cross-sectional area of the inlets and outlets perpendicular to the direction under consideration and so is shown as a vector;

$\boldsymbol{F}$: the net force exerted on the fluid within the control volume by the surrounding/control volume walls, *e.g.* frictional forces;

$m_T \boldsymbol{g}$: the body force exerted on the fluid within the control volume by gravitational acceleration.

Thus

$$G(\boldsymbol{u}_1 - \boldsymbol{u}_2) + (p_1 \boldsymbol{A}_1 - p_2 \boldsymbol{A}_2) + \boldsymbol{F} + m_T \boldsymbol{g} = 0 \qquad (3.12)$$

[See *Derivation of head loss in a sudden expansion* later in this section for an example of the use of the momentum equation.]

## 3.5.5   Steady-state Energy Balances

### 3.5.5.1   *Steady Flow Energy Equation*

$$g(z_2 - z_1) + \left(\frac{u_2^2}{2} - \frac{u_1^2}{2}\right) + (h_2 - h_1) = q - w_x \qquad (3.13)$$

where $h$ is the specific enthalpy of the fluid.

[Units: specific energy, energy per unit mass of fluid flowing [J kg$^{-1}$, m$^2$ s$^{-2}$]. To get rates [J s$^{-1}$, W] simply multiply each term by mass flow rate, $G$.]

The Steady Flow Energy Equation relates the changes in potential and kinetic energy, and enthalpy of streams flowing in and out of a control volume, to the rate of heat transfer to, and the rate of shaft work from, the control volume. Here the generation term is zero. If a chemical reaction were occurring within the control volume, the heat of reaction would need to be accounted for in a generation term. Note that if we are using the Chemists' convention for shaft work, the right-hand side of eqn (3.13) would become $q + w_x$.

### 3.5.5.2   *Mechanical Energy Balance Equation*

Assuming no heat transfer to or from the surroundings ($q = 0$) and that frictional losses end up being dissipated as heat, the Steady Flow Energy Equation for a constant density fluid becomes:

$$\frac{p_2 - p_1}{\rho} + \left(\frac{u_2^2}{2} - \frac{u_1^2}{2}\right) + g(z_2 - z_1) + g\Delta h_f + w_x = 0 \qquad (3.14)$$

and is known as the *Mechanical Energy Balance Equation*.

[Units: energy per unit mass of fluid flowing [J kg$^{-1}$, m$^2$ s$^{-2}$]. To get rates [J s$^{-1}$, W] multiply terms by mass flow rate, $G$.]

Here $\Delta h_f$ is the *head loss due to friction* and is directly related to the pressure loss due to friction $\Delta p_f$ by $\Delta p_f = \rho g \Delta h_f$.

The Mechanical Energy Balance eqn (3.14), is the key equation when calculating pressure changes through piping systems. See Section 3.10 for examples of the use of the Mechanical Energy Balance Equation in such calculations.

### 3.5.5.3   *Bernoulli's Equation*

Bernoulli's Equation is obtained by integrating a momentum (force) balance along a flow streamline assuming steady state, with an in-compressible ($\rho = $ constant) and inviscid ($\mu = 0$) fluid. As we integrate a momentum or force balance along a streamline, we get an energy balance: $\int F dx$ = force × distance moved = work done = energy exerted. Bernoulli's Equation is therefore an energy balance.

Bernoulli's Equation gives us the total energy in a flowing fluid and sums the energy due to fluid pressure, kinetic energy and potential energy.

$$\frac{p}{\rho g} \quad + \quad \frac{u^2}{2g} \quad + \quad z \quad = \quad \text{constant along a streamline}$$

| pressure head | + | velocity head | + | gravity or static head | = | total head of fluid |

(3.15)

Each term on the left-hand side has the same dimensions, length (energy per unit weight of fluid) in this version of Bernoulli's Equation. Each term can be thought of as a *head* of fluid, and the sum of these terms is called the *total head* of fluid.

The use of the term head is simply an extension of the practice of expressing pressures, $p$, as heights or heads of fluid, $h$, where $p = \rho g h$, *e.g.* 1 atm = 760 mm Hg = 406.8 in $H_2O$, *etc.*

[Note: eqn (3.15) is just one of several versions of Bernoulli's Equation where the terms are expressed in different but self-consistent sets of units (energy per unit mass with (3.15) multiplied by $\rho$ or energy per unit volume [pressure] with (3.15) multiplied by $\rho g$).]

Bernoulli's Equation is normally applied between two points, 1 and 2, along a streamline.

$$\frac{p_1}{\rho g} + \frac{u_1^2}{2g} + z_1 = \frac{p_2}{\rho g} + \frac{u_2^2}{2g} + z_2$$

(3.16)

Eqn (3.16) is similar to (3.14), the Mechanical Energy Balance Equation, but with $g\Delta h_f$ and $w_x$ both equal to zero.

[See *Venturi Meter* in Section 3.7 below for an example of the use of Bernoulli's Equation.]

## 3.5.6 Example of use of the Conservation Equations

### 3.5.6.1 *Derivation of Head Loss in a Sudden Expansion*

This example illustrates the use of all three conservation equations to derive an expression for the head loss through a sudden expansion in a pipe or as a pipe enters a large vessel, an *exit loss*, and is illustrated in Figure 3.6.

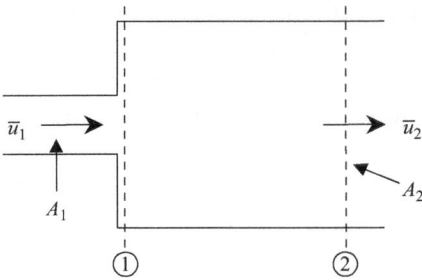

**Figure 3.6**  Flow through a sudden expansion.

In Figure 3.6, Section 1 is immediately downstream of the expansion where radial acceleration of fluid is small so pressure is uniform across the cross-section (and area is $A_2$ not $A_1$). Section 2 is sufficiently far downstream for flow to be parallel and pressure constant across the cross-section.

The Momentum eqn (3.12) is:

$$G(\pmb{u}_1 - \pmb{u}_2) + (p_1 A_1 - p_2 A_2) + \pmb{F} + m_T \pmb{g} = 0$$

Assume that the sudden expansion is horizontal, $z_1 = z_2$ and $\pmb{g} = 0$, and that there is negligible wall friction, $\pmb{F} = 0$, as length is short. In addition, let us assume that the velocity is constant across each cross-section (*i.e.* turbulent flow, flat profile).

$$(p_1 - p_2)A_2 = G(\bar{u}_2 - \bar{u}_1) = \rho Q(\bar{u}_2 - \bar{u}_1) = \rho \bar{u}_2 A_2(\bar{u}_2 - \bar{u}_1)$$

giving

$$\frac{(p_1 - p_2)}{\rho} = \bar{u}_2(\bar{u}_2 - \bar{u}_1) \tag{3.17}$$

The Mechanical Energy Balance eqn (3.14) for a constant density fluid is:

$$\frac{p_2 - p_1}{\rho} + \left(\frac{\bar{u}_2^2}{2} - \frac{\bar{u}_1^2}{2}\right) + g(z_2 - z_1) + g\Delta h_f + w_x = 0$$

or, as $z_1 = z_2$ and $w_x = 0$,

$$\Delta h_f = \frac{1}{g}\left[\frac{p_1 - p_2}{\rho} + \frac{\bar{u}_1^2 - \bar{u}_2^2}{2}\right]$$

Using eqn (3.17)

$$\Delta h_f = \frac{1}{g}\left[\bar{u}_2(\bar{u}_2 - \bar{u}_1) + \frac{\bar{u}_1^2 - \bar{u}_2^2}{2}\right] = \frac{1}{2g}(\bar{u}_1 - \bar{u}_2)^2 \tag{3.18}$$

The continuity eqn (3.19), assuming constant density, $\rho$, gives:

$$\bar{u}_1 A_1 = \bar{u}_2 A_2$$

and eqn (3.18) then becomes:

$$\Delta h_f = \frac{\bar{u}_1^2}{2g}\left(1 - \frac{A_1}{A_2}\right)^2 = \frac{\bar{u}_2^2}{2g}\left(\frac{A_2}{A_1} - 1\right)^2 = K\frac{\bar{u}_1^2}{2g} \tag{3.19}$$

where $K = (1 - (A_1/A_2))^2$, and as $\bar{u}_1^2/2g$ is one velocity head. $K$ is known as the number of velocity heads for the fitting loss.

If $A_2 \gg A_1$, *e.g.* flow from a pipe into a large vessel or reservoir, $K \approx 1$. Thus the exit loss from a pipe into a large vessel is 1 velocity head, *i.e.* all kinetic energy in the flowing fluid is lost when it comes to rest after entering a large vessel. This is the value reported later in Table 3.5 for an Exit Loss. Losses in fittings are described in more detail in the next section.

## 3.6 Pipe Flow

With turbulent flows we normally have to resort to the correlation of experimental data. Friction between fluids and surfaces are normally characterised by friction factors. Different flow situations give rise to different friction factors.

### 3.6.1 Friction Factors

#### 3.6.1.1 Fanning Friction Factor

The Fanning friction factor for pipe flow, $c_f$, is defined by:

$$\tau_w = c_f \frac{1}{2}\rho u^2 \tag{3.20}$$

where $\tau_w$ is the pipe wall shear stress, the frictional shear force per unit area of pipe wall. Applying a momentum (force) balance to a pipe of diameter, $d$, and length, $l$, gives $\Delta p_f \times (\pi d^2/4) = \tau_w \times \pi dl$ and so the relation between Fanning friction factor and pipe pressure drop may be obtained:

$$\Delta p_f = 2c_f \frac{l}{d}\rho u^2 \quad \text{or} \quad c_f = \frac{\Delta p_f}{2\rho u^2}\frac{d}{l} \tag{3.21}$$

Recall that, using dimensional analysis for pipe flow, we saw (see Section 3.2):

$$\frac{\Delta p_f}{l}\frac{d}{\rho \bar{u}^2} = f\left(\frac{\rho \bar{u}d}{\mu}, \frac{e}{d}\right)$$

or,

$$c_f = f\left(\mathrm{Re}, \frac{e}{d}\right) \tag{3.22}$$

### 3.6.1.2  Other Friction Factors

Beware, there are other definitions of friction factor.

- *Blasius or Darcy* friction factor, $\lambda$ or $f$: $\Delta h_f = \lambda$(or $f$)$(l/d)(\bar{u}^2/2g)$, giving $\lambda = f = 4c_f$
- *Coulson and Richardson* friction factor,[1] $f$: $\tau_w = f\rho\bar{u}^2$, giving $f = c_f/2$

The form of the $c_f$ – Re relationship depends upon the flow regime.

### 3.6.1.3  Laminar Flow, Re ≤ 2000

A momentum balance on a differential fluid element yields the laminar velocity profile (see Table 3.2) and

$$\frac{\Delta p_f}{l} = \frac{32\mu\bar{u}}{d^2} \tag{3.23}$$

which is known as the Hagen–Poiseuille Equation, and indicates that for laminar flow,

$$\Delta p_f \propto \mu\bar{u}$$

Rearranging (3.23) yields

$$c_f = \frac{16}{\mathrm{Re}} \tag{3.24}$$

Note: roughness has no effect on laminar flow in pipes.

### 3.6.1.4  Transition, 2000 < Re < 5000

Flow fluctuates between laminar and turbulent. Pressure drops fluctuate and are uncertain. Avoid the transition region.

### 3.6.1.5  Turbulent flow, Re ≥ 5000

In turbulent flow, one has to rely on correlation of experimental data

*Smooth pipes, $e = 0$*
  Blasius:

$$c_f = 0.0791 \, \text{Re}^{-1/4}$$

$$\text{for } 3000 < \text{Re} < 10^5 \tag{3.25}$$

giving

$$\Delta p_f \propto \rho \bar{u}^{1.75}$$

Nikuradse:

$$\frac{1}{\sqrt{c_f}} = 4.0 \, \log_{10}(\text{Re}\sqrt{c_f}) - 0.40$$

$$\text{for } 3000 < \text{Re} < 3 \times 10^6 \tag{3.26}$$

*Rough pipes, $e > 0$*
  Nikuradse:

$$\frac{1}{\sqrt{c_f}} = 4.0 \, \log_{10}\left(\frac{r_0}{e}\right) + 4.6$$

$$\text{for } \left(\frac{r_0}{e}\right)(\text{Re}\sqrt{c_f}) > 0.005, \text{ fully rough pipes} \tag{3.27}$$

Moody:

$$c_f = 0.001375\left[1 + \left(20\,000\frac{e}{d} + \frac{10^6}{\text{Re}}\right)^{1/3}\right]$$

$$\text{for } 3000 < \text{Re} < 10^7 \text{ and } e/d < 0.1 \tag{3.28}$$

giving $\Delta p_f \propto \rho \bar{u}^2$ *for fully rough pipes*
  Typical values of pipe roughness, $e$, are given in Table 3.4.
  Figure 3.7 shows the form of $c_f$ as a function of Re and $e/d$ as predicted by eqn (3.24)–(3.26) and (3.28).

**Table 3.4** Typical values of pipe roughness.[1,4,21,22]

| Pipe | $e$, Absolute roughness (mm) |
|------|------------------------------|
| Drawn tubing | 0.0015 |
| Commercial steel & wrought iron | 0.05 |
| Cast iron | 0.25 |
| Concrete | 0.3–3 |
| Riveted steel | 1–10 |

**Figure 3.7**  Fanning friction factor chart for pipe flow. (■) piping system example (*see Section 3.10*) Re = 363 000, *e/d* = 0.0005, $c_f$ = 0.00459.

With knowledge of $c_f$ from the above correlations and using eqn (3.21) we can calculate frictional pressure drops in straight pipes. However, we also need to be able to evaluate frictional losses in pipe fittings.

### 3.6.2 Losses in Fittings

The flow in pipe fittings, *e.g.* bends and valves, is generally too complex to determine theoretically. For *turbulent flow*, these minor losses are approximately equal to the square of the flow velocity. Thus, we define a loss coefficient, $K$,

$$\Delta p_{loss} = K \frac{\rho \bar{u}^2}{2}, \quad \text{or} \quad \Delta h_{loss} = K \frac{\bar{u}^2}{2g} \tag{3.29}$$

where $\bar{u}^2/2g$ is *one velocity head* (see Section 3.5). So $K$ is known as the *number of velocity heads* for a fitting loss.

These minor fitting losses can also be expressed in terms of an equivalent length of straight horizontal pipe, $l_e$, giving the same frictional loss.

$$\Delta h_{loss} = 4c_f \frac{l_e}{d} \frac{\bar{u}^2}{2g} \equiv K \frac{\bar{u}^2}{2g} \tag{3.30}$$

thus

$$K = 4c_f \frac{l_e}{d} = 4c_f n \tag{3.31}$$

where $n = l_e/d$, the *number of pipe diameters* of straight pipe that gives an equivalent frictional head or pressure loss.

Therefore, the total pressure or head loss due to friction in pipework due to the pipe and fittings is given by

$$\Delta p_{f_T} = 2c_f \frac{(l + l_e)}{d} \rho \bar{u}^2 = 2c_f \frac{l_T}{d} \rho \bar{u}^2$$

$$\text{or} \quad \Delta h_{f_T} = 4c_f \frac{(l + l_e)}{d} \frac{\bar{u}^2}{2g} = 4c_f \frac{l_T}{d} \frac{\bar{u}^2}{2g} \tag{3.32}$$

Table 3.5 shows typical losses in a range of fittings expressed as values of $K$ and $n$.

### 3.6.3 Economic Velocities

Economic velocities for turbulent flow (the most common industrially), taking into account piping capital costs (piping, fittings, pumps, *etc.*) and running costs (pressure drops), are primarily a function of

**Table 3.5**  Typical losses in fittings.[1,4]

| Fitting | $n$, Number of pipe diameters | $K$, Number of velocity heads |
|---|---|---|
| 45° elbows: | | |
| Standard radius | 15 | 0.3 |
| Long radius | 10 | 0.2 |
| | | |
| 90° elbows: | | |
| Standard radius | 35 | 0.7 |
| Long radius | 22 | 0.45 |
| Square or mitre | 60 | 1.2 |
| | | |
| T-piece: | | |
| Entry from leg | 60 | 1.2 |
| Entry into leg | 90 | 1.8 |
| Unions and couplings | Small, 2 | Small, 0.04 |
| | | |
| Globe valve: | | |
| Fully open | 300 | 6 |
| 1/2 open | 450 | 9 |
| | | |
| Gate valve: | | |
| Fully open | 7 | 0.15 |
| 3/4 open | 40 | 1 |
| 1/2 open | 200 | 4 |
| 1/4 open | 800 | 16 |
| | | |
| Check valve: | | |
| Swing | 100 | 2 |
| Disk | 500 | 10 |
| Ball | 3500 | 70 |
| Entrance loss | 25 | 0.5 |
| Exit loss | 50 | 1 |
| Orifice meter | $\Delta p_{loss} \sim 60\%$ of $(p_1 - p_2)$ | |
| Venturi meter | $\Delta p_{loss} \sim 10\%$ of $(p_1 - p_2)$ | |

fluid density $[\Delta p_{f_T} \propto \rho \bar{u}^2]$. Details are given in the literature.[4] Order of magnitude economic velocities are:

- for low viscosity liquids, $\bar{u}_{economic} \approx 1 \text{ m s}^{-1}$
- for gases, $\bar{u}_{economic} \approx 10 \text{ m s}^{-1}$

# 3.7 Flow Measurement

The details of design and construction of various flow meters are shown in many standard chemical engineering and fluid mechanics texts. Some reviews and advice in choosing flowmeters for particular duties are available in the literature.[10–12]

## 3.7.1 Variable Head Meters[13,14]

### 3.7.1.1 *Venturi Meter*

Figure 3.8 illustrates a Venturi meter. Fluid is accelerated through a gradual contraction, half angle approximately 20°, through a throat, and then decelerated though an even more gradual expansion, approximately 7°, for good pressure recovery.

The Continuity and Bernoulli Equations may be used to derive equations relating flow rate to measured pressure difference for the Venturi meter (and the orifice plate meter discussed below).

Applying Bernoulli's eqn (3.16), along a streamline between points 1 and 2 and assuming the meter is horizontal, $z_1 = z_2$, and that the velocities are constant across a cross-section, gives:

$$\frac{p_1 - p_2}{\rho g} = \frac{\bar{u}_2^2 - \bar{u}_1^2}{2g} \tag{3.33}$$

Continuity, assuming constant density $\rho$, eqn (3.10), gives $\bar{u}_2 = \bar{u}_1 (A_1/A_2)$. Substituting for $\bar{u}_2$ into eqn (3.33) and rearranging gives:

$$\bar{u}_1 = \left[ \frac{2}{\rho} \frac{(p_1 - p_2)}{(A_1/A_2)^2 - 1} \right]^{1/2} \tag{3.34}$$

**Figure 3.8** Venturi meter.

or

$$G = \rho \bar{u}_1 A_1 = A_1 \left[ \frac{2\rho(p_1 - p_2)}{(A_1/A_2)^2 - 1} \right]^{1/2} \tag{3.35}$$

Adding $C_d$, a *discharge coefficient*, to account for frictional losses (assumed zero in Bernoulli's Equation) and other non-idealities, we get eqn (3.36), the operating equation for the Venturi meter.

$$\bar{u}_1 = C_d \left[ \frac{2}{\rho} \frac{(p_1 - p_2)}{(A_1/A_2)^2 - 1} \right]^{1/2} \tag{3.36}$$

or

$$G = C_d \rho \bar{u}_1 A_1 = C_d A_1 \left[ \frac{2\rho(p_1 - p_2)}{(A_1/A_2)^2 - 1} \right]^{1/2} \tag{3.37}$$

$C_d$ for a well-designed Venturi meter is normally in the range 0.97–0.99. The appropriate British Standards[13,14] give correlations of $C_d$ values for standard geometry meters.

### 3.7.1.2  Orifice Plate Meter

Figure 3.9 illustrates an orifice plate meter. Here fluid is suddenly accelerated as it passes through an orifice in a plate, which is normally held between two flanges. Pressure is measured upstream (point 1) and downstream (point 2) of the orifice. There are a number of standard positions for the pressure tapings, "*D* & *D*/2", *i.e.* one pipe diameter upstream and one half diameter downstream of the orifice, or "flange tapings", or "corner tapings".

Eqn (3.37) is also used for the orifice plate meter. The values of $C_d$ are lower for orifice plate meters. They can vary significantly with flow rate and are a function of tapping location, as the measured pressure difference is a function of tapping position (see the pressure profile illustrated in Figure 3.9). For well-designed meters operating in their optimum range, $C_d$ values typically range between 0.6 and 0.7. The appropriate British Standards[13,14] give correlations of $C_d$ values for standard geometry meters.

## 3.7.2  Variable Area Meters

### 3.7.2.1  Rotameter

The rotameter consists of a vertical tapered tube, the inside diameter of which increases with height, and is normally made of glass. Inside

**Figure 3.9** Orifice plate meter showing various tapping locations and pressure profile.

the tube is a spherical or cylindrical "float". At a steady flow rate the float settles at a height where the weight of the float is balanced by the drag and buoyancy forces acting on it. As the flow rate increases, the float rises in the tapered tube (so giving a greater annular area for flow between the float and the tube walls) until the weight of the float is again balanced by the drag and buoyancy forces. The height of the float is related to the flow rate.

### 3.7.2.2   Variable Area Orifices

Variable area orifices consist of an orifice and a concentric obstruction. Fluid flows through the orifice around the concentric obstruction. As the flow rate increases, the area between the orifice and the obstruction changes, either by displacement of a spring loaded obstruction relative to a fixed orifice or by displacement of a spring loaded orifice relative to a fixed obstruction. The displacement is related to the flow rate.

## 3.7.3   Some Other Flowmeters

### 3.7.3.1   Vortex Shedding Flowmeter

At Reynolds numbers greater than $\sim 1000$, vortices are shed behind bluff bodies giving an oscillating wake. The rate of oscillation is proportional to the flow rate and is measured by the meter and the flow rate determined. See Chapter 7 "An Introduction to Particle Systems" for further information on oscillating wakes.

### 3.7.3.2  Magnetic Flowmeter

When a conductor moves though a magnetic field, an electromagnetic force (emf), is generated. The magnetic flowmeter uses this principle to measure flow rates in electrically conducting liquids. A magnetic field is applied across a section of non-conducting pipe and, if the liquid is electrically conducting, an emf will be generated that is proportional to liquid velocity. The emf is picked up by electrodes in the pipe wall and converted into a flow rate.

# 3.8  Pumps and Pumping

## 3.8.1  Positive Displacement Pumps

In positive displacement pumps, a discrete "parcel" of fluid is taken from the pump inlet to the pump outlet where it is discharged. This continues even if there is substantial resistance to flow at the outlet. Positive displacement pumps may develop very high pressures and pressure relief devices should be fitted.

### 3.8.1.1  Reciprocating Pumps

The pump may deliver fluid on every other stroke of the reciprocating piston, known as "single-acting", or on each stroke known as "double-acting", see Figure 3.10(a) and (b), respectively. Flow through the pump is controlled by inlet and outlet valves. Delivery of fluid is pulsating.

### 3.8.1.2  Diaphragm Pump

The movement of the fluid is generated by a flexible diaphragm, driven by a reciprocating piston or by compressed air/gas (Figure 3.11). The diaphragm isolates the moving parts of the pump (except the inlet and outlet valves) from the fluid being pumped. Delivery of fluid is pulsating.

### 3.8.1.3  Other Positive Displacement Pumps

There is a wide range of other types of positive displacement pumps, including gear pumps (internal and external), lobe pumps and screw

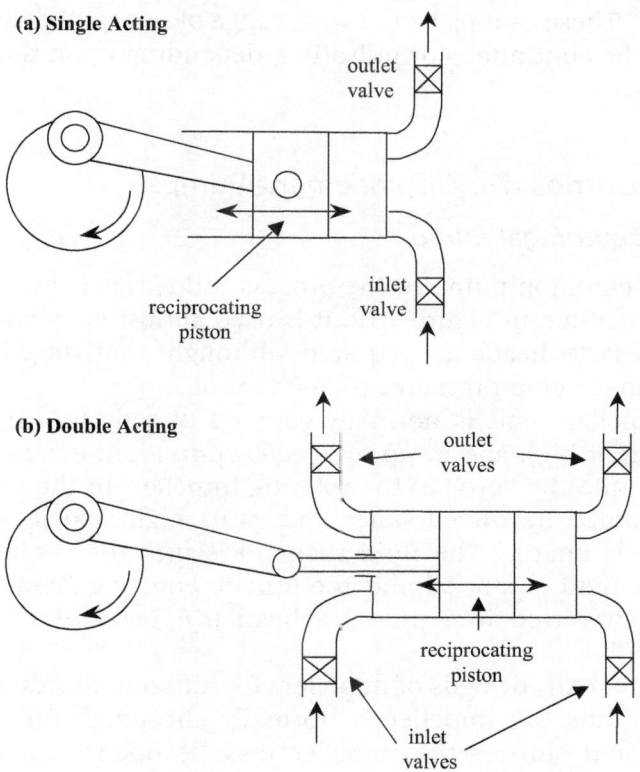

**Figure 3.10** Reciprocating piston pumps.

**Figure 3.11** Diaphragm pump.

pumps.[5,15] These pumps have a wide range of uses and delivery of the fluid may be continuous or pulsating, depending upon the design of the pump.

## 3.8.2   Non-positive Displacement Pumps

### 3.8.2.1   *Centrifugal Pump*

The most common pump in the process industries is the centrifugal pump, illustrated in Figure 3.12. It is used almost everywhere, except when very large heads are required, although multistage centrifugal pumps can develop pressures of 200 atm or more.

The impeller shaft is normally coupled directly to a motor drive and driven at high speed, up to ~4000 rpm. Fluid enters the pump and flows into the "eye" of the rotating impeller. In the impeller the fluid is guided by curved vanes and gains high velocity and hence high kinetic energy. The fluid then flows into the "volute casing" where the fluid is slowed and the kinetic energy gained in the impeller is converted into pressure head (*c.f. Bernoulli's* eqn (3.16), above).

There are many designs of impellers for different duties, *e.g.* for low viscosity fluids the impeller is normally shrouded; for solids suspensions and slurries the impeller may be open-faced to prevent blockage.

The delivery is continuous. Inlet and outlet valves are not required, but a non-return, or "check" valve is normally installed on the pump outlet line to prevent reverse flow when the pump is switched off. The outlet may be blocked (for short periods) without damage to the pump.

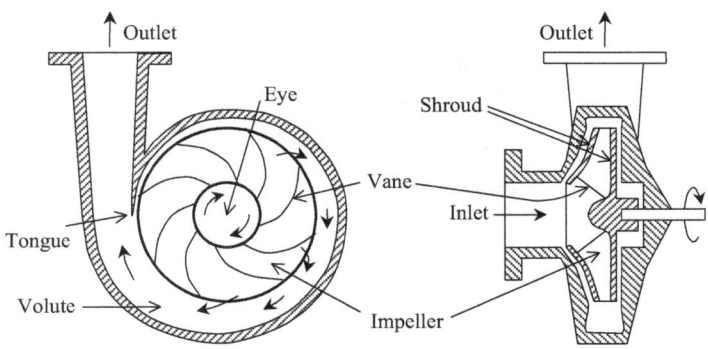

**Figure 3.12**   Centrifugal pump (motor drive not shown).

**Figure 3.13** Characteristic curves.

### 3.8.2.2 Characteristic Curves

The pressure head *versus* flow relationship for a centrifugal pump, called the *characteristic curve*, and illustrated in Figure 3.13, depends very much upon the design of the impeller, its vanes and the volute casing.

Maximum head is attained at zero flow, when the maximum kinetic energy gained by the fluid within the impeller is converted into pressure. As the flow through the pump is increased, the fluid leaving the outlet takes with it kinetic energy. Therefore, less of the kinetic energy gained within the impeller gets converted into pressure and hence the head generated falls.

In a well-designed system, a centrifugal pump will normally operate at or close to its maximum efficiency point. Pump efficiencies, $\eta$, are typically in the range 60–80%, larger pumps having the higher efficiencies.

## 3.8.3 Matching Centrifugal Pumps to Flow Requirements

### 3.8.3.1 Pump Characteristic and System Head

The pump characteristic curve needs to be matched with the head loss through a piping system, which is known as the *system head*.

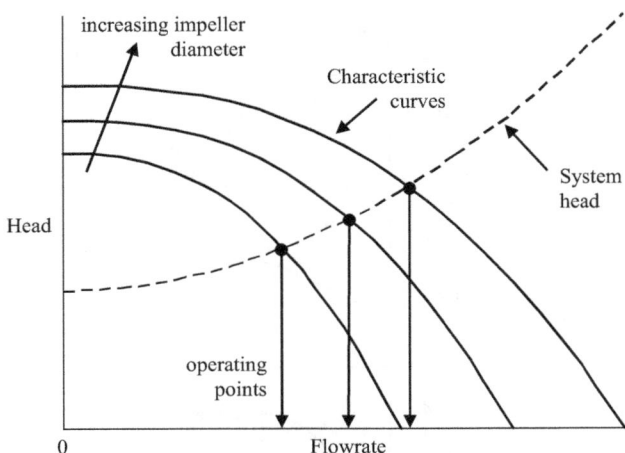

**Figure 3.14** Characteristic curves and system head.

The system head increases approximately in proportion to the square of the flow rate ($\Delta p \propto u^2$). An example of a system head calculation is given in Section 3.10.

The intersection of the characteristic curve (pump head) and the system head, illustrated in Figure 3.14, gives the *operating point* (flow rate) of the pump and pipe system. At this point the head generated by the pump is just balanced by the system head.

Different pumps with different characteristic curves will intersect the system head at different flow rates and give different operating points.

### 3.8.3.2  Control Valves

A flow system will normally require control and a control valve would normally be inserted into the pipeline.

The required head loss (pressure drop), across the control valve, $\Delta h_{cv}$, at any desired operating point (op point) can easily be determined, see Figure 3.15. An operating range of head losses required across a control valve, $\Delta h_{cv\,min}$ to $\Delta h_{cv\,max}$, say, may also be determined.

### 3.8.3.3  Net Positive Suction Head – NPSH

The pressure of liquid inside a pump must always exceed the liquid's vapour pressure, otherwise the liquid will tend to vaporise within the pump, causing cavitation. As the fluid passes on through the pump its

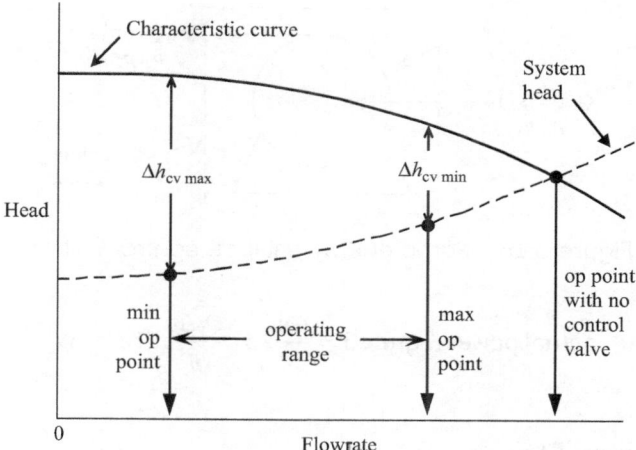

**Figure 3.15**  Effect of control valves on operating point.

pressure increases and any vapour bubbles collapse rapidly causing mechanical damage to the pump, which is often severe. Pumps may have their operating life reduced dramatically as a result.

The point of minimum pressure is not at the pump inlet, but will be somewhere within the impeller of the pump. Therefore, the absolute pressure (or head) of the liquid at the pump inlet, the suction side, must always be greater than the fluid vapour pressure by an amount, called the *Net Positive Suction Head* or *NPSH*. Manufacturers will quote the NPSH requirements for their pumps.

The Mechanical Energy Balance eqn (3.14), may be used to determine the fluid pressure at the inlet to the pump (see Section 3.10). This must exceed the liquid's vapour pressure by the NPSH requirement of the pump, otherwise the pipework should be redesigned or a different pump with a smaller NPSH requirement installed. This explains why pumps are normally located at the lowest part of the pipework (maximum head) and close to the source (minimum inlet pipework pressure drop).

### 3.8.4  Pump Power Requirements

Assuming $\rho = $ constant and $A_1 = A_2$, then $u_1 = u_2$, as $Q_v = $ constant, the mechanical energy balance over a pump (see Figure 3.16) gives the work done on the fluid as: $-w_x = \int_{p_1}^{p_2} \mathrm{d}p/\rho = -\Delta p/\rho$ [J kg$^{-1}$].

$$\text{Theoretical power required} = Gw_x = \rho Q_v \frac{\Delta p}{\rho} = Q_v \Delta p \text{ [J s}^{-1}, \text{ W]} \qquad (3.38)$$

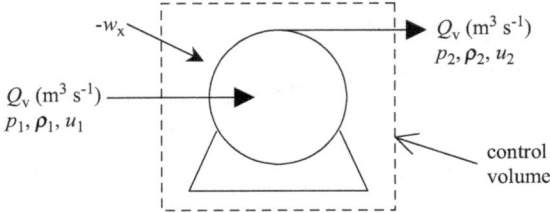

**Figure 3.16**  Pump energy balance control volume.

$$\text{Actual power required} = \frac{Gw_x}{\eta} = \frac{Q_v \Delta p}{\eta} \ [\text{J s}^{-1}, \text{W}] \tag{3.39}$$

## 3.9  Other Flows

### 3.9.1  Equivalent Hydraulic Diameter

When the pipe or duct is non-circular or non-full, the equivalent hydraulic diameter, $d_e$, is often used in place of $d$, the pipe diameter, in *turbulent flow* equations (although not for very narrow or slotted flow cross-sections).

$$d_e = 4 \times \frac{\text{flow cross-sectional area}}{\text{wetted perimeter}} \tag{3.40}$$

and the head loss due to friction becomes

$$\Delta h_f = 2c_f \frac{l}{d_e} \frac{\bar{u}^2}{g} \tag{3.41}$$

#### 3.9.1.1  Example of Equivalent Hydraulic Diameters

Rectangular duct, sides $a$ and $b$:

$$d_e = 4 \times \frac{ab}{2(a+b)} = \frac{2ab}{(a+b)} \tag{3.42}$$

Parallel plates, width $b$, distance $a$ apart, $a \ll b$:

$$d_e = 4 \times \frac{ab}{2(a+b)} = 2a \tag{3.43}$$

Equilateral triangular duct, side $a$:

$$d_e = 4 \times \frac{(1/2)a^2 \sin 60°}{3a} = 0.577a \tag{3.44}$$

Open channel flow, non-full sloping duct, depth $a$ width $b$:

$$d_e = 4 \times \frac{ab}{2a+b} = \frac{4ab}{2a+b} \tag{3.45}$$

With open channel flow the driving force is gravity; there is no net pressure force as the free surface is uniformly exposed to the atmosphere. Thus, in the Mechanical Energy Balance eqn (3.14), the $(p_2 - p_1)/\rho$ and $w_x$ terms are both zero and the $g(z_2 - z_1)$ term quantifies the driving force and is given by $gl \sin\theta$, where $l$ is the length of the duct which is at an angle $\theta$ to the horizontal.

### 3.9.2 Flow Around Bodies

#### 3.9.2.1 Spheres and Cylinders

The drag force $F$ exerted by a fluid on a body would be expected to be a function of the body size (area), relative velocity with fluid, fluid density and viscosity, and body shape (see Figure 3.17). Thus $F = f(A, u_\infty, \rho, \mu, \text{shape})$.

Dimensional analysis yields, $C_D = f(\text{Re}, \text{shape})$, where

- Drag coefficient:

$$C_D = \frac{F}{\frac{1}{2}\rho u_\infty^2 A_p} \equiv \frac{\text{drag force}}{\text{inertial forces}}$$

- Reynolds No:

$$\text{Re} = \frac{\rho u_\infty d}{\mu} \equiv \frac{\text{inertial forces}}{\text{viscous forces}}$$

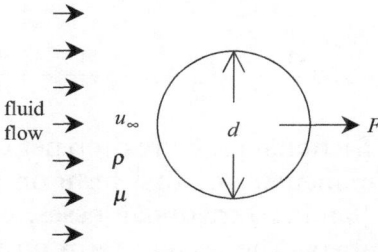

**Figure 3.17** Drag on a sphere or cylinder.

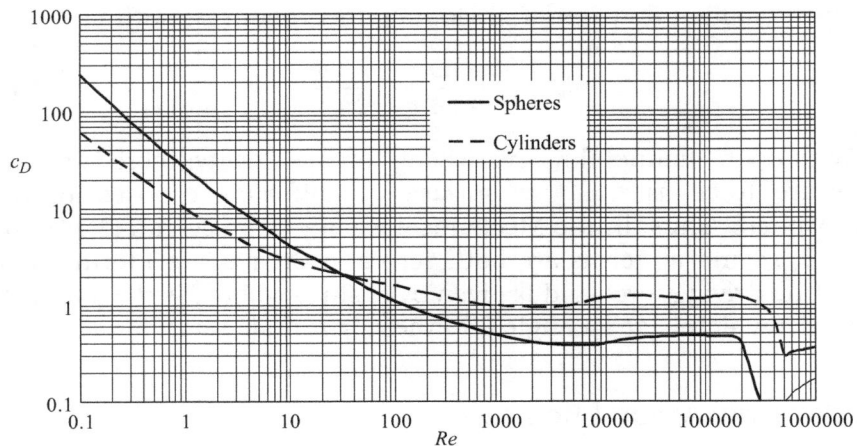

**Figure 3.18**  Drag coefficients.

$A_\mathrm{p}$ is the body's projected area in the direction of flow. For a sphere, $A_\mathrm{p} = \pi d^2/4$, and for a cylinder $A_\mathrm{p} = dL$ ($L$ = length of the cylinder).

$C_\mathrm{D}$ is a complex function of Re and geometry. The curves for cylinders and spheres show many similarities (see Figure 3.18), and may be divided into the following flow regimes:

- laminar: $\mathrm{Re} \lesssim 0.5$;
- transition: $0.5 \lesssim \mathrm{Re} \lesssim 1000$;
- turbulent, with laminar boundary layer: $1000 \lesssim \mathrm{Re} \lesssim 20\,000$;
- turbulent, with turbulent boundary layer: $\mathrm{Re} \gtrsim 200\,000$.

Further discussion on flow around and drag forces on bodies is available in the literature.[1,2,4]

### 3.9.2.2  *Flow Through a Packed Bed of Spheres*

The Ergun eqn (3.46), gives the pressure drop through a packed bed of spheres:[2]

$$\frac{\Delta p}{l} = 150 \frac{(1-\varepsilon)^2}{\varepsilon^3} \frac{\mu u}{d_\mathrm{p}^2} + 1.75 \frac{(1-\varepsilon)}{\varepsilon^3} \frac{\rho u^2}{d_\mathrm{p}} \tag{3.46}$$

We see that $\Delta p/l$, the frictional pressure drop per unit depth of bed, is made up of two components. The first term on the right-hand-side accounts for viscous (laminar) frictional losses, $\propto \mu u$, and dominates at low Reynolds numbers. The second term on the right-hand-side accounts for the inertial (turbulent) frictional losses, $\propto \rho u^2$, and dominates at high Reynolds numbers. For further information about flow through packed beds, see Chapter 7 "An Introduction to Particle Systems".

### 3.9.3  Stirred Tanks[16–20]

The power $P$ to turn an impeller is given by $2\pi N \times M$, (angular velocity $\times$ torque). This power would be expected to be a function of impeller diameter and speed, and geometry. Gravitational acceleration should also be included in the analysis as there is normally a free liquid surface present and gravity affects its shape and the flow within the vessel.

Thus

$$P = f(D,\ N,\ \rho,\ \mu,\ g,\ \text{geometry}).$$

Assuming geometric similarity, dimensional analysis yields, Po $= f(\text{Re},\ \text{Fr})$, where

- Power No:

$$\text{Po} = \frac{P}{\rho N^3 D^5} \equiv \frac{\text{drag force on impeller}}{\text{inertial forces}}$$

- Impeller Reynolds No:

$$\text{Re}_\text{I} = \frac{\rho N D^2}{\mu} \equiv \frac{\text{inertial forces}}{\text{viscous forces}}$$

- Froude No:

$$\text{Fr} = \frac{N^2 D}{g} \equiv \frac{\text{inertial forces}}{\text{gravitational forces}}$$

Po is a complex function of $\text{Re}_\text{I}$ and geometry. Figure 3.20 shows the Po *versus* $\text{Re}_\text{I}$ relationships for a number of impellers in tanks fitted with baffles, which suppress the formation of a vortex (see dashed lines in Figure 3.19), and hence where Fr has little effect. The curves may be divided into the following flow regimes:

- laminar: $\text{Re}_\text{I} \lesssim 10$;
- transition: $10 \lesssim \text{Re}_\text{I} \lesssim 5000$;
- turbulent: $\text{Re}_\text{I} \gtrsim 5000$.

**Figure 3.19**   Stirred Tank.

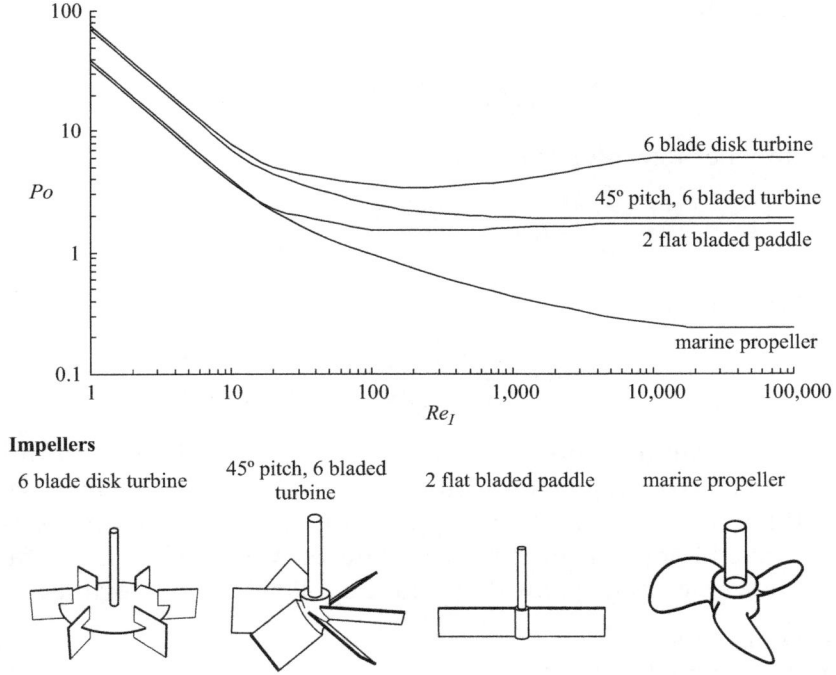

**Figure 3.20**   Po *versus* Re₁ curves for baffled tanks for different impellers.

# 3.10 Example Calculations for the Pipe-flow System

This section illustrates the use of material presented earlier to determine the pressure drop in a pipe system, mating of a pump to the pipe system and the head loss over a control valve.

10 kg s$^{-1}$ of water at 80 °C is pumped from a holding vessel through a pipe system to a process vessel, which is illustrated in Figure 3.21. The pipe is of made of commercial steel and has an internal diameter of 100 mm. The pipework has a total length of straight pipe of 35 m.

We are required to specify a suitable centrifugal pump to achieve the required flow.

## 3.10.1 Data

### 3.10.1.1 Datum and Reference Points

| | |
|---|---|
| Elevation datum | Level with the inlet to the pump, but can be any convenient level |
| Point ①, subscript 1 | At the minimum level of liquid in the holding vessel |
| Point ②, subscript 2 | Inside the pressure vessel, level with the pipe inlet |
| Point ③, subscript 3 | At the inlet to the pump |

**Figure 3.21** Example pipe-flow system.

### 3.10.1.2   Fluid Properties, Water at 80 °C

| | |
|---|---|
| Density | $\rho = 972$ kg m$^{-3}$ |
| Viscosity | $\mu = 0.351 \times 10^{-3}$ Pa s |
| Vapour pressure | $p_v = 0.4748 \times 10^5$ Pa |

### 3.10.1.3   Pipe Properties

| | |
|---|---|
| Total length of straight pipe | $l = 35$ m |
| Internal diameter | $d = 0.1$ m |
| Roughness | $e = 0.05 \times 10^{-3}$ m (see Table 3.4) |
| Elevation of ① | $z_1 = 4$ m |
| Elevation of ② | $z_2 = 6$ m |
| Elevation of ③ | $z_3 = 0$ m |
| Relative roughness | $e/d = 0.05 \times 10^{-3}/0.1 = 0.0005$ |
| Pipe cross-sectional area | $A = \pi\, d^2/4 = \pi\, 0.1^2/4 = 7.85 \times 10^{-3}$ m$^2$ |

### 3.10.1.4   Flow Properties

| | |
|---|---|
| Mass flow rate | $G = 10$ kg s$^{-1}$ |
| Volumetric flow rate | $Q_v = G/\rho = 10.3 \times 10^{-3}$ m$^3$ s$^{-1}$ |
| Mean velocity | $\bar{u} = Q_v/A = 1.31$ m s$^{-1}$ |
| Reynolds number | $\mathrm{Re} = \dfrac{\rho \bar{u} d}{\mu} = \dfrac{972 \times 1.31 \times 0.1}{0.000351} = 363\,000,$ *i.e.* turbulent |

### 3.10.1.5   Pipe Fittings

Orifice meter and control valve will be considered separately, see Table 3.5.

| No. fitting | Each fitting $n$ | Subtotal $n$ |
|---|---|---|
| 1 × entrance loss from holding vessel | 25 | 25 |
| 2 × shut off valves (open gate valves) | 7 | 2 × 7 |
| 1 × check valve (swing) | 100 | 100 |
| 3 × 90° standard radius bends | 35 | 3 × 35 |
| 1 × exit loss into process vessel | 50 | 50 |
| | | Total: $n = 294$ |

Equivalent length of straight pipe: $l_e = n \times d = 294 \times 0.1 = 29.4$ m.

## 3.10.2 Friction Losses

By using eqn (3.32), total frictional head loss:

$$\Delta h_{f_T} = 4c_f \frac{(l + l_e)\,\bar{u}^2}{d}\frac{\bar{u}^2}{2g} = 4c_f \frac{l_T}{d}\frac{\bar{u}^2}{2g}$$

Fanning friction factor, from eqn (3.28),

$$c_f = 0.001375\left[1 + \left(20\,000\,\frac{e}{d} + \frac{10^6}{Re}\right)^{1/3}\right]$$

$$= 0.001375\left[1 + \left(20\,000 \times 0.0005 + \frac{10^6}{36\,3000}\right)^{1/3}\right] = 0.00459$$

or look up $c_f$ for appropriate values of Re and $e/d$ in Figure 3.7, see also point ■.

Total equivalent length of straight pipe $l_T = l + l_e = 35 + 29.4 = 64.4$ m. Substituting

$$\Delta h_{f_T} = 4c_f \frac{l_T}{d}\frac{\bar{u}^2}{2g} = 4 \times 0.00459 \times \frac{64.4}{0.1} \times \frac{1.31^2}{2 \times 9.81} = 1.03\,\text{m H}_2\text{O}$$

## 3.10.3 Orifice Meter

Eqn (3.37):

$$G = C_d A_1\left[\frac{2\rho(p_1 - p_2)_{\text{orifice}}}{(A_1/A_2)^2 - 1}\right]^{1/2} = C_d A_1\left[\frac{2\rho\Delta p_{\text{orifice}}}{(A_1/A_2)^2 - 1}\right]^{1/2}$$

typical discharge coefficient for orifice $C_d = 0.65$ (say)
pipe csa. $A_1 = \pi\,d^2/4 = 7.85 \times 10^{-3}$ m$^2$
orifice csa. $A_2 = 1/2\,A_1$ say $= 3.92 \times 10^{-3}$ m$^2$ (a typical ratio of areas).

From eqn (3.37)

$$10 = 0.65 \times 7.85 \times 10^{-3}\left[\frac{2 \times 972 \times \Delta p_{\text{orifice}}}{(2)^2 - 1}\right]^{1/2}$$

$\therefore \Delta p_{\text{orifice}} = 5930$ Pa $= 0.62$ m H$_2$O. This is the pressure difference across the flowmeter that would be measured and from which the flow rate would be determined for control purposes.

Overall frictional loss across orifice meter, see Table 3.5 $\Delta p_{\text{loss}} \approx 60\% \times \Delta p_{\text{orifice}} \approx 3560$ Pa $\therefore \Delta h_{\text{loss}} = \Delta p_{\text{loss}}/\rho\, g = 0.37$ m H$_2$O.

## 3.10.4 Pump Shaft Work

Use the Mechanical Energy Balance eqn (3.14):

$$\frac{p_2 - p_1}{\rho} + \left(\frac{u_2^2}{2} - \frac{u_1^2}{2}\right) + g(z_2 - z_1) + g\Delta h_f + w_x = 0$$

and apply between two points.

Point 1: chosen at the liquid surface in the holding vessel ① $u_1 \approx 0$, $p_1 = p_{\text{atm}} = 1.013$ bar, $z_1 = 4$ m.

Point 2: chosen inside the pressure vessel, level with the pipe inlet ② $u_2 \approx 0$, $p_2 = 3$ bar, $z_2 = 6$ m.

Substituting into eqn (3.14):

$$\frac{(3 - 1.013) \times 10^5}{972} + \left(\frac{0^2}{2} - \frac{0^2}{2}\right) + 9.81(6-4) + 9.81(1.03 + 0.37) + w_x = 0$$

$$204.4 \quad + \quad 0 \quad + \quad 19.6 \quad + \quad 13.7 \quad + w_x = 0$$

and solving yields theoretical pump power, $w_x = -237.7$ J kg$^{-1}$

$$P_{\text{ideal}} = -Gw_x = -10 \times -237.7 = 2.38 \text{ kW}$$

Actual pump power, assuming 70% efficiency, $P_{\text{actual}} = 2.38/0.7 = 3.4$ kW $= 4.6$ hp.

## 3.10.5 System Head

The required shaft work for the pump, $w_x$, when expressed as a head, is known as the system head, $\Delta h_{\text{system}}$. This is the head required from the pump to achieve the required flow through the pipe system.

System head

$$\Delta h_{\text{system}} = -\frac{w_x}{g} = -\frac{-237.7}{9.81} = 24.2 \text{ m H}_2\text{O}$$

## 3.10.6 Pump Characteristics

Figure 3.22 shows characteristic curves and NPSH requirements for a family of different sized centrifugal pumps adapted from curves given by Perry.[23] Superimposed is the system head curve for the example piping system. The system head curve is generated by repeating the above calculations for $\Delta h_{\text{system}}$ over the required range of flow rates.

**Figure 3.22** Pump characteristics curves and system head. Adapted from Perry.[23]

It can be seen that a pump with an impeller diameter of 138 mm is suitable for the required duty, in that it will deliver a head which comfortably exceeds the required system head of 24.2 m at the required flow rate of 0.0103 m³ s⁻¹. A smaller diameter impeller, *e.g.* 125 mm, will not simultaneously deliver the required combination of head and flow rate and is therefore not suitable for this duty.

### 3.10.7 Control Valve

Note that at the required flow rate, the head produced by the pump exceeds the required system head. The excess head, ~6 m H₂O, will be taken up by the control valve, $\Delta h_{cv}$. Also by changing the head across the control valve, $\Delta h_{cv}$, the operating point of the system, *i.e.* the flow rate, will change. Figure 3.22 may be used to determine this effect.

### 3.10.8 Net Positive Suction Head

The absolute pressure (or head) of the liquid at the pump inlet, the suction side, must always be greater than the fluid vapour pressure by the NPSH. Let us use the Mechanical Energy Balance eqn (3.14), to find the pressure at the pump inlet.

Point 1: no change, at the liquid surface in the holding vessel ①: $u_1 \approx 0$, $p_1 = p_{atm} = 1.013$ bar, $z_1 = 4$ m.

Point 2: now chosen to be the inlet to the pump ③ where we need to determine the pressure: $u_3 = \bar{u}$, $z_3 = 0$, $p_3$ is the unknown pressure at pump inlet.

Note: $\Delta h_f$ has to be recalculated based on the shorter length of pipe from the holding vessel to the pump inlet, 8 m, and the fewer fittings in this section of pipework. In this case $l_e = 6.7$ m, giving $\Delta h_f = 0.24$ m $H_2O$.

$$\frac{p_3 - p_1}{\rho} + \left(\frac{u_3^2}{2} - \frac{u_1^2}{2}\right) + g(z_3 - z_1) + g\Delta h_f + w_x = 0$$

$$\frac{p_3 - 1.013 \times 10^5}{972} + \left(\frac{1.31^2}{2} - \frac{0^2}{2}\right) + 9.81(0 - 4) + 9.81 \times 0.24 + 0 = 0$$

$$\frac{p_3}{972} - 104.2 + 0.86 - 39.2 + 2.3 + 0 = 0$$

$$p_3 = 1.36 \times 10^5 \, \text{Pa} = 1.36 \, \text{bar}$$

NPSH available is the difference between absolute pressure ($p_3$) of the liquid at the pump and the fluid vapour pressure ($p_v$), expressed as a head.

$$\text{NPSH}_{\text{available}} = \frac{p_3 - p_v}{\rho g} = \frac{(1.36 - 0.475) \times 10^5}{972 \times 9.81} = 9.4 \, \text{m} \, H_2O$$

From Figure 3.22 it is seen that the NPSH required by the pump is less than 3 m.

Therefore, as the NPSH available exceeds that required by the selected pump, the pump is suitable. Our task is now complete.

## Nomenclature

| Symbol | Description | Dimensions | SI unit |
|--------|-------------|------------|---------|
| $A$ | Area, cross-sectional area | $L^2$ | $m^2$ |
| $a, b$ | Dimensions, lengths | L | m |
| $C_D$ | Drag coefficient | — | — |
| $C_d$ | Discharge coefficient | — | — |
| $c_f$ | Fanning friction factor | — | — |

| Symbol | Description | Dimensions | SI unit |
| --- | --- | --- | --- |
| $d, D$ | Diameter, pipe diameter, impeller diameter | L | m |
| $e$ | Absolute roughness | L | mm |
| $F$ | Force | $M\,L\,T^{-2}$ | N |
| $f$ | Function | — | — |
| $f$ | Friction factor | — | — |
| Fr | Froude number | — | |
| $g$ | Gravitational acceleration | $L\,T^{-2}$ | $m\,s^{-2}$ |
| $G$ | Mass flow rate | $M\,T^{-1}$ | $kg\,s^{-1}$ |
| $h$ | Head of fluid | L | m *of fluid* |
| | Specific enthalpy | $L^2\,T^{-2}$ | $J\,kg^{-1}$ |
| $K$ | Loss coefficient, number of velocity heads | — | |
| $l, L$ | (Pipe) length, cylinder length | L | m |
| $m$ | Number of variables | — | — |
| $m_T$ | Total mass within control volume | M | kg |
| $n$ | Number of fundamental dimensions, Number of pipe diameters | — | — |
| $N$ | Rate of rotation | $T^{-1}$ | Hz |
| $p$ | Pressure | $M\,L^{-1}\,T^{-2}$ | Pa |
| $p_v$ | Equilibrium vapour pressure | $M\,L^{-1}\,T^{-2}$ | Pa |
| $P$ | Power | $M\,L^2\,T^{-3}$ | W |
| Po | Power number | — | |
| $q$ | Heat input to control volume per unit mass | $L^2\,T^{-2}$ | $J\,kg^{-1}$ |
| $Q$ | Quantity, variable | *n/a* | *n/a* |
| $Q_v$ | Volumetric flow rate | $L^3\,T^{-1}$ | $m^3\,s^{-1}$ |
| $r, r_0$ | Radial co-ordinate, pipe radius | $LT^{-1}$ | $m\,s^{-1}$ |
| Re | Reynolds number | — | — |
| $u, U, u_{max}$ | Velocity in $x$-direction, velocity, maximum velocity | $L\,T^{-1}$ | $m\,s^{-1}$ |
| $\bar{u}, \tilde{u}$ | Average velocity across a cross-section, mean at a point | $L\,T^{-1}$ | $m\,s^{-1}$ |
| $u_\infty$ | Velocity relative to infinity or to undisturbed flow | $L\,T^{-1}$ | $m\,s^{-1}$ |
| $u^+$ | Dimensionless velocity | — | |
| $w_x$ | Shaft work out from control volume per unit mass | $L^2\,T^{-2}$ | $J\,kg^{-1}$ |

| Symbol | Description | Dimensions | SI unit |
|---|---|---|---|
| $W_x$ | Rate shaft work out from control volume | $M\,L^2\,T^{-3}$ | W |
| $x$ | Distance in $x$-direction, distance along pipe | L | m |
| $y$ | Distance in $y$-direction, distance from (pipe) wall | L | m |
| $y^+$ | Dimensionless distance in $y$-direction | — | |
| $Y$ | Length | L | m |
| $z$ | Distance in $z$-direction, elevation | L | m |
| $\varepsilon$ | Voidage | — | — |
| $\eta$ | Efficiency | — | — |
| $\lambda, \phi$ | Friction factors | — | — |
| $\mu$ | Dynamic viscosity | $M\,L^{-1}\,T^{-1}$ | Pa s, N s $m^{-2}$ |
| $\nu$ | Kinematic viscosity | $L^2\,T^{-1}$ | $m^2\,s^{-1}$ |
| $\Pi$ | Dimensionless group | — | — |
| $\rho$ | Fluid density | $M\,L^{-3}$ | kg m$^{-3}$ |
| $\tau, \tau_w$ | Shear stress, wall shear stress | $M\,L^{-1}\,T^{-2}$ | Pa, N m$^{-2}$ |

## Subscripts

| | |
|---|---|
| 1, 2, 3 | at point 1, 2, 3 |
| e | Equivalent hydraulic |
| f | Frictional |
| I | Impeller |
| p | Projected, particle |
| T | Total |

## Symbols

| | |
|---|---|
| $\Delta$ | "the change in" |
| $[=]$ | "has the dimensions of" |

# References

1. J. M. Coulson, J. F. Richardson, with J. R. Backhurst and J. H. Harker, *Coulson & Richardson's Chemical Engineering*, 6th edn, Butterworth-Heinemann, Oxford, 1999, vol. 1.
2. J. M. Kay and R. M. Neddermann, *Fluid Mechanics and Transfer Processes*, Cambridge University Press, Cambridge, 1985, ch. 1–12.
3. B. S. Massey, revised by J. Ward-Smith, *Mechanics of Fluids*, Taylor & Francis, Abingdon, Oxford, 8th edn, 2006, ch. 5.
4. R. H. Perry and D. Green, *Perry's Chemical Engineers' Handbook*, McGraw Hill, New York, 7th edn, 1997, Section 6.
5. R. H. Perry and D. Green, *Perry's Chemical Engineers' Handbook*, McGraw Hill, New York, 7th edn, 1997, Section 10.
6. C. Schaschke, *Fluid Mechanics Worked Examples for Engineers*, IChemE, Rugby, 1998.
7. R. B. Bird, W. E. Stewart and E. N. Lightfoot, *Transport Phenomena*, John Wiley & Sons Inc., New York, 2nd edn, 2001.
8. C. R. Duhne, Viscosity–temperature correlations for liquids, *Chem. Eng.*, 1979, **86**, 83–91.
9. R. P. Chhabra and J. F. Richardson, *Non-Newtonian Flow in the Process Industries*, Butterworth-Heinemann, Oxford, 1999.
10. J. E. Edwards, Flow measurement practice, *Chem. Eng.*, 1979, **344**, 325–333.
11. D. Ginesi, Choices abound in flow measurement, *Chem. Eng.*, 1991, **98**, 88–100.
12. D. Ginesi, A raft of flowmeters on tap, *Chem. Eng.*, 1991, **98**, 146–155.
13. BS 1042, Measurement of fluid flow in closed conduits, BSI.
14. BS EN ISO 5167-1 ... -4, Measurement of fluid flow by means of pressure differential devices inserted in circular cross-section conduits running full, BSI.
15. I. J. Karassik, J. P. Messina, P. Cooper and C. C. Heald, *Pump Handbook*, McGraw-Hill, New York, 3rd edn, 2001.
16. V. W. Uhl and J. B. Gray, *Mixing: Theory and Practice*, Academic Press, New York, 1966, 1967 and 1986, vol. I–III.
17. S. Nagata, *Mixing: Principles and Applications*, Kodansha (Wiley), New York, 1975.
18. N. Harnby, M. F. Edwards and A. W. Nienow, *Mixing in the Process Industries*, Butterworth-Heinemann, Oxford, 2nd edn, 1992.
19. G. B. Tatterson, *Fluid Mixing and Gas Dispersion in Agitated Tanks*, McGraw Hill, New York, 1991.

20. J. J. Ulbrecht and G. K. Patterson, *Mixing of Liquids by Mechanical Agitation*, Gordon and Breach, New York, 1985.
21. L. F. Moody, *Trans. Am. Soc. Mech. Eng.*, 1944, **66**, 671–684; *Mech. Eng.*, 1947, **69**, 1005–1006.
22. E. E. Ludwig, *Applied Process Design for Chemical and Petro-chemical Plants*, Gulf Publishing Co, Houston, Texas, 1964, ch. 2, vol. 1.
23. R. H. Perry and D. Green, *Perry's Chemical Engineers' Handbook*, 7th edn, McGraw Hill, New York, 1997, Figure 10–31.

# 4 An Introduction to Heat Transfer

Tim Elson and Paola Lettieri

## 4.1 Introduction and Objectives

Heat is transferred from regions of high temperature to regions of low temperature and may take place by three different mechanisms or modes.

### 4.1.1 Modes of Heat Transfer

#### 4.1.1.1 Conduction

Conduction of heat results from the greater motion of molecules and atoms at higher energy (temperature) levels imparting energy to adjacent molecules and atoms at lower energy (temperature) levels. There is no overall motion of the material and it occurs in gases, liquids and solids. The physical mechanisms of heat conduction are complex, but in gases and liquids comprise of molecular collisions and lattice vibrations in solids and flow of free electrons in metals.

#### 4.1.1.2 Convection

Heat is transferred by moving fluids and therefore only occurs in gases and liquids. A moving fluid takes thermal energy with it, and

Concepts of Chemical Engineering for Chemists: 2nd Edition
Edited by Stefaan Simons
© The Royal Society of Chemistry 2017
Published by the Royal Society of Chemistry, www.rsc.org

when it contacts another fluid or a surface at a different temperature, heat is transferred. There are two types of convection:

- *Forced Convection*, where the flow is induced by external forces, *e.g.* from a pump or a fan.
- *Natural Convection*, where the flow is induced by buoyancy forces, which result from the thermal expansion of hotter fluids that rise relative to cooler, denser fluids.

### 4.1.1.3  Radiation

All bodies emit and absorb electromagnetic radiation, such as infrared, ultraviolet and visible light. The predominant frequency and the energy of the emitted radiation increase as the temperature of the body increases. Heat may therefore be transferred from hotter bodies to cooler bodies by electromagnetic radiation. This will only occur through materials that are transparent or partially transparent to electromagnetic radiation.

## 4.1.2  Scope and Objectives

We will look at these three modes of heat transfer in more detail and consider the laws that govern the rates of heat transfer by these mechanisms. We will look at some of the more common types of heat exchanger and how we may use these laws to specify and design a heat exchanger.

Heat transfer is covered in many chemical engineering texts. Kern,[1] although first published in 1950, still remains a standard reference. Coulson and Richardson's Chemical Engineering series covers heat transfer in general in volume 1[2] and the detailed design of heat exchangers in volume 6.[3] Perry[7] also gives a good coverage of the subject.

## 4.2  Modes of Heat Transfer

## 4.2.1  Conduction

### 4.2.1.1  Fourier's Law of Conduction

Experiments have shown that at steady state, the heat flux $q_y$, which is the rate of heat transfer, $Q$, per unit area, $A$, through a material due to conduction, is proportional to the temperature gradient in the direction of heat flow, $y$ in this case.

$$q_y = \frac{Q}{A} = -k\frac{dT}{dy} \tag{4.1}$$

The proportionality constant $k$ is called the *thermal conductivity* of the material. Heat is transferred from regions of high temperatures to low temperatures – hence the negative sign in front of $k$. Eqn (4.1) holds for liquids and gases, as well as for solids, as long as convection and radiation are prevented.

### 4.2.1.2  Thermal Conductivity

Figure 4.1 shows the temperature dependence of a range of materials, and it can be seen that the values of thermal conductivities of materials vary widely. Typical values are as follows:

- Gases: $0.01 \sim 0.1 \ \mathrm{W\,m^{-1}K^{-1}}$, with higher values for $H_2$ and He
- Liquids: $0.1 \sim 1 \ \mathrm{W\,m^{-1}K^{-1}}$, with much higher values for liquid metals
- Solids: $1 \sim 100 \ \mathrm{W\,m^{-1}K^{-1}}$, with higher value for good conductors, *e.g.* Cu and Al, and much lower values for good insulators.

Some fibrous or lamellar materials have non-isotropic thermal conductivities. Examples are pine wood, with $k \approx 0.15 \ \mathrm{W\,m^{-1}K^{-1}}$ across the grain and $\approx 0.35 \ \mathrm{W\,m^{-1}K^{-1}}$ parallel to the grain, and graphite at 300 K, $k \approx 10 \ \mathrm{W\,m^{-1}K^{-1}}$ parallel to the basal plane and $\approx 2000 \ \mathrm{W\,m^{-1}K^{-1}}$ perpendicular to the basal plane.[7]

### 4.2.1.3  Rate of Heat Transfer

Fourier's Law may be integrated and solved for a number of geometries to relate the rate of heat transfer by conduction to the temperature driving force. Equations are given below that allow the calculation of steady-state heat flux and temperature profiles for a number of geometries.

#### 4.2.1.3.1  Large, Flat Plate

The heat flux through a flat plate, with thickness $b$ in the $y$ direction, is given by

$$q_y = \frac{Q}{A} = k\frac{(T_0 - T_b)}{b} \tag{4.2}$$

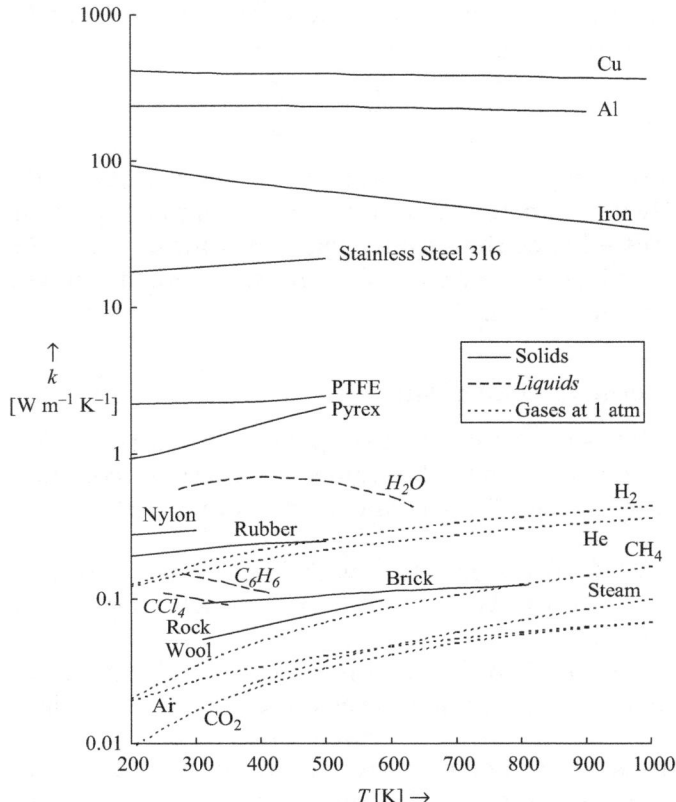

**Figure 4.1**  Temperature variation of thermal conductivity for a number of materials.[7,8]

where the surfaces are being held at $T_0$ at $y = 0$ and at $T_b$ at $y = b$. The temperature profile, $T(y)$, is found to be linear with distance through the plate, $y$:

$$T(y) = T_0 - \frac{y}{b}(T_0 - T_b) \qquad (4.3)$$

#### 4.2.1.3.2  Composite Plate

The heat flux through a flat plate comprising of, for example, three layers of material of different thickness $b_1$, $b_2$ and $b_3$, each with its own thermal conductivity, $k_1$, $k_2$ and $k_3$, respectively, and the outside surfaces being held at $T_0$ and $T_3$, is given by

$$q_y = \frac{T_0 - T_3}{(b_1/k_1) + (b_2/k_2) + (b_3/k_3)} = \frac{T_0 - T_3}{\Sigma(b/k)} = \frac{\Delta T}{\Sigma(b/k)} \qquad (4.4)$$

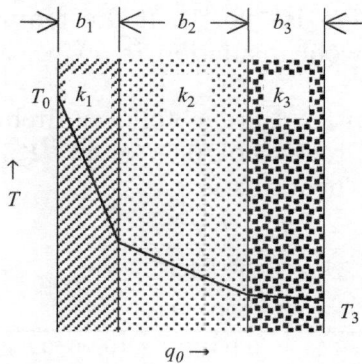

**Figure 4.2** Conduction through a composite plate (illustrated with $k_1 < k_2 < k_3$).

**Figure 4.3** Electrical resistances in series.

The temperature profile, $T(y)$, in each layer is given by eqn (4.3) and illustrated in Figure 4.2.

Eqn (4.4) is analogous to electrical current flowing through electrical resistances in series. Figure 4.3 illustrates such an electrical circuit, where an electrical current, $i$, flows through a series of three resistances, $R_1$, $R_2$ and $R_3$. The overall voltage difference of $V_0 - V_3$, is given by

$$i = \frac{V_0 - V_3}{R_1 + R_2 + R_3} = \frac{V_0 - V_3}{\Sigma R} = \frac{\Delta V}{\Sigma R} \tag{4.5}$$

Comparing eqn (4.4) with eqn (4.5), $i \equiv q_y$, $\Delta V \equiv \Delta T$ and $R_n \equiv (b_n/k_n)$. Thus, we can think of the $(b/k)$ values as thermal resistances, the resistance of the material layer to heat transfer by conduction. From eqn (4.4), it can be seen that the heat flux is equal to the overall temperature difference $\Delta T$, divided by the sum of the thermal resistance of each layer $\Sigma(b/k)$. The concept of thermal resistance of a layer is very useful and we will return to it later.

### 4.2.1.3.3 Example Calculation of Heat Loss through a Composite Wall

A furnace wall is constructed of a 3 cm thick, flat steel plate, with a firebrick insulation 30 cm thick on the inside, and rock wool insulation 6 cm thick on the outside. The inside surface temperature of the firebrick insulation is 700 °C. If the temperature of the outer surface of the rock wool insulation is 50 °C, what is the heat flux through the wall?

Thermal conductivities[7,8] are as follows: firebrick, 0.1 W m$^{-1}$ K$^{-1}$; steel, 40 W m$^{-1}$ K$^{-1}$; rock wool, 0.04 W m$^{-1}$ K$^{-1}$.

From eqn (4.4) the heat flux is

$$q_y = \frac{T_0 - T_3}{(b_1/k_1) + (b_2/k_2) + (b_3/k_3)}$$

$$= \frac{700 - 50}{(0.3/0.1) + (0.03/40) + (0.06/0.04)}$$

$$= \frac{650}{3.000 + 0.001 + 1.5} = \frac{650}{4.501} = 144.4 \text{ W m}^{-2}$$

What is the temperature of the steel? From eqn (4.2)

$$T_1 = T_0 - q_y \frac{b_1}{k_1} = 700 - 144.4 \frac{0.3}{0.1} = 266.74 \,°\text{C}$$

This is the inner surface temperature of the steel. The steel outer surface temperature is then given by

$$T_2 = T_1 - q_y \frac{b_2}{k_2} = 266.74 - 144.43 \frac{0.03}{40} = 266.63 \,°\text{C}$$

We see that there is only a small temperature difference across the steel wall. This is due to the relatively small thermal resistance offered by this layer $(b_2/k_2)$, approximately 0.02% of the total resistance. The wall is effectively at a uniform temperature of 267 °C.

#### 4.2.1.3.4   Hollow Cylinder

Radial heat flux, $q_r$, through the wall of a long cylinder of length $L$ and inner radius $r_i$ held at temperature $T_i$, and outer radius $r_o$ held at temperature $T_o$, e.g. the wall of a pipe or tube in a heat exchanger, is given by

$$q_r = \frac{Q}{2\pi r L} = \frac{k(T_i - T_o)}{r \ln(r_o/r_i)} \tag{4.6}$$

The temperature profile, $T(r)$ is given by

$$T(r) = \frac{T_i \ln(r_o/r) + T_o \ln(r/r_i)}{\ln(r_o/r_i)} \tag{4.7}$$

#### 4.2.1.3.5   Spherical Shell

Radial heat flux, $q_r$, at any radius $r$ through a spherical shell of inner radius $r_i$ held at temperature $T_i$, and outer radius $r_o$ held at temperature $T_o$, *e.g.* heat loss through the wall of a storage sphere, is given by

$$q_r = \frac{Q}{4\pi r^2} = \frac{k(T_i - T_o)}{r^2(1/r_i - 1/r_o)}$$

(4.8)

The temperature profile, $T(r)$ is given by

$$T(r) = T_i - \frac{(T_i - T_o)(r - r_i)}{rr_i(1/r_i - 1/r_o)}$$

(4.9)

## 4.2.2 Convection

Convective heat transfer results from the motion of fluids of different temperature relative to each other and is therefore intimately related to fluid flow. In general, we would expect that the convective heat flux $q$ to be proportional to the temperature difference driving force $\Delta T$.

$$q = \frac{Q}{A} = h\Delta T$$

(4.10)

This proportionality constant is known as the convective heat transfer coefficient and is given by the symbol $h$. Eqn (4.10) is the defining equation for the convective heat transfer coefficient. Table 4.1 lists typical ranges of heat transfer coefficients.

Fluid flow is often turbulent, and so heat transfer by convection is often complex and normally we have to resort to correlations of experimental data. Dimensional analysis will give us insight into the pertinent dimensionless groups; see Chapter 6, Scale-Up in Chemical Engineering, Section 6.7.4.

We have noted earlier that fluid flow may be either natural or forced and different variables will come into play in each.

**Table 4.1** Typical ranges of convective heat transfer coefficients.[3]

| Typical values of $h$ | $h$ (W m$^{-2}$ K$^{-1}$) |
| --- | --- |
| Gases | 10–200 |
| Liquids | 100–1000 |
| Low viscosity liquids | 1400–1800 |
| Boiling liquids | 1000–3000 |
| Boiling water | 1000–2000 |
| Condensing vapours | 1000–3000 |
| Condensing steam | 4000–10 000 |

### 4.2.2.1　Forced Convection

In forced convection, flow is generated by some external means, *e.g.* a pump, and heat transfer coefficients would be expected to be a function of

- fluid flow properties, density, $\rho$ and viscosity, $\mu$
- fluid thermal properties, specific heat capacity, $c_p$ and thermal conductivity, $k$
- a characteristic length, $d$
- a characteristic velocity, $u$, generated by the external means

So, $h = f(\rho, \mu, c_p, k, d, u)$.

Dimensional analysis gives the following dimensionless groups:

$$\mathrm{Nu} = \frac{hd}{k} = f\left(\frac{\rho u d}{\mu}, \frac{c_p \mu}{k}\right) = f(\mathrm{Re}, \mathrm{Pr}) \qquad (4.11)$$

As we will see in Chapter 6, dimensionless groups may be considered as ratios. The Nusselt number Nu is a ratio of heat transfer resistances:

$$\mathrm{Nu} = \frac{hd}{k} = \frac{d/k}{1/h} \equiv \frac{\text{conductive resistance}}{\text{convective resistance}}$$

The Reynolds number is a ratio of forces:

$$\mathrm{Re} = \frac{\rho u d}{\mu} = \frac{\rho u^2 d^2}{\mu u d} \equiv \frac{\text{inertial forces}}{\text{viscous forces}}$$

and the Prandtl number is a ratio of diffusivities:

$$\mathrm{Pr} = \frac{c_p \mu}{k} = \frac{\mu/\rho}{k/\rho c_p} \equiv \frac{\text{momentum diffusivity}}{\text{thermal diffusivity}}$$

### 4.2.2.2　Natural Convection

Natural convection arises from the differential thermal expansion of fluids at different temperatures. The resulting density differences give rise to buoyancy driven flows. Heat transfer coefficients would be expected to be a function of

- fluid flow properties, density, $\rho$ and viscosity, $\mu$
- fluid thermal properties, specific heat capacity, $c_p$ and thermal conductivity, $k$
- a characteristic length, $L$

- buoyancy forces that may be characterised by
  - gravitational acceleration, $g$
  - temperature difference between surface and bulk fluid, $\Delta T$
  - coefficient of thermal expansion, $\beta$, which is defined by $\rho = \rho_0(1 + \beta\Delta T)$ and for an ideal gas, $\beta = 1/T(K)$

Now, from the above, $h = f(\rho, \mu, c_p, k, L, g, \Delta T, \beta)$ and dimensional analysis gives

$$\text{Nu} = \frac{hL}{k} = f\left(\frac{\beta g \rho^2 L^3 \Delta T}{\mu^2}, \frac{c_p \mu}{k}\right) = f(\text{Gr}, \text{Pr}) \tag{4.12}$$

For natural convection, the Grashof number, Gr, is found instead of the Reynolds number for forced convection correlations. The Grashof number may be shown to be equivalent to a ratio of forces:

$$\frac{\beta g \rho^2 L^3 \Delta T}{\mu^2} \propto \frac{(\rho_0 - \rho)L^3 g \times \rho u^2 L^2}{(\mu u L)^2} \propto \frac{\text{buoyancy forces} \times \text{inertial forces}}{(\text{viscous forces})^2}$$

Table 4.2 lists heat transfer correlations for forced and natural convection for a number of geometries.

**Table 4.2** Some heat transfer coefficient correlations.[1,2,7]

| Flow and geometry | Conditions | Correlation |
|---|---|---|
| *Forced convection* | | |
| Laminar inside a pipe | $(\text{Re Pr } d/L) > 10$ | $\text{Nu} = 1.86 \, (\text{Re Pr } d/L)^{0.33} (\mu_b/\mu_w)^{0.14}$ |
| Turbulent inside a pipe | $\text{Re} > 10^4$, $0.7 < \text{Pr} < 17\,000$, $L/d > 60$ | $\text{Nu} = 0.023 \, \text{Re}^{0.8} \, \text{Pr}^{0.33} (\mu_b/\mu_w)^{0.14}$ |
| Outside a horizontal cylinder | $1000 < \text{Re} < 1\,00\,000$ | $\text{Nu} = 0.26 \, \text{Re}^{0.6} \, \text{Pr}^{0.3}$ |
| Across a bank of cylinders | $\text{Re} > 3000$ (characteristic length: $d_e$) | $\text{Nu} = 0.36 \, \text{Re}^{0.55} \, \text{Pr}^{0.33} (\mu_b/\mu_w)^{0.14}$ |
| Around a sphere | $1 < \text{Re} < 7 \times 10^4$, $0.6 < \text{Pr} < 400$ | $\text{Nu} = 2 + 0.6 \, \text{Re}^{0.5} \, \text{Pr}^{0.33}$ |
| *Natural convection* | | |
| Outside a horizontal cylinder | $10^4 < \text{Gr Pr} < 10^9$ | $\text{Nu} = 0.525 \, (\text{Gr Pr})^{0.25}$ |
| Around a sphere | $\text{Gr}^{0.25} \, \text{Pr}^{0.33} < 2000$ | $\text{Nu} = 2 + 0.6 \, \text{Gr}^{0.5} \, \text{Pr}^{0.33}$ |
| Vertical flat plate | $10^4 < \text{Gr Pr} < 10^9$ | $\text{Nu} = 0.59 \, (\text{Gr Pr})^{0.25}$ |
| | $10^9 < \text{Gr Pr} < 10^{13}$ | $\text{Nu} = 0.13 \, (\text{Gr Pr})^{0.33}$ |
| Horizontal flat plate | $10^5 < \text{Gr Pr} < 2 \times 10^7$ | $\text{Nu} = 0.54 \, (\text{Gr Pr})^{0.25}$ |
| *Facing upwards* | $2 \times 10^7 < \text{Gr Pr} < 3 \times 10^{10}$ | $\text{Nu} = 0.14 \, (\text{Gr Pr})^{0.33}$ |
| *Facing downwards* | $3 \times 10^5 < \text{Gr Pr} < 3 \times 10^{10}$ | $\text{Nu} = 0.27 \, (\text{Gr Pr})^{0.25}$ |

## 4.2.3  Radiation

All bodies emit and absorb electromagnetic radiation resulting in a transfer of energy. All materials are characterised by their ability to absorb, emit, transmit or reflect electromagnetic radiation.

The absorptivity, $\alpha$, is the ratio of the absorbed energy to the incident energy from the source. A perfect absorber is called a *black body*. Non-perfect absorbers, $1 \geq \alpha \geq 0$, are known as *grey bodies*.

The transmittance, $\tau$, is the ratio of the transmitted energy to the incident energy. The fraction of energy reflected will be $1 - \alpha - \tau$. For a black body, $\tau = 0$.

The emissivity, $\varepsilon$, is the ratio of the emitted energy to that of a perfect emitter. A perfect emitter is also a perfect absorber, a black body. For grey bodies, $\varepsilon = \alpha = $ constant is often assumed. Values of $\varepsilon$ for many materials are given in the literature, *e.g.* ref. 7.

The energy emitted by a body is given by Stefan's law:

$$q = \frac{Q}{A} = \sigma \varepsilon T^4 \tag{4.13}$$

where $\sigma = 5.67 \times 10^{-8}$ $\mathrm{W\,m^{-2}\,K^{-4}}$ and is known as the Stefan–Boltzmann constant.

It can be seen that the energy emitted is a very strong function of temperature. All other things being equal, a hot surface will radiate more energy than a cold surface and therefore there is a net heat transfer from hot to cold surfaces. The surfaces do not need to be in physical contact, but must be in *thermal contact*, *i.e.* they must be able to "see" each other, and material in between the surfaces must be transparent to the radiation.

The net rate of heat transfer from surface 1 to surface 2 is

$$Q_{1 \to 2} = \sigma \varepsilon_1 A_1 (T_1^4 - T_2^4) \tag{4.14}$$

This assumes that all energy emitted by surface 1 reaches surface 2. However, normally only a fraction of the energy emitted by surface 1 will reach surface 2. This fraction is called the *view factor*, $F_{12}$. Thus

$$Q_{1 \to 2} = \sigma \varepsilon_1 A_1 F_{12} (T_1^4 - T_2^4) \tag{4.15}$$

Since for two black bodies at equal temperatures the absorbed energy must equal the emitted energy, it can be shown that $A_1 F_{12} = A_2 F_{21}$.

The detailed analysis of radiative heat transfer can easily become extremely complicated when transmission, reflection and complex geometries are taken into account.[4] An important conclusion which may be reached is that the heat flux corresponding to the surrounding black body is the maximum radiative heat flux that may be achieved, *i.e.* for $F_{12} = 1$ and $\varepsilon = 1$:

$$q_{\text{RadB}} = \frac{Q_{\text{RadB}}}{A_1} = \sigma(T_1^4 - T_2^4) \tag{4.16}$$

As a rough approximation, one can define a radiative heat transfer coefficient and describe the heat flux using

$$q = \frac{Q}{A_1} = h_{\text{Rad}}(T_1 - T_2) \tag{4.17}$$

where

$$h_{\text{Rad}} = \frac{\sigma(T_1^4 - T_2^4)}{T_1 - T_2} \tag{4.18}$$

Figure 4.4 shows the predictions of eqn (4.18) for various values of $T_1$ and $T_2$.

**Figure 4.4** Maximum obtainable values for $h_{\text{Rad}}$.

## 4.3   The Overall Heat Transfer Coefficient, *U*

When considering heat transfer by conduction, we have neglected the mechanism by which heat is transferred to and from the outer solid surfaces. In convective heat transfer, heat is transferred to and from flowing fluids.

When a fluid is present in contact with each solid wall, there will be an additional resistance to heat transfer in each fluid boundary layer or "film". The combined mechanism of heat transfer from a hot fluid through a dividing wall to a cold fluid has many similarities to conduction through a composite slab reviewed earlier.

Figure 4.5 shows hot and cold fluids separated by a flat, conducting, dividing wall. Overall, heat is transferred from the bulk of the hot fluid to the bulk of the cold fluid. In detail, hot fluid in the turbulent bulk will be at a uniform temperature $T_h$. Next to the wall will be a boundary layer through which heat is transferred to the wall. Heat is then transferred through the dividing wall by conduction to the cold fluid, and temperature drops from $T_1$ to $T_2$. Next to the wall on the cold side will be another boundary layer, through which heat is transferred into the cold side turbulent bulk, temperature $T_c$.

The boundary layers are often called *"films"*, and heat transfer by conduction predominates in them giving rise to significant resistances to heat transfer. We assume that all the resistance to heat transfer on the hot side is confined to the hot film and all that on the cold side is confined to the cold film.

The associated heat transfer coefficients for these films are known as *"film heat transfer coefficients"*, $h_h$ and $h_c$ for the hot and cold side, respectively. The heat flux through the hot film is given by $q = h_h(T_h - T_1)$, through the wall by $q = (k_w/b_w)(T_1 - T_2)$ and through

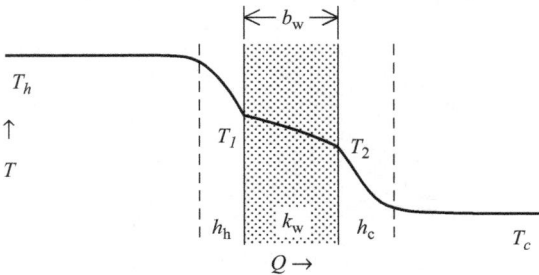

**Figure 4.5**   Heat transfer from a hot fluid, though a wall, to a cold fluid.

the cold film by $q = h_c(T_2 - T_c)$. At steady state, the heat fluxes are constant and it is easy to show that

$$q = \frac{Q}{A} = \frac{(T_h - T_c)}{(1/h_h) + (b_w/k_w) + (1/h_c)} = U(T_h - T_c) \tag{4.19}$$

$U$ is known as the *"overall heat transfer coefficient"* and relates the heat flux to the overall temperature difference $T_h - T_c$.

The overall heat transfer coefficient is related to the individual film coefficients by

$$\frac{1}{U} = \frac{1}{h_h} + \frac{b_w}{k_w} + \frac{1}{h_c} \tag{4.20}$$

Eqn (4.20) may be compared to eqn (4.4) for the heat flux through a composite plate. As we have seen, $(b_w/k_w)$ is the thermal resistance of the wall to heat transfer by conduction. Similarly, $(1/h_h)$ and $(1/h_c)$ are the resistances to heat transfer offered by the hot and cold films, respectively.

The reciprocal of the overall heat transfer coefficient, $1/U$, is equivalent to the overall resistance to heat transfer, being equal to the sum of the resistances in the hot film, the wall and the cold film.

## 4.3.1  Example Calculation of $U$

Water flows through a mild steel pipe: $k = 45$ W m$^{-1}$ K$^{-1}$ and wall thickness 1 mm. Air flows around the outside of the pipe. If $h_{air} = 50$ W m$^{-2}$ K$^{-1}$ and $h_{water} = 1000$ W m$^{-2}$ K$^{-1}$, estimate the overall heat transfer coefficient $U$.

$$U = \frac{1}{(1/h_{air}) + (b_w/k_w) + (1/h_{water})}$$

$$= \frac{1}{(1/50) + (0.001/45) + (1/1000)}$$

$$= \frac{1}{0.02 + 0.000022 + 0.001} = 47.6 \text{ W m}^{-2} \text{ K}^{-1}$$

It can be seen that in this example the largest resistance, the controlling film, is that of the air, $1/U \approx 1/h_{air}$, *i.e.* $U \approx h_{air}$, and so the air film controls the overall heat transfer process. If we wish to increase the overall rate of heat transfer by increasing $U$, we should look at the dominating resistance, the air film, and try to reduce it, *e.g.* by increasing the air-side velocity. Reducing the water film resistance has very little effect upon $U$.

### 4.3.2 Overall Heat Transfer Coefficients with Curvature, for Example, Transfer through a Pipe Wall

Here we will consider the problem of heat transfer through a cylindrical wall, for example, the wall of a pipe or tube in a shell-and-tube heat exchanger (Figure 4.6).

The area for heat transfer varies with radius; therefore, we define the overall heat transfer coefficient based upon the outer area of the cylinder, $A_o = 2\pi r_o L$, where $L$ is the length of the cylinder, giving

$$Q = U_o A_o (T_h - T_c) \tag{4.21}$$

$$\frac{1}{U_o} = \frac{1}{h_o} + \frac{r_o \ln(r_o/r_i)}{k_w} + \frac{r_o}{r_i h_i} \tag{4.22}$$

We again see that the reciprocal of the overall heat transfer coefficient is equivalent to the overall resistance to heat transfer, which is equal to the sum of resistances in the hot film $(1/h_o)$, the wall $[r_o \ln(r_o/r_i)]/k_w$ and the cold film $(r_o/r_i h_i)$, as shown in eqn (4.22). However in this case, the expressions for the resistances in the wall and the cold film are modified to take the curvature into account. No such modification is required for the outside film resistance, as the overall heat transfer coefficient is defined based on the outside surface area of the cylinder.

### 4.3.3 Overall Heat Transfer Coefficients with Convection and Radiation

For processes in parallel, the conductance is additive; so for combined convective and radiative heat transfer at a surface

$$h_{Tot} = h_{Conv} + h_{Rad} \tag{4.23}$$

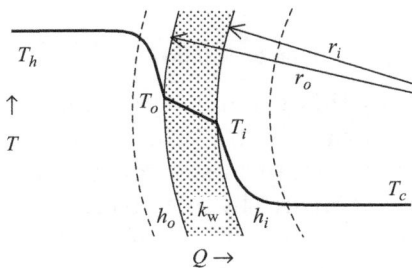

**Figure 4.6** Heat transfer through a cylindrical pipe wall.

For example, for heat transfer through a flat wall, where heat is transferred by convection and radiation from the wall, the overall heat transfer coefficient is related to the individual coefficients by

$$\frac{1}{U} = \frac{1}{h_{\mathrm{h}}} + \frac{b_{\mathrm{w}}}{k_{\mathrm{w}}} + \frac{1}{h_{\mathrm{Conv}} + h_{\mathrm{Rad}}} \tag{4.24}$$

Since the two contributions are additive, we need to consider radiation only when the two coefficients are comparable. Therefore, it is often not necessary to carry out the detailed analysis of radiative heat transfer, since it is possible to obtain the value of $h_{\mathrm{Conv}}$ from correlations and compare it to the maximum value of $h_{\mathrm{Rad}}$, as shown in Figure 4.4.

## 4.4 Transient Heat Transfer

When manufacturing goods or sterilising lines, one often needs to know how long it will take for a certain part of a body to reach a specified temperature. In order to address this issue, it is necessary to consider the dynamic behaviour of heat transfer and write the balance equations introducing time. In the case of conduction in a solid with plate geometry, this becomes

$$\frac{\partial T}{\partial t} = \frac{k}{\rho c_{\mathrm{p}}} \frac{\partial^2 T}{\partial x^2} \tag{4.25}$$

or for conduction in three dimensions

$$\frac{\partial T}{\partial t} = \frac{k}{\rho c_{\mathrm{p}}} \left( \frac{\partial^2 T}{\partial x^2} + \frac{\partial^2 T}{\partial y^2} + \frac{\partial^2 T}{\partial z^2} \right) \tag{4.26}$$

These partial differential equations need to be solved with the associated initial and boundary conditions. A comprehensive list of solutions is available in Carslaw and Jaeger.[5]

Williamson and Adams[6] presented data, shown in Figure 4.7, for the dimensionless centre temperature $(T_{\mathrm{c}} - T_{\mathrm{surf}})/(T_0 - T_{\mathrm{surf}})$ of variously shaped bodies as a function of dimensionless time Fo. Here $T_{\mathrm{c}}$ is the centre point temperature, $T_0$ the initial body temperature and $T_{\mathrm{surf}}$ the surface temperature. Fo is the Fourier number, which is given by $\mathrm{Fo} = \alpha t/\delta^2$, where $\alpha$ is the thermal diffusivity, $t$ is the time and $\delta$ is the characteristic length for conduction, the distance of the centre point or centreline of the body to the nearest part of the surface. The data given in Figure 4.7 neglects the thermal resistance at the surface.

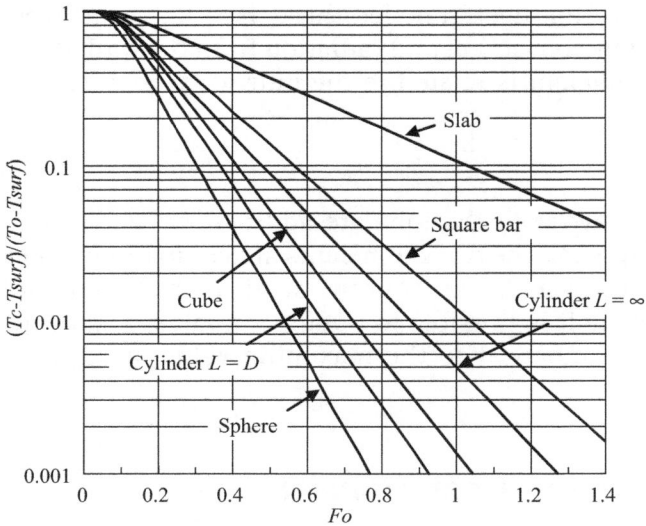

**Figure 4.7**  Central dimensionless temperature variation with dimensionless temperature for a number of bodies, with negligible surface resistance.[6]

When the heat transfer resistance is concentrated at the surface of a body of area $A$, the *thermally thin body* approximation can be applied, and the temperature throughout the solid, $T$, is assumed to be independent of position. In this case, the heat balance becomes

$$m_{solid}c_{p_{solid}} \frac{dT}{dt} = h_{fluid}A(T_{surf} - T) \tag{4.27}$$

or

$$\frac{T - T_{surf}}{T_0 - T_{surf}} = \exp\left(-\frac{h_{fluid}A}{m_{solid}c_{p_{solid}}}t\right) \tag{4.28}$$

To determine whether the thin body approximation may be used, one should compare the surface heat transfer coefficient, $h_{fluid}$ and the thermal conductance of the solid, $k_{solid}/\delta$. Their ratio is the Biot number,

$$Bi = \frac{h_{fluid}\delta}{k_{solid}} = \frac{\delta/k_{solid}}{1/h_{fluid}} = \frac{\text{conductive resistance of solid}}{\text{convective resistance of fluid}} \tag{4.29}$$

When Bi is very small, the thin body approximation can be used, *e.g.* a small metal sphere cooled by air ($k_{solid}$ large, $\delta$ and $h_{fluid}$ small).

When Bi is very large, the external resistance is negligible, *e.g.* large body with low thermal conductivity immersed in a flowing liquid ($k_{solid}$ small, $\delta$ and $h_{fluid}$ large).

## 4.4.1 Example Calculation, Small Bi Case: Cooling of a Copper Sphere in Air

A copper sphere 1 cm in diameter initially at 100 °C is cooled in a stream of air at 20 °C, flowing at 5 m s$^{-1}$. Estimate the time taken for the sphere to reach 25 °C.

### 4.4.1.1 Physical Properties:[7,8]

|  | Copper | Air |
|---|---|---|
| Density ($\rho$, kg m$^{-3}$) | 8300 | 1.3 |
| Specific heat ($c_p$, J kg$^{-1}$ K$^{-1}$) | 419 | 812 |
| Thermal conductivity ($k$, W m$^{-1}$ K$^{-1}$) | 372 | 0.0234 |
| Viscosity ($\mu$, Pa s) | — | 0.000018 |

### 4.4.1.2 Estimate h and, hence, Bi

$$\text{Re}_{air} = \frac{\rho u d}{\mu} = \frac{1.3 \times 5 \times 0.01}{0.000018} = 3611$$

$$\text{Pr}_{air} = \frac{c_p \mu}{k} = \frac{812 \times 0.000018}{0.0234} = 0.62$$

From Table 4.2 for forced convection to a sphere under these conditions,

$$\text{Nu}_{air} = 2 + 0.6\text{Re}^{0.5}\text{Pr}^{0.33} = 2 + 0.6 \times 3611^{0.5} \times 0.62^{0.33} = 32.9$$

Heat transfer coefficient,

$$h_{air} = \frac{\text{Nu}_{air} \times k_{air}}{d} = \frac{32.9 \times 0.0234}{0.01} = 76.9 \text{ W m}^{-2}\text{ K}^{-1}$$

Biot number,

$$\text{Bi} = \frac{h_{air}\delta}{k_{Cu}} = \frac{76.9 \times 0.01/2}{372} = 0.001$$

Bi is small and so the thermal resistance of the air film dominates the overall heat transfer process.

### 4.4.1.3  Estimate Time

Rearranging eqn (4.28) and solving for time $t$ when $T = 25\ °C$

$$t = -\frac{m_{Cu}c_{p_{Cu}}}{h_{air}A}\ln\left(\frac{T - T_{surf}}{T_0 - T_{surf}}\right)$$

$$= -\frac{0.004346 \times 419}{76.9 \times 0.000314}\ln\left(\frac{25 - 20}{100 - 20}\right) = 209\,s = 3.5\,min$$

## 4.4.2  Example Calculation, Large Bi Case: Cooling of a Perspex Plate in Water

A vertical Perspex plate, 3 cm thick, 10 cm high and initially at 100 °C, is cooled by immersion in a large water bath at 20 °C. Estimate the time taken for the centre of the plate to reach 25 °C.

### 4.4.2.1  Physical Properties[7,8]

|                                                 | Perspex | Water    |
|-------------------------------------------------|---------|----------|
| Density ($\rho$, kg m$^{-3}$)                    | 1180    | 1000     |
| Specific heat ($c_p$, J kg$^{-1}$ K$^{-1}$)     | 1440    | 4182     |
| Thermal conductivity ($k$, W m$^{-1}$ K$^{-1}$) | 0.184   | 0.6      |
| Viscosity ($\mu$, Pa s)                          | —       | 0.001    |
| Coefficient of thermal expansion ($\beta$, K$^{-1}$) | —  | 0.000888 |

### 4.4.2.2  Estimate h and, hence, Bi

As $\Delta T$ changes with time from 80 to 5 °C, an average $\Delta T$ of 25 °C is taken to estimate Gr.

$$Gr_{Water} = \frac{\beta g \rho^2 L^3 \Delta T}{\mu^2}$$

$$= \frac{0.000888 \times 9.81 \times 1000^2 \times 0.1^3 \times 25}{0.001^2} = 2.2 \times 10^8$$

$$Pr_{Water} = \frac{c_p \mu}{k} = \frac{4182 \times 0.001}{0.6} = 6.97$$

From Table 4.2, for natural convection from a vertical plate with Gr Pr $= 1.533 \times 10^9$,

$$Nu_{water} = 0.13(Gr\ Pr)^{0.33} = 0.13 \times (1.533 \times 10^9)^{0.33} = 139$$

Heat transfer coefficient,

$$h_{\text{water}} = \frac{\text{Nu}_{\text{water}} \times k_{\text{water}}}{L} = \frac{139 \times 0.6}{0.1} = 836 \, \text{W m}^{-2} \, \text{K}^{-1}$$

Biot number,

$$\text{Bi} = \frac{h_{\text{water}} \delta}{k_{\text{perspex}}} = \frac{836 \times 0.03/2}{0.184} = 68$$

Bi is large, and so the thermal resistance of the water film may be neglected.

### 4.4.2.3 Estimate Time

Dimensionless final temperature,

$$\frac{T - T_{\text{surf}}}{T_0 - T_{\text{surf}}} = \frac{25 - 20}{100 - 20} = 0.0625$$

From Figure 4.7, for a slab the equivalent Fourier number Fo at this dimensionless temperature is 1.22.

Now, $\text{Fo} = \alpha t/\delta^2$. Rearranging and solving for time $t$, when $T = 25 \, °\text{C}$,

$$t = \frac{\text{Fo} \delta^2}{\alpha_{\text{perspex}}} = \frac{\text{Fo} \delta^2}{(k_{\text{perspex}}/\rho_{\text{perspex}} c_{p_{\text{perspex}}})}$$

$$= \frac{1.22 \times (0.03/2)^2}{0.184/(1180 \times 1440)} = 2535 \, \text{s} = 42.2 \, \text{min}$$

## 4.5 Heat Exchangers

A heat exchanger is any device that effects transfer of thermal energy between two fluids that are at different temperatures. The two fluids do not come in direct contact but are separated by a solid surface or tube wall. Common heat exchangers include

- Shell-and-tube
- Flat plate
- Finned tubes

Applications of heat exchangers include air conditioning systems and car/truck radiators, but they are also widely employed in the process industry whenever there is a need for cooling or heating up a process fluid.

In practice, engineers often have to do the following:

- *Select* a heat exchanger that will achieve a specific temperature change in a fluid stream of known mass flow rate.
- *Design* a heat exchanger, *i.e.* determine the surface area to transfer heat at a given rate for given fluid temperatures and flow rates.
- *Predict* the outlet temperatures of hot and cold fluid streams in a specified heat exchanger.

## 4.5.1 Double Pipe Heat Exchanger

The simplest form of a heat exchanger consists of two concentric cylindrical tubes, the *double pipe heat exchanger*, as schematically represented in Figure 4.8:

In the schematic in Figure 4.8, a hot fluid "A" flows in the inner tube and exchanges heat through an inner wall with a cooler fluid "B", which flows in the outer tube. Heat transfer involves convection in each fluid (with $h_o$ and $h_i$ being the convective heat transfer co-efficients of the fluid in the outer and inner tube, respectively) and conduction through the wall of thermal conductivity $k_w$ separating the two fluids.

Two different types of heat exchangers can be identified. In the *co-current-flow* heat exchanger, both fluids flow in the same direction, as shown in Figure 4.9. In this case, the two fluids enter the heat exchanger from the same end.

In a *counter-current-flow* heat exchanger, the fluids flow in opposite directions, as shown in Figure 4.10. In this case, the hot and cold

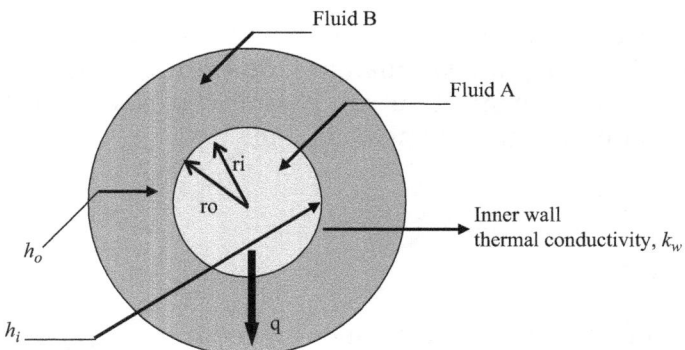

Direction of heat flow if fluid "A" is hotter than fluid "B"

**Figure 4.8** Schematic representation of a double pipe heat exchanger.

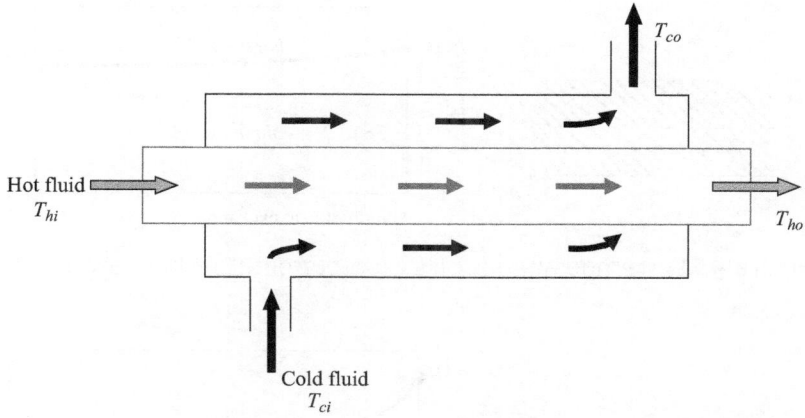

**Figure 4.9** Schematic representation of a co-current-flow heat exchanger.

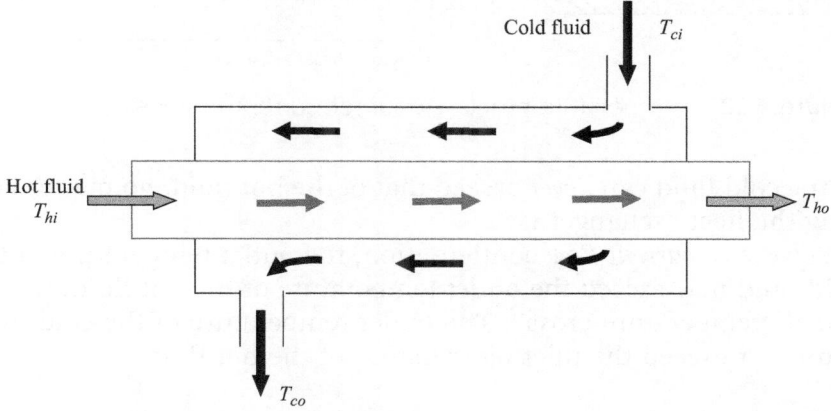

**Figure 4.10** Schematic representation of a counter-current-flow heat exchanger.

fluids enter the heat exchanger from opposite ends. In both the parallel and counter-flow configurations, the two fluids are forced to flow along the heat exchanger by either pumps or fans.

In the schematics depicted in Figures 4.9 and 4.10, the inlet and outlet temperatures of the hot fluid $(T_{hi}, T_{ho})$ and cold fluid $(T_{ci}, T_{co})$ are also shown. The temperature profiles for the two configurations are shown in Figures 4.11 and 4.12.

For the *co-current-flow* configuration, the temperature difference $\Delta T$ is large at the inlet but decreases exponentially towards the outlet. The temperature of the hot fluid decreases and the temperature of the cold fluid increases along the heat exchanger. The outlet temperature

**Figure 4.11**   Temperature profile – co-current-flow heat exchanger.

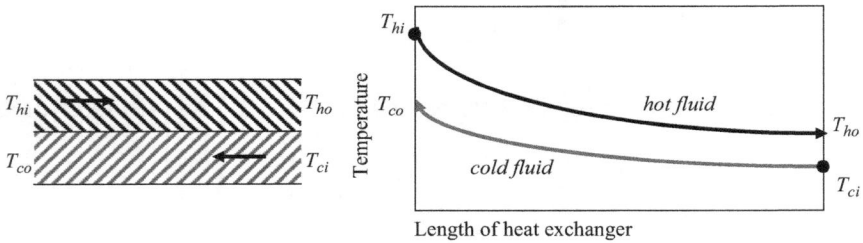

**Figure 4.12**   Temperature profile – counter-current-flow heat exchanger.

of the cold fluid can *never* exceed that of the hot fluid, no matter how long the heat exchanger is.

For the *co-current-flow* configuration, the outlet temperature of the cold fluid may exceed the outlet temperature of the hot fluid, the so-called "temperature cross". The outlet temperature of the cold fluid can never exceed the inlet temperature of the hot fluid.

### 4.5.2   Heat Transfer in Heat Exchangers

Heat exchangers operate for long periods with no change in the operating conditions; thus, they can be modelled as *steady-flow* devices, for which the following assumptions are valid:

- The overall heat transfer coefficient, $U$, is constant with time and is independent of the position along the exchanger.
- The mass flow rate of each fluid remains constant.
- The specific heats of the fluids are constant.
- The temperature of the two fluids is constant over a specific cross-section.
- The outer surface is perfectly insulated, so that any heat transfer occurs between the two fluids only, and there are no heat losses to the surroundings.

Under these assumptions, it follows that the *rate of heat transfer, q,* from the hot fluid is equal to the rate of heat transfer to the cold one. The basic design equations for heat exchangers are therefore the energy balances for each fluid:

*Energy given up by hot fluid:*

$$q = - G_h c_{ph} (T_{ho} - T_{hi})$$ (4.30)

*Energy gained by cold fluid:*

$$q = G_c c_{pc} (T_{co} - T_{ci})$$ (4.31)

where $G$ (kg s$^{-1}$) is the mass flow rate and $c_p$ (kJ kg$^{-1}$ K$^{-1}$) is the specific heat, and the defining equation for $U$ (eqn (4.19)) which is reiterated here:

$$q = UA\Delta T$$ (4.19)

### 4.5.2.1  Log Mean Temperature Difference, $\Delta T_{lm}$

The temperature difference between the hot and cold fluids, $\Delta T$, varies along the length of the heat exchanger. Therefore, we have to express $\Delta T$ in terms of a suitable mean temperature difference, the log mean temperature difference, $\Delta T_{lm}$. This can be determined by simply applying the design eqn (4.30), (4.31) and (4.19) to a cross-section of the heat exchanger, as represented in Figure 4.13 for the case of *co-current-flow*.

Based on the assumptions set out, the heat transfer rate for a differential length is:

$$\delta q = U(T_h - T_c)\delta A$$ (4.32)

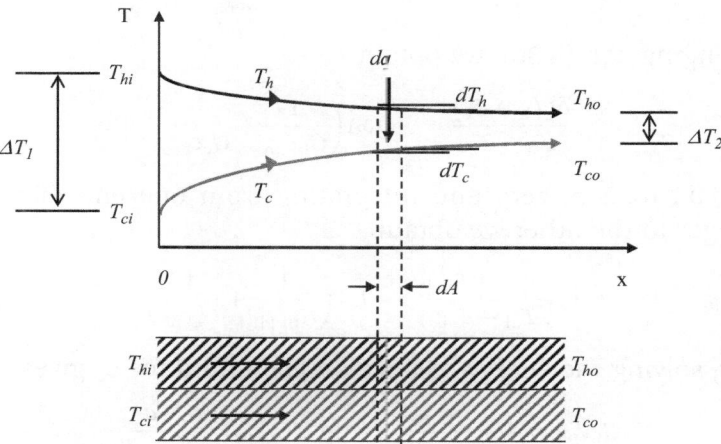

**Figure 4.13** Temperature profile along a co-current-flow heat exchanger.

Having assumed that the outer surface of the exchanger is perfectly insulated, the energy gained by the cold fluid is equal to that given up by the hot fluid. Thus, an energy balance on each fluid in the differential section is as follows:

*Energy given up by hot fluid:*

$$\delta q = -G_h c_{ph} \delta T_h \tag{4.33}$$

*Energy gained by cold fluid:*

$$\delta q = G_c c_{pc} \delta T_c \tag{4.34}$$

Note that the temperature change of the hot fluid, $\delta T_h$, is a negative quantity to make the heat transfer a positive quantity. Solving the above equations for $\delta T_h$ and $\delta T_c$ gives

$$\delta T_h = -\frac{\delta q}{G_h c_{ph}} \quad \text{and} \quad \delta T_c = \frac{\delta q}{G_c c_{pc}}$$

Taking their difference, we get

$$\delta T_h - \delta T_c = \delta(T_h - T_c) = -\delta q \left( \frac{1}{G_h c_{ph}} + \frac{1}{G_c c_{pc}} \right) \tag{4.35}$$

The rate of heat transfer in the differential section of the exchanger is also given by eqn (4.32), reported below for convenience:

$$\delta q = U(T_h - T_c) \delta A \tag{4.32}$$

Substituting this into eqn (4.35) gives

$$\delta(T_h - T_c) = -U\delta A(T_h - T_c) \left( \frac{1}{G_h c_{ph}} + \frac{1}{G_c c_{pc}} \right) \tag{4.36}$$

Re-arranging eqn (4.36), we obtain

$$\frac{\delta(T_h - T_c)}{T_h - T_c} = -U\delta A \left( \frac{1}{G_h c_{ph}} + \frac{1}{G_c c_{pc}} \right) \tag{4.37}$$

Letting $\delta A$ tend to zero and integrating from one end of the heat exchanger to the other we obtain

$$\ln \frac{(T_{ho} - T_{co})}{T_{hi} - T_{ci}} = -UA \left( \frac{1}{G_h c_{ph}} + \frac{1}{G_c c_{pc}} \right) \tag{4.38}$$

Finally, solving the energy balance on each fluid for $c_p$ gives

$$G_h c_{ph} = -\frac{q}{T_{ho} - T_{hi}} \quad \text{and} \quad G_c c_{pc} = \frac{q}{T_{co} - T_{ci}} \tag{4.39}$$

Substituting eqn (4.39) into eqn (4.38), we obtain

$$\ln \frac{(T_{ho} - T_{co})}{(T_{hi} - T_{ci})} = -UA \frac{(T_{hi} - T_{ho}) + (T_{co} - T_{ci})}{q} \tag{4.40}$$

Or, in terms of the differences in end temperatures (see Figure 4.14)

$$\ln \frac{\Delta T_2}{\Delta T_1} = UA \frac{\Delta T_2 - \Delta T_1}{q} \tag{4.41}$$

Thus, the heat transfer rate is equal to

$$q = UA \frac{\Delta T_2 - \Delta T_1}{\ln(\Delta T_2 / \Delta T_1)} = UA \, \Delta T_{1m} \tag{4.42}$$

where the *log mean temperature difference*, $\Delta T_{1m}$, is the mean temperature along the heat exchanger:

$$\Delta T_{1m} = \frac{\Delta T_2 - \Delta T_1}{\ln(\Delta T_2 / \Delta T_1)} \tag{4.43}$$

The $\Delta T_1$ and $\Delta T_2$ expressions in *co-current-flow* are given by the difference between the inner and outer temperatures of the two fluids, as shown in Figure 4.14.

The procedure to obtain $\Delta T_{1m}$ is the same for a *counter-current-flow* heat exchanger. It is again given by eqn (4.43); however in this case, $\Delta T_1$ is the temperature difference between the inlet and outlet temperature of the hot fluid and the cold fluid respectively at one end of the exchanger, and $\Delta T_2$ is the temperature difference between the hot fluid and the cold fluid at the other end, as shown in Figure 4.15.

Therefore, whether considering *co-current-flow* or *counter-current-flow*, the procedure to obtain $\Delta T_{1m}$ is the same.

**Figure 4.14** Temperature profile showing $\Delta T_1$ and $\Delta T_2$ for a co-current-flow heat exchanger.

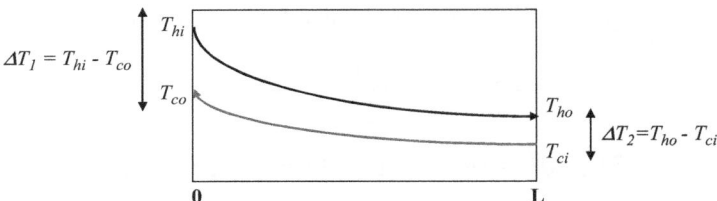

**Figure 4.15**  Temperature profile showing $\Delta T_1$ and $\Delta T_2$ for a counter-current-flow heat exchanger.

### 4.5.2.2  *Comparison between Co-current-flow and Counter-current-flow Heat Exchangers*

The worked examples reported below will demonstrate the different efficiencies of the two heat exchanger configurations in terms of the heat transfer area required in both cases for the same *U*. Let us consider two fluids between which heat is being exchanged. For a given amount of heat exchanged, *q*, the log mean temperature difference, $\Delta T_{1m}$, will be calculated for both *co-current-flow* and *counter-current-flow* and the heat transfer area required in both cases compared, assuming that *U* is unchanged. Both inlet and outlet temperatures of each fluid are given:

| Hot fluid | Cold fluid |
|---|---|
| $T_{hi} = 200\ ^\circ C$ | $T_{ci} = 80\ ^\circ C$ |
| $T_{ho} = 150\ ^\circ C$ | $T_{co} = 120\ ^\circ C$ |

Using the given data, $\Delta T_{1m}$ for *co-current-flow* is given by

$$\Delta T_{1m} = \frac{\Delta T_2 - \Delta T_1}{\ln(\Delta T_2/\Delta T_1)} = \frac{(150 - 120) - (200 - 80)}{\ln[(150 - 120)/(200 - 80)]} = 64.9\,^\circ C$$

For *counter-current-flow*

$$\Delta T_{1m} = \frac{\Delta T_2 - \Delta T_1}{\ln(\Delta T_2/\Delta T_1)} = \frac{(150 - 80) - (200 - 120)}{\ln[(150 - 80)/(200 - 120)]} = 74.9\,^\circ C$$

For a given duty, *q*, and assuming that *U* is the same for parallel and counter-flow exchangers; the heat transfer area required is given by eqn (4.42):

$$A = \frac{q}{U\Delta T_{1m}}$$

From which

$$\frac{A_{\text{parallel}}}{A_{\text{counter}}} = \frac{\Delta T_{1m\,\text{counter}}}{\Delta T_{1m\,\text{parallel}}} = \frac{74.9}{64.9} = 1.15$$

The ratio between the heat transfer areas required for parallel and counter flow shows that the heat transfer area required with the parallel flow configuration is larger by 15% than with counter flow. This demonstrates the higher efficiency of the counter-flow heat exchanger compared to the parallel flow configuration, which would require a larger and, thus, more expensive heat exchanger to perform the same duty.

## 4.5.3 Performance of Heat Exchangers – Fouling

The performance of a heat exchanger depends upon the transfer surfaces being clean and uncorroded. The performance deteriorates with time due to accumulation of deposits on the heat transfer surfaces. The layer of deposits represents additional resistance to heat transfer. In the design of heat exchangers, this added resistance is accounted for by a *fouling factor* or *dirt factor*, $R_d$.

The most common type of fouling is the precipitation of solid deposits in a fluid on the heat transfer surfaces. A layer of calcium-based deposits forms after prolonged use on the surfaces at which boiling occurs, similar to what can be observed on the inner surface of a kettle. To avoid this potential problem, water in process plants is treated to remove its solid content.

Another form of fouling is corrosion or chemical fouling due to accumulation of the products of chemical reactions. This can be avoided by coating the metal surfaces with glass or by using plastic pipes. Heat exchangers may also be fouled by the growth of algae in warm fluids, biological fouling that can be prevented by chemical treatment.

Fouling has to be considered in the design and selection of heat exchangers. Larger and more expensive heat exchangers may be necessary to ensure the heat transfer requirements even after fouling occurs. Periodic cleaning of exchangers and the resulting down time are additional penalties associated with fouling.

Thermal resistance due to fouling has to be included in the calculation of the overall heat transfer coefficient. By definition, the overall heat transfer coefficient is the reciprocal of the overall thermal resistance. Thus, including the thermal resistance due to fouling, it follows

$$\frac{1}{U} = \frac{1}{(1/h_0) + R_{do} + (b_w/k_w) + R_{di} + (1/h_i)} \tag{4.44}$$

Typical values for $R_d$ are given in the literature.[1,3,7]

### 4.5.4 Types of Heat Exchangers

#### 4.5.4.1 Shell-and-tube

The most widely used types of heat exchangers in the process industry are shell-and-tube heat exchangers. This can be one-shell pass and one-tube pass (see Figure 4.16) or multi-pass shell-and-tube heat exchangers (as shown in Figure 4.17 for the case of one-shell pass and two-tube passes). Baffles are installed in the shell in order to promote mixing and cross-flow, thereby giving higher heat transfer coefficients.

The flow arrangement in shell-and-tube heat exchangers can involve both *co-current-flow* and *counter-current-flow*, as shown in the schematic in Figure 4.17.

**Figure 4.16** Schematic of a shell-and-tube heat exchanger (one-shell pass and one-tube pass).

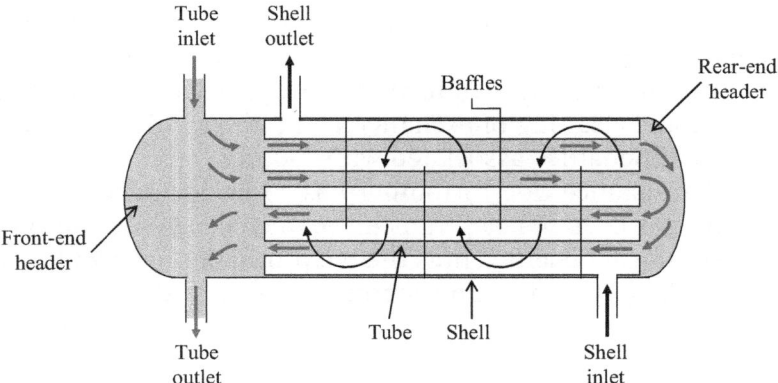

**Figure 4.17** Schematic of a shell-and-tube heat exchanger (one-shell pass and two-tube pass).

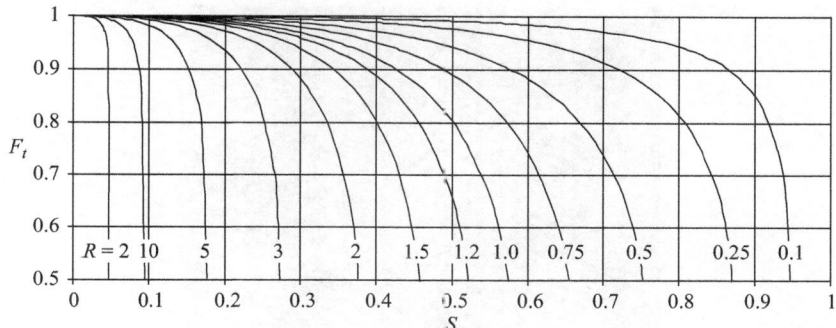

**Figure 4.18**   LMTD correction factor $F_t$, for 1 shell pass and 2 or more tube passes.[1]

Determination of the average temperature difference, $\Delta T_{lm}$, is very complex for these types of heat exchangers. In the top half of the exchanger illustrated in Figure 4.17, there is parallel flow and in the bottom half, counter flow. For this reason, it is common practice to introduce a *correction factor*, $F_t$, into eqn (4.42). The heat transfer rate is therefore given by

$$q = UAF_t\Delta T_{lm} \tag{4.45}$$

where $\Delta T_{lm}$ is that for the counter-flow double-pipe heat exchanger with the same fluid inlet and outlet temperatures as in the more complex design shown in Figure 4.17. $F_t$ values for one-shell pass and two or more tube passes are given in Figure 4.18, where

$$R = \frac{T_{s1} - T_{s2}}{T_{t2} - T_{t1}} \quad \text{and} \quad S = \frac{T_{t2} - T_{t1}}{T_{s1} - T_{t1}} \tag{4.46}$$

Derivations of $F_t$ for various geometries are given by Kern,[1] and charts for $F_t$ are given in the literature.[1,3,7]

### 4.5.4.2   Plate Heat Exchangers

Plate heat exchangers are built up from individual plates separated by gaskets. These plates are assembled into a pack and clamped in a frame, as shown in Figure 4.19. They are applied in the energy recovery section of many processes because of their *high efficiency* and *low costs*. In this type of heat exchanger, fluids are in *co-current-flow* and *counter-current-flow* in alternate channels, as shown in Figure 4.20.

**Figure 4.19**  Plate heat exchanger (Picture courtesy of WCR Inc. of Dayton, Ohio).

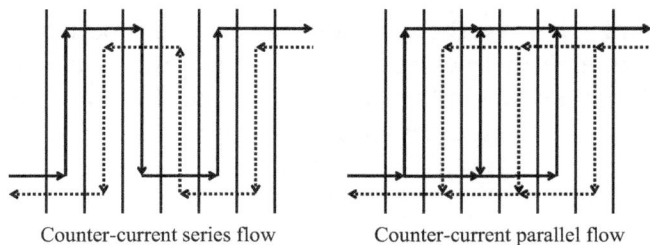

Counter-current series flow          Counter-current parallel flow

**Figure 4.20**  Flow channels in a plate heat exchanger.

Plates are made of thin sheet material, resulting in economic units, particularly when expensive material is involved. Plates are especially corrugated to promote turbulence even at low Re, resulting in

- very high heat transfer coefficients
- reduced fouling
- easy chemical cleaning

They are applied to various processes, such as the energy recovery sections of crystallisation and desalination plants.

### 4.5.4.3 Finned Heat Exchangers

Finned tube heat exchangers are used when heat is to be transferred between gases and liquids. The finned tubes provide a large heat transfer area, since gases, which are good thermal insulators, transfer smaller amounts of heat than liquids.

There are various types of finned tubes depending on the application. They are applied, for example, in large air conditioning systems, for radiators of vehicles or where cooling water is in short supply.

## 4.5.5 Numerical Examples

### 4.5.5.1 Cooling Hot Process Fluid in a Counter-flow Heat Exchanger

A counter-flow double-pipe heat exchanger is used to cool a hot process fluid using water. The process fluid flows at 18 kg s$^{-1}$ and is cooled from 105 °C to 45 °C. The water flows counter-currently to the process fluid, entering at 25 °C and leaving at 50 °C. Assuming no heat losses, calculate the required flow rate for the cooling water. Neglecting the tube wall curvature, calculate the required area for heat exchange.

The specific heat for water is 4.2 kJ kg$^{-1}$ K$^{-1}$ and that of the process fluid is 3.4 kJ kg$^{-1}$ K$^{-1}$. The process fluid side film heat transfer coefficient is 2500 W m$^{-2}$ K$^{-1}$ and the cooling water side heat transfer coefficient is 1200 W m$^{-2}$ K$^{-1}$. The tube wall thickness is 3 mm and the thermal conductivity is 220 W m$^{-1}$ K$^{-1}$.

#### 4.5.5.1.1 Problem Solution

Calculate the required flow rate for the cooling water: The heat given up by hot process fluid is equal to

$$Q_f = G_f c_{pf}(T_{hi} - T_{ho}) = 18 \times 3.4 \times 10^3 \times (105 - 45)$$

$$= 3\,672\,000 \text{ W} = 3672 \text{ kW}$$

From the energy balance eqn (4.30) and (4.32), the heat given up by hot process fluid is equal to the heat gained by the cooling water; thus, we can obtain the flow rate required for the cooling water:

$$G_w = \frac{Q_w}{c_{pw}(T_{co} - T_{ci})} = \frac{3672 \times 10^3}{4.2 \times 10^3 \times (50 - 25)} = 34.97 \text{ kg s}^{-1}$$

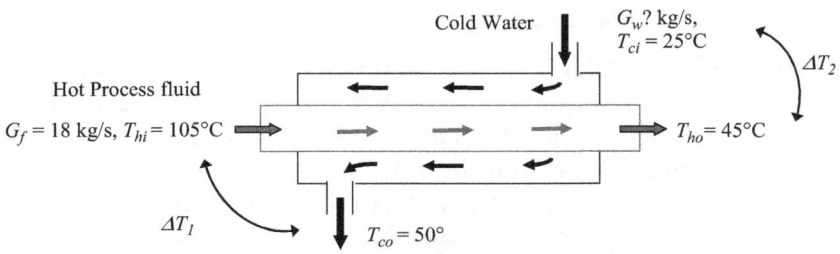

**Figure 4.21**  Problem double-pipe heat exchanger.

Calculate the required area for heat exchange: The heat exchanged is given by eqn (4.42), where the log mean temperature difference is given by (see Figure 4.21):

$$\Delta T_{1m} = \frac{\Delta T_2 - \Delta T_1}{\ln(\Delta T_2 / \Delta T_1)} = \frac{(T_{ho} - T_{ci}) - (T_{hi} - T_{co})}{\ln[(T_{ho} - T_{ci}) / (T_{hi} - T_{co})]}$$

$$= \frac{(45 - 25) - (105 - 50)}{\ln[(45 - 25) / (105 - 50)]} = \frac{-35}{-1.01} = 34.6 \, ^\circ C$$

Neglecting the effect of fouling, the overall heat transfer coefficient is

$$U = \frac{1}{R_{tot}} = \frac{1}{(1/h_f) + (x/k) + (1 / h_w)}$$

$$= \frac{1}{(1/2500) + (0.003/220) + (1/1200)}$$

$$= 802 \, W \, m^{-2} \, K^{-1}$$

Thus, the heat transfer area is therefore equal to

$$A = \frac{q}{U\Delta T_{1m}} = \frac{3672 \times 1000}{802 \times 34.6} = 132.3 \, m^2$$

### 4.5.5.2  Shell-and-tube Heat Exchanger Rating

A 20 000 kg h$^{-1}$ of light hydrocarbon liquid stream is to be cooled from 200 °C to 100 °C by 65 000 kg h$^{-1}$ of a heavy hydrocarbon liquid stream with a feed temperature of 35 °C.

A shell-and-tube heat exchanger is available with one-shell pass and two-tube passes (illustrated schematically in Figure 4.17) and with dimensions listed below. Is it suitable for the duty?

*Heat Exchanger Geometry:*
Shell:

- Inside diameter, $D_{iS}$, 0.54 m
- Number of baffles, $N_B$, 38
- Number of shell passes, $N_{PS}$, 1

Tubes:

- Outside diameter, $D_{oT}$, 0.025 m
- Inside diameter, $D_{iT}$, 0.020 m
- Length, $L_T$, 4.4 m
- Number of tubes, $N_T$, 158
- Number of tube passes, $N_{PT}$, 4
- Pitch, $P_T$, 0.032 m
- Layout pattern, square

### 4.5.5.2.1 Heat Balance

The first task is to calculate the required duty of the exchanger and the outlet temperature of the heavy hydrocarbon stream. Heat lost by light hydrocarbon stream:

$$Q_{duty} = G \times c_p \times \Delta T = G \times c_p \times (T_{in} - T_{out})$$

$$= \frac{20\,000}{3600} \times 2470 \times (200 - 100) = 1372\,\text{kW}$$

Outlet temperature of heavy hydrocarbon stream:

$$T_{out} = T_{in} + \Delta T = T_{in} + \frac{Q_{duty}}{G \times c_p} = 35 + \frac{1372 \times 10^3}{(65\,000/3600) \times 2050} = 72\,°\text{C}$$

It is proposed to put the heavy hydrocarbon stream on the tube-side as it has the higher fouling factor. The tube-side is easier to clean than the shell-side.

*Fluid Physical Properties at Mean Bulk Temperatures*[1,3]

| | Light $h/c$ Shell-side | Heavy $h/c$ Tube-side |
|---|---|---|
| Mean bulk temperature ($T_b$, °C) | 150 | 53.5 |
| Density ($\rho$, kg m$^{-3}$) | 730 | 820 |
| Viscosity ($\mu$, Pa s) | 0.0004 | 0.0025 |
| Specific heat ($c_p$, J kg$^{-1}$ K$^{-1}$) | 2470 | 2050 |
| Thermal conductivity ($k$, W m$^{-1}$ K$^{-1}$) | 0.1329 | 0.1333 |
| Fouling factor ($R_d$, m$^2$ K W$^{-1}$) | 0.0003 | 0.0005 |

The calculation procedure shown is based on Kern's method,[1] with modifications to include the heat transfer correlations presented earlier in this chapter. The individual film heat transfer coefficients will be estimated first and then combined to give the overall heat transfer coefficient and the overall rate of heat transfer.[1,3]

### 4.5.5.2.2  Tube-side Calculations
The tube-side velocity is given by

$$v_T = \frac{G}{\rho} \times \frac{1}{\text{flow area}} = \frac{G}{\rho} \times \frac{4}{\pi D_{iT}^2} \times \frac{N_{PT}}{N_T}$$

$$= \frac{65\,000}{3600 \times 820} \times \frac{4}{\pi(0.020)^2} \times \frac{4}{158} = 1.77 \text{ m s}^{-1}$$

It is seen that the flow area through the tubes is the cross-sectional area for flow of one tube multiplied by the total number of tubes and divided by the number of tube passes.

Reynolds Number eqn (4.11):

$$Re_T = \frac{\rho v_T D_{iT}}{\mu} = \frac{820 \times 1.77 \times 0.02}{0.0025} = 11\,640$$

Prandtl Number eqn (4.11):

$$Pr_T = \frac{c_p \mu}{k} = \frac{2050 \times 0.0025}{0.1333} = 38.5$$

Nusselt Number:

$$Nu_T = 0.023 \ Re^{0.8} \ Pr^{0.33}(\mu_b/\mu_w)^{0.14}$$

See Table 4.2 (note all conditions for this correlation are met: $Re_T > 10^4$; $0.7 < Pr_T < 17\,000$ and $L_T/D_{iT} = 244 > 60$). $Nu_T = 0.023 \times 11640^{0.8} \times 8.5^{0.33} = 137$, neglecting the viscosity correction $(\mu_b/\mu_w)^{0.14}$ as the wall temperature is not yet known.

The uncorrected tube-side heat transfer coefficient in eqn (4.11) is

$$h_T = \frac{Nu_T k}{D_{iT}} = \frac{139 \times 0.1333}{0.02} = 915 \text{ W m}^2 \text{ K}^{-1}$$

### 4.5.5.2.3  Shell-side Calculations
In the shell, kerosene flows between the baffles over the tubes. The area for cross-flow between the tubes is given by the length of tube between baffles, $L_T/(N_b + 1)$, multiplied by the shell internal diameter,

$D_{iS}$, adjusted by the fractional available area for flow between the tubes, $(P_T - D_{oT})/P_T$. Thus, the shell-side cross-flow area is given by

$$A_S = \frac{L_T}{N_B + 1} \times D_{iS} \times \frac{P_T - D_{oT}}{P_T}$$

$$= \frac{4.4}{38 + 1} \times 0.54 \times \frac{0.032 - 0.025}{0.032} = 0.0133 \, \text{m}^2$$

Hence, the shell-side velocity becomes

$$v_S = \frac{G}{\rho A_S} = \frac{20\,000}{3600 \times 730 \times 0.0133} = 0.57 \, \text{m s}^{-1}$$

To estimate a Reynolds number for the flow on the shell-side, we use the equivalent hydraulic diameter

$$D_{eS} = \frac{4 \times \text{flow area}}{\text{wetted perimeter}}$$

(see Chapter 3). For a square tube layout, consider a square section of the pipe bundle, whose four corners are the centre-lines of four adjacent tubes, as illustrated in Figure 4.22.

The flow area, shaded in Figure 4.22, is given by the area of a square of side equal to the tube pitch, minus four quarter pipe cross-sectional areas. The wetted perimeter is four-quarter pipe perimeters.

$$D_{eS} = \frac{4 \times \text{flow area}}{\text{wetted perimeter}} = \frac{4[P_T^2 - (\pi D_{oT}^2/4)]}{\pi D_{oT}}$$

$$= \frac{4[0.0320^2 - (\pi \times 0.025^2/4)]}{\pi \times 0.025} = 0.0272 \, \text{m}$$

Reynolds Number eqn (4.11):

$$\text{Re}_S = \frac{\rho v_S D_{eS}}{\mu} = \frac{730 \times 0.57 \times 0.0272}{0.0004} = 28\,297$$

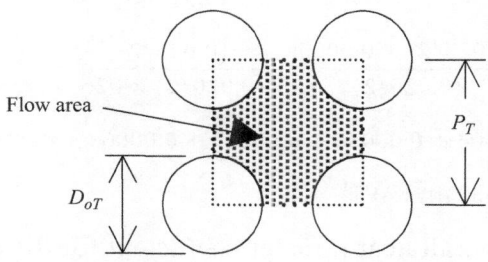

Flow area

$D_{oT}$

$P_T$

**Figure 4.22** Flow area for a square tube layout.

Prandtl Number eqn (4.11):

$$Pr_S = \frac{c_p\mu}{k} = \frac{2470 \times 0.0004}{0.1329} = 7.44$$

Nusselt Number:

$$Nu_S = 0.36\ Re^{0.55}Pr^{0.33}(\mu_b/\mu_w)^{0.14}$$

See Table 4.2 (note flow is turbulent, $Re \gg 3000$)

$Nu_S = 0.36 \times 28,\ 297^{0.55} \times 7.44^{0.33} = 196$, again neglecting the viscosity correction.

Shell-side heat transfer coefficient:

$$h_S = \frac{Nu_S k}{D_{eS}} = \frac{196 \times 0.1329}{0.0272} = 959\,\mathrm{W\,m^2\,K^{-1}}$$

### 4.5.5.2.4   Overall Heat Transfer Coefficients

Overall resistance to heat transfer based upon the outside area for the clean exchanger:

$$\frac{1}{U_o} = \frac{1}{h_o} + \frac{r_o \ln(r_o/r_i)}{k_w} + \frac{r_o}{r_i h_i}$$

$$\frac{1}{U_o} = \frac{1}{959} + \frac{(0.025/2) \times \ln(0.025/0.020)}{42} + \frac{(0.025/2)}{(0.020/2) \times 926} \qquad (4.22a)$$

$$= 0.00104 + 0.00007 + 0.00135 = 0.00246\,\mathrm{m^2 K\,W^{-1}}$$

Therefore, the overall heat transfer coefficient for the clean exchanger is $U_{oc} = 407\ \mathrm{W\ m^{-2}\ K^{-1}}$.

The overall resistance to heat transfer based upon the outside area, including the fouling coefficients for the dirty exchanger, is

$$\frac{1}{U_o} = \frac{1}{h_o} + R_{do} + \frac{r_o \ln(r_o/r_i)}{k_w} + \frac{r_o R_{di}}{r_i} + \frac{r_o}{r_i h_i}$$

$$\frac{1}{U_o} = \frac{1}{959} + 0.0003 + \frac{(0.025/2) \times \ln(0.025/0.020)}{42}$$

$$+ \frac{(0.025/2) \times 0.0005}{(0.020/2)} + \frac{(0.025/2)}{(0.020/2) \times 926} \qquad (4.44)$$

$$= 0.00104 + 0.00030 + 0.00007 + 0.00063 + 0.00135$$

$$= 0.00338\,\mathrm{m^2\ K\,W^{-1}}$$

Therefore, the overall heat transfer coefficient for the dirty exchanger is $U_{oc} = 295\ \mathrm{W\,m^{-2}\,K^{-1}}$.

### 4.5.5.2.5 Viscosity Correction for the Dirty Exchanger

These overall coefficients are based upon the uncorrected film heat transfer coefficients, *i.e.* those without the viscosity correction parameter $(\mu_b/\mu_w)^{0.14}$. As the temperature difference across a layer is proportional to the layer's thermal resistance (see eqn (4.2) and (4.10)), the relative resistances, above, allow estimation of the mean tube wall temperature, and hence evaluation of the viscosity corrections $(\mu_b/\mu_w)^{0.14}$ to the film coefficients. Mean bulk temperature difference:

$$\Delta T_b = T_{bS} - T_{bT} = 150.0 - 53.5 = 96.5 \; ^{\circ}C$$

Mean tube-side wall temperature:

$$T_{wT} = T_{bT} + \frac{\text{tube-side film resistance}}{\text{total resistance}} \times \Delta T_b$$

$$= 53.5 + \frac{0.00135}{0.00338} \times 96.5 = 92.0 \,^{\circ}C$$

Mean shell-side wall temperature:

$$T_{wS} = T_{bS} - \frac{\text{shell-side film resistance}}{\text{total resistance}} \times \Delta T_b$$

$$= 150.0 - \frac{0.00104}{0.00338} \times 96.5 = 120.3 \,^{\circ}C$$

Note, in the above analysis for the dirty exchanger the fouling has been included as part of the wall. Neglecting fouling, *i.e.* for a clean exchanger, these temperatures become 106.5 °C and 109.2 °C, respectively.

|  | Shell-side | Tube-side |
|---|---|---|
| Mean wall temperature ($T_w$, °C) | 120.3 | 92.0 |
| Viscosity[1] ($\mu_w$, Pa s) | 0.0006 | 0.001 |
| Viscosity correction factor $(\mu_b/\mu_w)^{0.14}$ | 0.95 | 1.14 |
| Corrected film coefficients ($h$, W m$^{-2}$ K$^{-1}$) | 913 | 1040 |

Corrected overall heat transfer coefficient for the dirty exchanger: $U_{od} = 304$ W m$^{-2}$ K$^{-1}$.

For the clean exchanger, the overall heat transfer coefficient: $U_{oc} = 423$ W m$^{-2}$ K$^{-1}$.

### 4.5.5.2.6  Calculation of Log Mean Temperature Difference

Log mean temperature difference:

$$\Delta T_{1m} = \frac{(200 - 72) - (100 - 35)}{\ln[(200 - 72)/(100 - 35)]} = 92.9\,\text{K}$$

$\Delta T_{lm}$ correction factor $F_t$, using $R = (200 - 100)/(72 - 35) = 2.70$ and $S = (72 - 35)/(200 - 35) = 0.225$, gives $F_t = 0.922$, from Figure 4.18.

### 4.5.5.2.7  Calculation of Rate of Heat Transfer

Heat transfer area:

$$A_o = N_T \times \pi \times D_{oT} \times L = 158 \times \pi \times 0.025 \times 4.4 = 54.6\,\text{m}^2$$

Clean exchanger:

$$Q_c = U_{oc} \times A_o \times F_t \times \Delta T_{lm} = 423 \times 54.6 \times 0.922 \times 92.9 = 1979\,\text{kW}$$

Dirty exchanger:

$$Q_d = U_{od} \times A_o \times F_t \times \Delta T_{lm} = 304 \times 54.6 \times 0.922 \times 92.9 = 1422\,\text{kW}$$

Thus, the clean exchanger comfortably exceeds the required duty. The dirty exchanger also exceeds the required duty and so this exchanger seems suitable.

### 4.5.5.2.8  Pressure Drop

All that remains is to check the pressure drops during flow through the exchanger. Kern[1] states that it is customary to allow a pressure drop of 30 to 70 kPa for an exchanger. If the pressure drop is too low, there will be maldistribution of the flow through the exchanger. If it is too high, the pumping/compression costs will become significant.

The flow on the shell-side is complex and the prediction of shell-side pressure drop is beyond the scope of this chapter.[1,3] However, we will estimate the tube-side pressure losses, considering only the losses during flow through the tubes and the tube entrance and exit losses.

It will be assumed that the tubes have an absolute roughness, $e$, of 0.0015 mm (drawn tubing; Chapter 3, Table 4). At $Re_T = 11\,640$ and $e/D_{iT} = 0.000075$, from Chapter 3 Equation (32), or from Chapter 3 Figure 9, $c_f = 0.0075$.

From Chapter 3, Equation (25), the frictional pressure loss in the tubes is

$$\Delta p_f = 2c_f \frac{N_{PT} \times L_T}{D_{iT}} \rho v_T^2$$

$$= 2 \times 0.0075 \times \frac{4 \times 4.4}{0.020} \times 730 \times 1.77^2 = 34.0\,\text{kPa}$$

For each tube pass there will be one entrance loss and one exit loss, 0.5 and 1.0 velocity heads, respectively, (Chapter 3, Table 5). From Chapter 3, Equation (33), these losses are

$$\Delta p_{\text{loss}} = N_{\text{PT}} \times (0.5 + 1.0) \times \frac{\rho v_{\text{T}}^2}{2} = 4 \times 1.5 \times \frac{730 \times 1.77^2}{2} = 7.7 \,\text{kPa}$$

Total tube-side pressure loss $= \Delta p_{\text{f}} + \Delta p_{\text{loss}} = 34.0 + 7.7 = 41.7$ kPa

This is a reasonable pressure loss, within the range recommended by Kern[1] and noted above.

The above, simplified analysis used example correlations given earlier in this chapter. Coefficients, assumed constant above, can change significantly with temperature, and especially if there is a change of phase, with boiling and/or condensing on one or both sides of the exchanger. Hence, coefficients can change with position within an exchanger. Often exchangers are divided into sections and heat transfer coefficients are evaluated in each section.

The design of heat exchangers, rather than the rating of an existing one as shown, is an iterative process:

- Guess a value for $U$ and estimate the required heat transfer area.
- Select an exchanger geometry that gives this area.
- Calculate film coefficients, $U$, and estimate pressure drops.
- Compare calculated $U$ with the assumed value.
- Iterate upon the assumed $U$ and geometry until converged and pressure drops are reasonable.

More detailed analyses and detailed design procedures for various types of heat exchangers are given in the literature.[1,3,7,8]

## 4.6 Conclusions

In this chapter, we have looked at the three modes of heat transfer, *viz.*, conduction, convection and radiation, and the laws relating heat fluxes to temperature driving forces for each mode. From these laws we have shown how to estimate heat transfer rates in a range of geometries and situations.

We have introduced the concept of thermal resistances and shown how these may be combined to give overall resistances to heat transfer and overall heat transfer coefficients.

Finally, we looked at a number of common types of heat exchanger along with some example design calculations.

# Nomenclature

| Symbol | Description | Dimensions | SI unit |
|---|---|---|---|
| ***Variables*** | | | |
| $A$ | Area, surface area for heat transfer | $L^2$ | $m^2$ |
| $A_o$ | Outside surface area for heat transfer | $L^2$ | $m^2$ |
| $b$ | Layer/pipe wall thickness | $L$ | $m$ |
| Bi | Biot number | — | — |
| $c_f$ | Fanning friction factor | — | — |
| $c_P$ | Heat capacity at constant pressure | $H\,M^{-1}\,\Theta^{-1}$ | $J\,kg^{-1}\,K^{-1}$ |
| $D$ | Diameter | $L$ | $m$ |
| $e$ | Absolute roughness | $L$ | $mm$ |
| $f$ | Function | n/a | n/a |
| $F_{12}$ | View factor from 1 to 2 | — | — |
| Fo | Fourier number | — | — |
| $F_t$ | LMTD correction factor | — | — |
| $g$ | Gravitational acceleration | $L\,T^{-2}$ | $m\,s^{-2}$ |
| $G$ | Mass flow rate | $M\,T^{-1}$ | $kg\,s^{-1}$ |
| Gr | Grashof number | — | — |
| $h$ | Heat transfer coefficient | $H\,L^{-2}\,T^{-1}\,\Theta^{-1}$ | $W\,m^{-2}\,K^{-1}$ |
| $i$ | Electrical current | $A$ | $A$ |
| $k$ | Thermal conductivity | $H\,L^{-1}\,T^{-1}\,\Theta^{-1}$ | $W\,m^{-1}\,K^{-1}$ |
| $L$ | Characteristic length scale, pipe/tube length | $L$ | $m$ |
| $L_T$ | Pipe/tube length | $L$ | $m$ |
| $N_B$ | Number of baffles | — | — |
| $N_{PT}$ | Number of tube passes | — | — |
| $N_T$ | Number of tubes | — | — |
| Nu | Nusselt number | — | — |
| $p$ | Pressure | $M\,L^{-1}\,T^{-2}$ | $Pa$ |
| Pr | Prandtl number | — | — |
| $P_T$ | Tube pitch | — | — |
| $q$ | Heat flux | $H\,L^{-2}\,T^{-1}$ | $W\,m^{-2}$ |
| $Q$ | Rate of heat flow | $H\,T^{-1}$ | $W$ |
| $R$ | Electrical resistance | $H\,A^{-2}$ | $\Omega$ |
| $r$ | Radius, radial co-ordinate | $L\,T^{-1}$ | $m\,s^{-1}$ |
| $R$ | Thermal resistance Parameter in calculation of $F_t$ | $H^{-1}\,L^2\,T\,\Theta$ | $m^2\,KW^{-1}$ |
| $R_d$ | Thermal resistance of dirt, fouling factor | $H^{-1}\,L^2\,T\,\Theta$ | $m^2\,KW^{-1}$ |

| Symbol | Description | Dimensions | SI unit |
|---|---|---|---|
| Re | Reynolds number | — | — |
| $S$ | Parameter in calculation of $F_t$ | | |
| $T$ | Temperature | $\Theta$ | K |
| $t$ | Time | T | s |
| $U$ | Overall heat transfer coefficient | $HL^{-2}\,T^{-1}\,\Theta^{-1}$ | $Wm^{-2}\,K^{-1}$ |
| $U_o$ | Overall heat transfer coefficient based upon outside area | $HL^{-2}\,T^{-1}\,\Theta^{-1}$ | $Wm^{-2}\,K^{-1}$ |
| $V$ | Electrical potential | $H\,A^{-1}$ | V |
| $v$ | Velocity | $L\,T^{-1}$ | $ms^{-1}$ |
| $V$ | Voltage | $L\,T^{-1}$ | volts |
| $y$ | Distance in $y$-direction, distance from (pipe) wall | $L$ | m |
| $\alpha$ | Absorptivity | — | — |
| | Thermal diffusivity | $L^2\,T^{-2}$ | $m^2\,s^{-1}$ |
| $\beta$ | Coefficient of thermal expansion | $\Theta^{-1}$ | $K^{-1}$ |
| $\delta$ | Distance, | $L$ | m |
| | Small difference | n/a | n/a |
| $\varepsilon$ | Emissivity | — | — |
| $\mu$ | Dynamic viscosity | $M\,L^{-1}\,T^{-1}$ | Pa s, N s m$^{-2}$ |
| $\rho$ | Fluid density | $M\,L^{-3}$ | $kg\,m^{-3}$ |
| $\sigma$ | Stefan–Boltzmann constant | $HL^{-2}\,T^{-1}\,\Theta^{-4}$ | $Wm^{-2}\,K^{-4}$ |
| $\tau$ | Transmittance | — | — |

**Subscripts**

| | |
|---|---|
| 0 | At time $= 0$ |
| 1, 2, *etc.* | At points 1, 2, *etc.* |
| b | At bulk conditions (temperature) |
| c | Cold-side, clean, centre |
| Conv | Convective |
| d | Dirt, dirty |
| e | Equivalent hydraulic |
| f | Frictional |
| h | Hot-side |
| i | Inside, inlet |
| lm | Log-mean |
| o | Outside, outlet |
| Rad | Radiatative |
| surf | Surface |
| S | Shell, shell-side |
| T | Tube, tube-side |
| Tot | Total |
| w | Of the wall, at wall conditions (temperature) |

| Symbol | Description | Dimensions | SI unit |
|---|---|---|---|
| x | in the *x*-direction | | |
| y | in the *y*-direction | | |
| z | in the *z*-direction | | |

*Symbols*

| Δ | "The change in" |
|---|---|
| δ | "Small difference in" |

*Fundamental Dimensions*

| M | Mass |
|---|---|
| L | Length |
| T | Time |
| Θ | Temperature |
| H | Heat |
| A | Electrical current |

# References

1. D. Q. Kern, *Process Heat Transfer*, McGraw-Hill, New York, 1965.
2. J. M. Coulson, J. F. Richardson, J. R. Backhurst and J. R. Harker, *Coulson & Richardson's Chemical Engineering*, Butterworth-Heinemann, Oxford, 6th edn, 1999, vol. 1.
3. R. K. Sinnott, *Coulson & Richardson's Chemical Engineering*, Butterworth-Heinemann, 3rd edn, Oxford, 1999, vol. 6.
4. J. P. Holman, *Heat Transfer*, McGraw-Hill, New York, 1989.
5. H. S. Carslaw and J. C. Jaeger, *Conduction of Heat in Solids*, Oxford University Press, Oxford, 2nd edn, 1959.
6. E. D. Williamson and H. A. Adams, *Phys. Rev.*, 1919, **14**, 99–114.
7. R. H. Perry and D. W. Green, *Perry's Chemical Engineers' Handbook*, McGraw-Hill, New York, 7th edn, 1997.
8. A. Bejan and A. D. Kraus, *Heat Transfer Handbook*, Wiley, Hoboken, New Jersey, 2003.

# 5 An Introduction to Mass-transfer Operations

Eva Sorensen

## 5.1 Introduction

A significant number of chemical engineering unit operations are concerned with the problem of changing the compositions of solutions and mixtures through methods not necessarily involving chemical reactions. Usually these operations are directed towards separating a substance into its component parts. For mixtures, such separations may be entirely mechanical, *e.g.* the filtration of a solid from a suspension in a liquid, or the separation of particles of a ground solid according to their density. On the other hand, if the operations involve changes in composition of *solutions*, they are knows as *mass-transfer operations* and it is these which concern us in this chapter.

The importance of mass-transfer operations in chemical engineering is profound. There is scarcely any chemical process which does not require either a preliminary purification of raw materials or a final separation of products from by-products, and for these, mass-transfer operations are commonly used. Frequently, the separations constitute the major part of the costs of a process.

The chemical engineer faced with the problem of separating the components of a solution must ordinarily choose from several possible methods. While the choice is usually limited by the specific physical characteristics of the materials to be handled, the necessity

Concepts of Chemical Engineering for Chemists: 2nd Edition
Edited by Stefaan Simons
© The Royal Society of Chemistry 2017
Published by the Royal Society of Chemistry, www.rsc.org

for making a decision nevertheless almost always exists. The principle basis for choice in any situation is cost; that method which costs the least is usually the one to be used. Occasionally, other factors also influence the decision, however. The simplest operation, while it may not be the least costly, is sometimes desired because it will be more trouble-free. Sometimes a method will be discarded because of imperfect knowledge of design methods or unavailability of data for design, as results cannot be guaranteed. Favourable previous experience with one method may be given strong consideration.

The choice of the most suitable separation method for a given separation depends on the mixture at hand and the required purity specifications. Although this is the most important decision in the design and operation of a separation process, it will therefore not be considered in this chapter.

In the following, the principles of mass-transfer separation processes will be outlined first. Details of mass-transfer calculations will be introduced next and examples will be given of both equilibrium-stage processes and diffusional rate processes. The chapter will then conclude with a detailed discussion of the two single most applied mass-transfer processes in the chemical industries, namely distillation and absorption.

## 5.2   Mechanisms of Separation

The separation of chemical mixtures into their constituents has been practised as an art for millennia. Separations, including enrichment, concentration, purification, refining and isolation, are important to both chemists and chemical engineers. Chemists use analytical separation methods, such as chromatography, to determine compositions of complex mixtures *quantitatively.* Chemists also use small-scale preparative separation techniques, often similar to analytical methods, to recover and purify chemicals. Chemical engineers, however, are more concerned with the *manufacture* of chemicals using economical, large-scale separation methods, which may differ considerably from laboratory techniques. For example, in a laboratory, chemists separate and analyse light-hydrocarbon mixtures by chromatography, while in a large manufacturing plant, a chemical engineer uses distillation to separate the same mixtures.

The mixing of chemicals to form a mixture is a spontaneous, natural process that is accompanied by an increase in entropy and randomness. The inverse process, the separation of that mixture into its

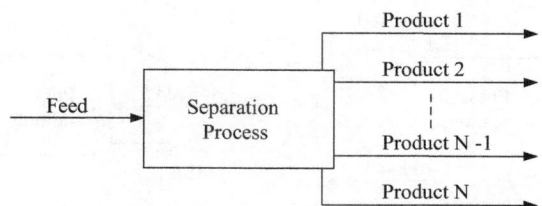

**Figure 5.1** General separation process.

constituent chemical species, is not a spontaneous process; it requires an expenditure of energy. A mixture to be separated usually originates as a single, homogeneous phase (solid, liquid or gas). If it exists as two or more immiscible phases, it is often best to first use some mechanical means, based on gravity, centrifugal force, pressure reduction or an electrical and/or magnetic field, to separate the phases. Then, appropriate separation techniques are applied to each phase.

A schematic diagram of a general separation process is shown in Figure 5.1. The feed mixture can be vapour, liquid or solid, while the two products may differ in composition and phase state from each other, as well as from the feed.

The separation is accomplished by forcing the different chemical species (components) in the feed into different spatial locations by any of five general separation techniques, or combinations thereof, as shown in Figure 5.2.[1]

The most common industrial technique involves the creation of a second phase (vapour, liquid or solid) that is immiscible with the feed phase (Figure 5.2a). The creation is accomplished by energy transfer (heat and/or shaft work) to or from the process or by pressure reduction.

In the second technique, separation by phase addition, the second phase (MSA) is introduced into the system in the form of a solvent that selectively dissolves some of the species in the feed mixture (Figure 5.2b).

Less common, but of growing importance, is the use of a barrier such as a membrane, which restricts and/or enhances the movement of certain chemical species with respect to other species (Figure 5.2c).

Also of growing importance are techniques that involve the addition of a solid agent, which acts directly or as an inert carrier for other substances so as to cause separation (Figure 5.2d).

Finally, external fields of various types are sometimes applied to specialised separations, although this is rare (Figure 5.2e).

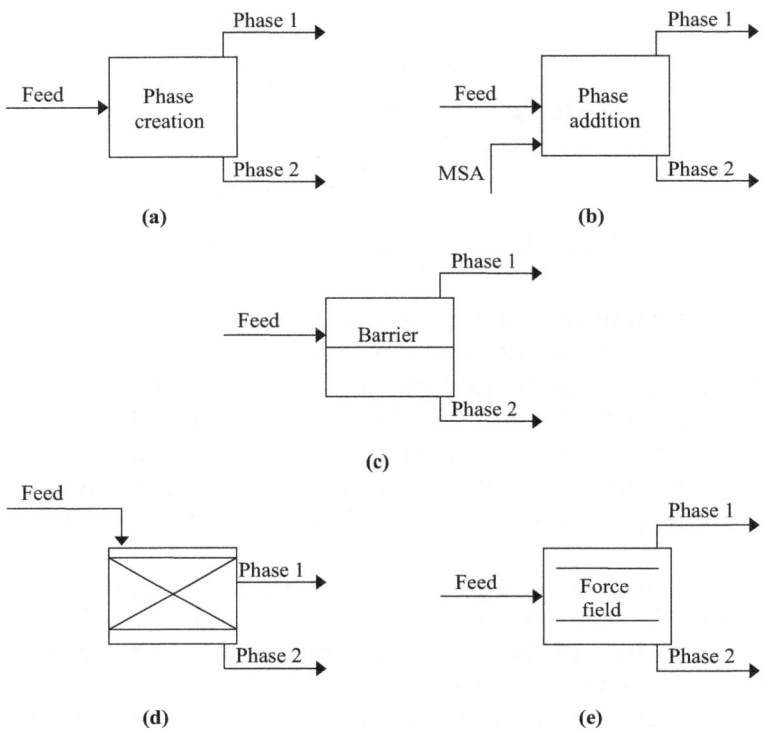

**Figure 5.2**  General separation techniques: (a) separation by phase creation; (b) separation by phase addition (MSA: mass separating
agent); (c) separation by barrier; (d) separation by solid agent;
(e) separation by force field or gradient.

For all of the general techniques of Figure 5.2, the separations are
achieved by enhancing the rate of mass transfer by diffusion of *certain*
species relative to mass transfer of *all* species by bulk movement
within a particular phase. The driving force and direction of mass
transfer by diffusion is governed by thermodynamics, with the usual
limitations of equilibrium. Thus, both transport and thermodynamic
considerations are crucial in separation operations. The *rate* of separation is governed by mass transfer, while the *extent* of separation is
limited by thermodynamic equilibrium. Fluid mechanics also plays
an important role, and applicable principles are included in other
chapters.

In laboratory analysis, the *extent* of separation is of major importance, *i.e.* is it thermodynamically possible to separate the components, however long this may take? In manufacturing, the extent of
separation will determine if the separation is possible, but the *rate* of

mass transfer will determine if the separation is fast enough to be economically viable. Chemical engineers strive towards designing separation processes in such a way as to maximise the rate of mass transfer, as it is generally not possible to change the thermodynamics, *i.e.* the extent of the separation.

The *extent of separation* achieved between or amongst the product phases for each of the chemical species present in the feed depends on the exploitation of differences in molecular, thermodynamic and transport properties of the species in the different phases present. Some properties of importance are as below.

(i) Molecular Properties: Molecular weight, van der Waals volume, polarisability, electrical charge.
(ii) Thermodynamic and Transport Properties: Vapour pressure, solubility, adsorptivity, diffusivity.

Values of these properties for many substances are available in handbooks, specialised reference books and journals. Some of these properties can also be estimated using commercial programmes. When the values are not available, the properties must be estimated or determined experimentally if a successful application of the appropriate separation operation(s) is to be designed.

## 5.3 General Separation Techniques

In the following, the first four separation techniques illustrated in Figure 5.2 are outlined with examples of separation methods which fall within each category. Most, if not all, of the methods mentioned are described in detail in standard chemical engineering textbooks.[1–6]

### 5.3.1 Separation by Phase Addition or Phase Creation

The first two separation techniques are very similar and will be considered together. If the feed mixture is a homogeneous, single-phase solution (gas, liquid or solid), a second immiscible phase must often be developed or added before separation of chemical species can be achieved (see Figure 5.2a and b). This second phase is created by an *energy-separating agent* and/or added as a *mass-separating agent.*

Application of an energy-separating agent involves heat transfer and/ or shaft work to or from the mixture to be separated. Alternatively, vapour may be created from a liquid phase by reducing the pressure.

A mass-separating agent may be partially immiscible with one or more of the species in the mixture. In this case, the mass-separating agent frequently remains the constituent with the highest concentration in the added phase. Alternatively, the mass-separating agent may be completely miscible with a liquid mixture to be separated, but may selectively alter the *partitioning* of species between vapour and liquid phases. This facilitates a more complete separation when used in conjunction with an energy-separating agent, such as in extractive distillation.

Table 5.1 is a compilation of the more common industrial separation operations based on inter-phase mass transfer between two phases, either created by an energy-separating agent or added as a mass-separating agent. A more comprehensive table is given by Seader and Henley.[1] In the following, the operations listed in Table 5.1 will be outlined briefly. The first two methods, distillation and absorption, will be discussed in more detail later.

*Distillation* is the most widely utilised industrial separation method. Distillation involves multiple contacts between counter-currently flowing liquid and vapour phases. Each contact consists of mixing the two phases to promote rapid partitioning of species by mass transfer, followed by phase separation. The contacts are often made on horizontal trays (referred to as stages) arranged in a vertical column. Vapour, while flowing up the column, is increasingly enriched with respect to the more volatile species. Correspondingly, liquid flowing down the column is increasingly enriched with respect to the less volatile species. Feed to the distillation column enters on a tray somewhere between the top and bottom trays, most often near the middle of the column. The portion of the column above the feed entry is called *the enriching or rectifying* section; and that below is the *stripping* section. Feed vapour will move up the column; feed liquid will move down. Liquid is required for making contacts with vapour above the feed tray, and vapour is required for making contacts with liquid below the feed tray. Normally, vapour from the top of the column is condensed in a condenser by cooling water or a refrigerant to provide contacting liquid, called *reflux*. Similarly, liquid at the bottom of the column passes through a reboiler, where it is heated by condensing steam or some other heating medium to provide contacting vapour, called *boilup*.

In *absorption*, a soluble vapour is absorbed by means of an added liquid mass-separating agent. The constituents of the vapour feed dissolve in the absorbent liquid to varying extents, depending on their solubilities. Vapourisation of a small fraction of the absorbent also

**Table 5.1** Common separation operations based on phase creation or addition (MSA: mass-separating agent; ESA: energy-separating agent).

| Separation operation | Initial or feed phase | Created or added phase | Separating agent | Industrial example |
|---|---|---|---|---|
| Distillation | Vapour and/or liquid | Vapour and liquid | ESA | Purification of styrene |
| Absorption | Vapour | Liquid | MSA | Separation of $CO_2$ from combustion products |
| Liquid–liquid extraction | Liquid | Liquid | MSA | Recovery of aromatics |
| Drying | Liquid and often solid | Vapour | Gas (MSA) and/ or heat transfer (ESA) | Removal of water from polyvinylchloride |
| Crystallisation | Liquid | Solid (and vapour) | ESA | Crystallisation of *p*-xylene from a mixture with *m*-xylene |
| Desublimation | Vapour | Solid | ESA | Recovery of phthalic anhydride from non-condensible gas |
| Leaching | Solid | Liquid | Liquid solvent | Extraction of sucrose from sugar beets with hot water |

generally occurs. The solute is subsequently recovered from the liquid absorbent by distillation, and the absorbing liquid can either be discarded or reused. The operation may not require an energy-separating agent and is frequently conducted at ambient temperature and high pressure.

*Liquid–liquid extraction*, using one or two solvents respectively, is widely used when distillation is impractical, especially when the mixture to be separated is temperature-sensitive and/or more than 100 distillation stages would be required. When one solvent is used, it selectively dissolves only one or a fraction of the components in the feed mixture. In a two-solvent extraction system, each solvent has its own specific selectivity for dissolving the components of the feed mixture. Additional separation operations are generally required to recover, for recycling, solvent from streams leaving the extraction operation.

Since many chemicals are produced wet but sold in dry, solid forms, one of the more common manufacturing steps is a *drying* operation, which involves removal of a liquid from a solid by vapourisation of the liquid. Although the only basic requirement in drying is that the vapour pressure of the liquid to be evaporated is higher than its partial pressure in the gas stream, the design and operation of dryers represents a complex problem in heat transfer, fluid mechanics and mass transfer.

*Crystallisation* is carried out in many organic, and almost all in-organic, chemical manufacturing plants, where the desired product is a finely divided solid. Since crystallisation is essentially a purification step, the conditions in the crystalliser must be such that impurities do not precipitate with the desired product.

Sublimation is the transfer of a substance from the solid to the gaseous state without formation of an intermediate liquid phase, usually at a relatively high vacuum. The reverse process, *desublimation*, is also practised, for example, in the recovery of phthalic an-hydride from gaseous reactor effluent.

Liquid–solid extraction, often referred to as *leaching*, is widely used in the metallurgical, natural product, and food industries under batch, semi-continuous or continuous operating conditions. The major problem in leaching is to promote diffusion of the solute out of the solid and into the liquid solvent. The most effective way of doing this is to reduce the dimensions of the solid to the smallest feasible particle size. For large-scale applications, large open tanks are used in counter-current operation. The major difference between solid–liquid and liquid–liquid systems centers around the difficulty in transporting the solid, or the solid slurry, from stage to stage. For this reason, the solid may be left in the same tank, with only the liquid transferred from tank to tank. In the pharmaceutical, food and natural-product industries, countercurrent solid transport is often provided by fairly complicated mechanical devices.

## 5.3.2 Separation by Barrier

The third main class of separation methods, the use of micro-porous and non-porous membranes as semi-permeable barriers (see Figure 5.2c) is rapidly gaining popularity in industrial separation processes for application to difficult and highly selective separations. Membranes are usually fabricated from natural fibres, synthetic polymers, ceramics or metals, but they may also consist of liquid films. Solid membranes are fabricated into flat sheets, tubes, hollow fibres or spiral-wound sheets. For the micro-porous membranes, separation is effected by differing rates of diffusion through the pores, while for non-porous membranes, separation occurs because of differences in both the solubility in the membrane and the rate of diffusion through the membrane. Table 5.2 is a compilation of the more common industrial separation operations based on the use of a barrier. A more comprehensive table is given by Seader and Henley.[1]

**Table 5.2** Common separation operations based on a barrier.

| Separation operation | Initial or feed phase | Separating agent | Industrial example |
|---|---|---|---|
| Reverse osmosis | Liquid | Non-porous membrane with pressure gradient | Desalination of sea water |
| Microfiltration | Liquid | Porous membrane with pressure gradient | Removal of bacteria from drinking water |
| Ultrafiltration | Liquid | Microporous membrane with pressure gradient | Separation of whey from cheese |
| Pervaporation | Liquid | Non-porous membrane with pressure gradient | Separation of azeotropic mixtures |
| Gas permeation | Vapour | Non-porous membrane with pressure gradient | Hydrogen enrichment |

Osmosis involves the transfer, by a concentration gradient, of a solvent through a membrane into a mixture of solute and solvent. The membrane is almost non-permeable to the solute. In *reverse osmosis*, transport of solvent in the opposite direction is effected by imposing a pressure, higher than the osmotic pressure, on the feed side. Using a non-porous membrane, reverse osmosis successfully desalts water.

Microporous membranes can be used in a manner similar to reverse osmosis to selectively allow small solute molecules and/or solvents to pass through the membrane and to prevent large dissolved molecules and suspended solids from passing through. *Microfiltration* refers to the retention of molecules typically in the size range from 0.05 to 10 μm. *Ultrafiltration* refers to the range from 1 to 100 nm. To retain even smaller molecules, reverse osmosis, sometimes called hyperfiltration, can be used down to less than 2 nm.

Although reverse osmosis can be used to separate organic and aqueous–organic liquid mixtures, very high pressures are required. Alternatively, *pervaporation* can be used in which the species being absorbed by, and transported through, the non-porous membrane are evaporated. This method, which uses much lower pressures than reverse osmosis, but where the heat of vapourisation must be supplied, is used to separate azeotropic mixtures.

The separation of gas mixtures by selective *gas permeation* through membranes using pressure as the driving force is a relatively simple process. Non-porous polymer membranes are being used commercially to enrich gas mixtures containing hydrogen, to recover hydrocarbons from gas streams and to produce nitrogen-enriched and oxygen-enriched air.

## 5.3.3 Separation by Solid Agent

Typical separation operations that use solid mass-separating agents are listed in Table 5.3 (see Figure 5.2d). The solid, usually in the form of a granular material or packing, acts as an inert support for a thin layer of absorbent, or alternatively, enters directly into the separation operation by selective adsorption of, or chemical reaction with, certain species in the feed mixture. Adsorption is confined to the surface of the solid adsorbent, unlike absorption, which occurs throughout the bulk of the absorbent. In all cases, the active separating agent eventually becomes saturated with solute and must be regenerated or replaced periodically. Such separations are often conducted batchwise or semi-continuously. However, equipment is available to

**Table 5.3** Common separation operations based on a solid agent.

| Separation operation | Initial or feed phase | Separating agent | Industrial example |
|---|---|---|---|
| Adsorption | Vapour or liquid | Solid adsorbent | Purification of *p*-xylene |
| Chromatography | Vapour or liquid | Solid adsorbent or liquid adsorbent on a solid support | Separation of xylene isomers and ethylbenzene |
| Ion exchange | Liquid | Resin with ion-active sites | Demineralisation of water |

simulate continuous operation, *e.g.* Simulated Moving Bed (SMB) chromatography.

*Adsorption* is used to remove components present in low concentrations in non-adsorbing solvents or gases and to separate the components in gas and liquid mixtures by selective adsorption on solids, followed by desorption to regenerate the adsorbent. Typical adsorbents are activated carbon, aluminium oxide, silica gel and synthetic sodium or calcium aluminiosilicate zeolite adsorbents (molecular sieves). Regeneration is accomplished by one of four methods: (1) vapourise the adsorbate with a hot purge gas (thermal swing adsorption), (2) reducing the pressure to vapourise the adsorbate (pressure-swing adsorption), (3) inert purge stripping without change in temperature or pressure and (4) displacement adsorption by a fluid containing a more strongly adsorbed species.

*Chromatography* is a method for separating the components of a feed gas or liquid mixture by passing the feed through a bed of packing. The feed may be volatilised into a carrier gas, and the bed may be a solid adsorbent (gas or liquid–solid chromatography) or a solid inert support that is coated with a very viscous liquid that acts as an absorbent (gas or liquid–liquid chromatography). Because of selective adsorption on the solid adsorbent surface or absorption into the liquid absorbent, followed by desorption, the different components of the feed mixture move through the bed at different rates, thus effecting the separation.

*Ion exchange* resembles adsorption in that solid particles are used and regeneration is necessary. However, a chemical reaction is involved. In water softening, a typical ion-exchange application, an organic or inorganic polymer in its sodium form removes calcium ions by exchanging calcium for sodium.

# 5.4  Mass Transfer Calculations

Mass-transfer calculations, such as the analysis or design of separation units, can be solved by two distinctly different methods, based on either the concept of: (1) Equilibrium stage processes or (2) Diffusional rate processes.

The choice of method depends on the kind of equipment in which the operation is carried out. Distillation, leaching and sometimes liquid extraction, are performed in equipment such as mixed-settler trains, diffusional batteries or plate towers, which contain a series of discrete processing units, and problems in these areas are commonly solved by equilibrium stage calculations. Gas absorption and other operations which are carried out in packed towers or similar equipment, are usually handled using the concept of a diffusional process.

Strictly speaking, all separation processes are diffusional, as one or more of the components in the feed are diffusing from one phase or region to another due to a concentration difference. However, chemical engineers often assume that certain regions of the mass-transfer equipment are at equilibrium. It is important to note that the two calculation methods are just two different ways of *describing* the mass-transfer taking place inside the separation units; the distinction is purely in terms of the way the calculations are carried out when analysing or designing these units. In the following, descriptions will be given of both types of calculations, although examples will only be provided for equilibrium-stage calculations.

## 5.4.1  Equilibrium-stage Processes

A large class of mass-transfer devices consists of assemblies of individual units, or stages, that are interconnected so that the materials being processesed pass through each stage in turn (see Figure 5.3). The two streams move through the assembly either counter-currently, co-currently or cross-currently. In each stage, they are brought into contact, mixed and then separated. Such multistage systems are called *cascades*. The counter-current arrangement is normally preferred, because it generally results in a higher degree of separation than the other two (Figure 5.3).

For mass transfer to take place, the streams entering each stage must not be in equilibrium with each other, for it is the *departure*, or *deviation*, from equilibrium conditions that provides the driving force for transfer. The leaving streams are usually not in equilibrium either

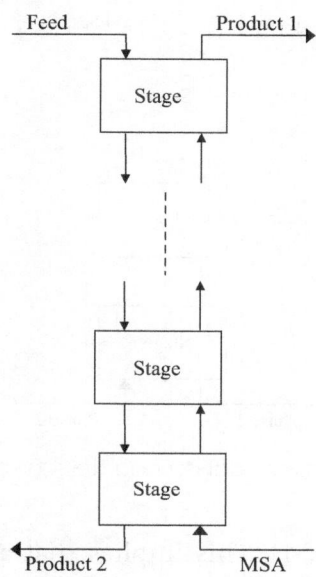

**Figure 5.3** Cascade of contacting stages (MSA).

but are much closer to being so than the entering streams are. The closeness of the approach to equilibrium depends on the effectiveness of mixing and mass transfer between the phases. To simplify the calculations for a cascade, the streams leaving each stage are often assumed to be in equilibrium, which, by definition, makes each stage ideal. A correction factor, or *efficiency* factor, is applied later to account for any actual departures from equilibrium.

The cascade shown in Figure 5.3 consists of a series of single sections or stages with streams entering and leaving only from the ends. Such cascades are used to recover components from a feed stream but are not generally useful for making a sharp separation between two selected feed components. To do this, it is best to provide a cascade consisting of two sections and the counter-current cascade of Figure 5.4 is often used. It consists of one section above the feed and one below. We will see later how this arrangement is used in distillation.

## 5.4.2 Stage Calculations

The solution to a multi-component, multi-phase, multi-stage separation problem is found in the simultaneous or iterative solution of the material balances, the energy balance and the phase equilibrium

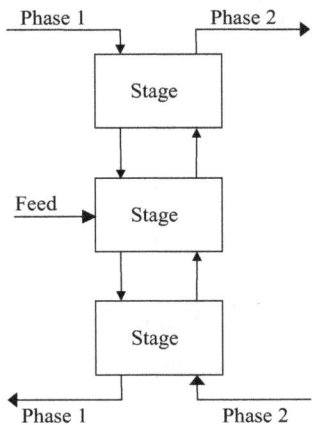

**Figure 5.4** Two section counter-current cascade.

equations (see Chapter 1). This implies that a sufficient number of design variables are specified so that the number of remaining unknown variables exactly equals the number of independent equations. When this is done, a separation process is said to be *specified.*

When considering multistage systems as a whole, attention is focused on the streams passing between the individual stages. An individual unit in a cascade normally receives two streams from the two units adjacent to it (*e.g.* one from the stage above and the other from the stage below), brings them into close contact, and delivers two streams respectively (*e.g.* one to the stage above and one to the stage below). The fact that the contact units may be arranged either one above the other, as in a distillation column, or side by side, as in a stage leaching plant, is important mechanically and may affect some of the details of the operation of individual units. However, the same balance equations are used for either arrangement.

The individual contact units in a cascade are numbered sequentially, starting from one end. In this chapter, the stages are numbered from the bottom and up. A general stage in the system is the $n$th stage, which is number $n$ counting from the bottom. The stage immediately ahead of stage $n$ in the sequence is stage $n-1$ and that immediately following it is stage $n+1$. To designate the streams pertaining to any one stage, all streams originating in that stage carry the number of the unit as a subscript. Thus, for a two-component system, $y_{n-1}$ is the mole fraction of the lightest component (*i.e.* the one with the lowest boiling point) in the *vapour* phase leaving stage $n-1$, and $L_n$ is the molar flow rate of the $L$ phase leaving the $n$th

**Figure 5.5** Equilibrium stage with heat addition.

stage. For a single equilibrium stage with two entering streams and two exit streams, as shown in Figure 5.5, the variables are those associated with the four streams plus the heat transfer rate to or from the stage. (Note that some textbooks number the stages from the top and down but the principle of numbering remains the same.)

The total input of material to stage $n$ is $L_{n+1} + V_{n-1}$ units (mass or moles) per time and the total output is $L_n + V_n$ units per time. Since, under steady flow, there is neither accumulation nor depletion, the input and the output are equal and the material balance can be written as:

$$L_{n+1} + V_{n-1} = L_n + V_n$$

This equation is the *total material balance* for the stage. Another balance can be written by equating input to output for component $i$ giving the *component i balance*:

$$L_{n+1}x_{i,n+1} + V_{n-1}y_{i,n-1} = L_n x_{i,n} = V_n y_{i,n}$$

Obviously, if all the component balances for a system were added up, the sum would be equal to the total material balance.

In most equilibrium stage processes, the general *energy balance* can be simplified by neglecting potential energy and kinetic energy. If, in addition, the process is workless and adiabatic, a simple enthalpy balance applies:

$$L_{n+1}H_{L,n+1} + V_{n-1}H_{V,n-1} + Q_n = L_n H_{L,n} + V_n H_{V,n}$$

where $H_L$ and $H_V$ are the enthalpies per unit (mass or moles) of the $L$ phase and $V$ phase, respectively, and $Q_n$ is the heat added (negative value if heat is removed).

Examples of stage calculations for the most common chemical engineering separation methods are given later in this chapter.

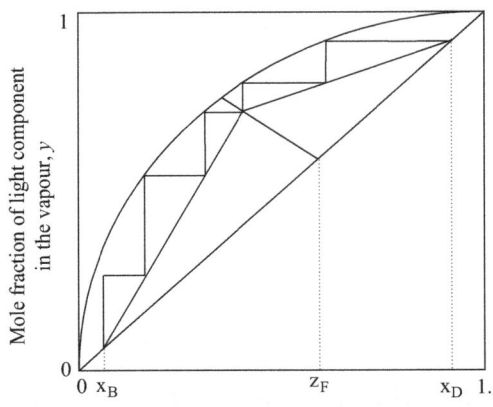

Figure 5.6   McCabe–Thiele diagram.

### 5.4.3   Graphical Methods

For systems containing only two components it is possible to solve certain simple mass-transfer problems graphically. The methods are based on material balances and equilibrium relationships; some more complex methods require enthalpy balances as well. A typical example is a McCabe–Thiele diagram used for staged distillation calculations as shown in Figure 5.6. This diagram can be used to find the number of ideal stages necessary for a distillation column to achieve a given separation, as well as the stage number where the feed is to enter, given the binary feed composition and column operating pressure. The method was published in 1925 and, although computer-aided methods are more accurate and easier to apply, the graphical construction of the McCabe–Thiele method greatly facilitates the visualisation of many of the important aspects of multistage distillation, and therefore the effort required to learn the method is well justified. A description of the method is given in all major chemical engineering textbooks.[1,2,4–6]

### 5.4.4   Diffusional Rate Processes

The second main class of mass-transfer operations take place in packed towers or similar equipment and are usually handled using the concept of a diffusional process. When considering this class of processes, the assumption of staging can no longer be made and the material and energy balances are instead derived for a differential

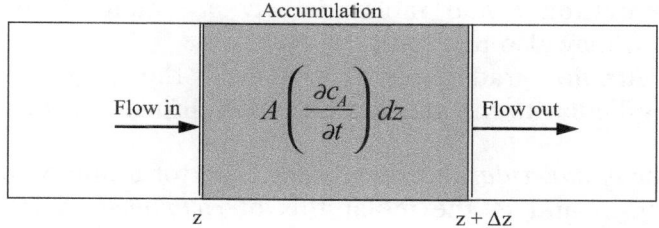

**Figure 5.7** Diffusion through a differential volume Adz.

element of the equipment as shown in Figure 5.7. The solution to the resulting equation set, which consists of several/many partial differential equations, is generally much more complex than those based on the concept of equilibrium stages, which is why chemical engineers normally simplify their calculations by assuming equilibrium stages.

*Diffusion* is the movement, under the influence of a physical stimulus, of an individual component through a mixture. The most common cause of diffusion is a concentration gradient of the diffusing component. A concentration gradient tends to move the component in such a direction as to equalise concentrations and destroy the gradient. When the gradient is maintained by constantly supplying the diffusing component to the high-concentration end of the gradient and removing the low-concentration end, the flow of the diffusing component is continuous. This movement is exploited in mass-transfer operations. Mass-transfer occurs by two mechanisms: (1) *molecular diffusion* by random and spontaneous movement of individual molecules in a gas, liquid or solid as a result of thermal motion; and (2) *eddy (turbulent) diffusion* by random macroscopic fluid motion.

## 5.4.5 Diffusion Calculations

In most calculations involving diffusion, attention is focused on diffusion in a direction *perpendicular* to the interface between the two phases and at a definite location in the equipment. Steady state is often assumed, and the concentrations at any point do not change with time. Five inter-related concepts are used in diffusion theory:

- Velocity $u$, defined as usual by length/time;
- Flux across a plane $N$, moles/area-time;
- Flux relative to a plane of zero velocity $J$, moles/area-time;

- Concentration $c$ and molar density $\rho_M$, moles/volume (mole fraction may also be used);
- Concentration gradient $dc/dz$, where $z$ is the length of the path perpendicular to the area across which diffusion is occurring.

*Fick's law of molecular diffusion* states that, for a binary mixture of components A and B, the molar flux of *component* A by ordinary molecular diffusion relative to the molar average velocity of *the mixture* in the positive $z$ direction, is proportional to the concentration gradient $dc_A/dz$, which is negative in the direction of ordinary molecular diffusion:

$$J_{Az} = -D_{AB} \frac{dc_A}{dz}$$

$$J_{Bz} = -D_{BA} \frac{dc_B}{dz}$$

Here, $D_{AB} = D_{BA}$ are the mutual diffusion coefficients of component A in component B and B in A, respectively.

In all mass-transfer operations, diffusion occurs in at least one phase and often in both phases. The material nature of diffusion and the resulting flow lead to three types of situations:

- *Unimolar Counter-Diffusion.* Only one component A of the mixture is transferred to or from the interface between the two phases, and the total flow is the same as the flow of A, thus:

$$N_B = 0 \quad \text{and} \quad N = N_A$$

  *Absorption* of a single component from a gas into a liquid is an example of this type.
- *Equimolar Counter-Diffusion.* The diffusion of component A in a mixture to and from the interface is balanced by an equal and opposite molar flow of component B, so that there is no net molar flow, thus:

$$N = N_A + N_B = 0 \quad \text{and} \quad J_A = -J_B$$

  This is generally the case in distillation, and it means there is no net volume flow in the gas phase. However, there is generally a net volume or mass flow in the liquid phase because of the difference in molar densities.
- *Uneven Counter-Diffusion.* Diffusion of A and B takes place in opposite directions, but the molar fluxes are unequal. This situation often occurs in diffusion of chemically reacting species to and from a catalyst surface.

In most common separation processes, the main mass transfer is across an interface between a gas and a liquid or between two liquid phases. At fluid–fluid interfaces, turbulence may persist to the interface. A simple theoretical model for turbulent mass transfer to or from a fluid-phase boundary was suggested in 1904 by Nernst, who postulated that the entire resistance to mass transfer in a given turbulent phase lies in a thin, stagnant region of that phase at the interface, called a film, hence the name *film theory*.[2,4,5] Other, more detailed, theories for describing the mass transfer through a fluid–fluid interface exist, such as the *penetration theory*.[1,4]

As mentioned earlier, calculations of diffusional rate processes are difficult as they involve the solution of partial differential equations. Even for processes which are clearly diffusional controlled, such as absorption, chemical engineers normally simplify the calculations by assuming equilibrium stages and may instead correct for possible deviations by using efficiency factors afterwards. Most commercial process design software, such as HYSYS, AspenPlus and ChemCAD, make the assumption of staged equilibrium processes.

## 5.5 Separation by Distillation

Distillation is the most mature separation method in industry and its design and operation procedures are well established. Only when vapour–liquid equilibrium or other data are uncertain, are laboratory and/or pilot plant studies necessary prior to the design of a commercial unit.

In distillation, a feed mixture of two or more components is separated into two or more products, including, and often limited to, an overhead distillate and a bottom product, whose compositions differ from that of the feed (see Figure 5.8). Most often, the feed is a liquid or a vapour–liquid mixture. The bottom product is always a liquid, but the distillate may be a liquid or a vapour or both. The separation requires that: (1) a second phase be formed so that both liquid and vapour phases are present and can contact each other; (2) the components have different volatilities so that they will partition between the two phases to different extents and (3) the two phases can be separated by gravity or other mechanical means.

Distillation differs from absorption in that the second fluid phase is created by thermal means (ESA), vapourisation and condensation, rather than by the introduction of a second phase that usually

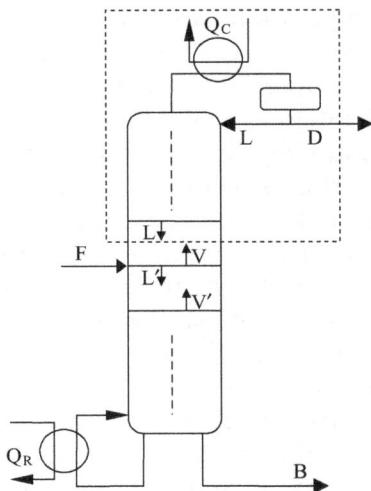

**Figure 5.8**  General distillation column.

contains an additional component or components not present in the feed mixture (MSA).

Throughout the twentieth century, multistage distillation has been by far the most widely used method for separating liquid mixtures of chemical components. Unfortunately, distillation is a very energy intensive technique, especially when the relative volatility of the components being separated is low (<1.5). The energy consumption for distillation alone in the US for 1976 totalled nearly 3% of the entire national energy consumption.[1] Approximately two-thirds of the distillation energy was consumed by petroleum refining, where distillation is widely used to separate crude oil into petroleum fractions, light hydrocarbons and aromatic chemicals.

## 5.5.1  Trays and Packing

Distillation has traditionally been carried out in trayed columns. However, more and more frequently, additional distillation capacity is being achieved with existing trayed columns by replacing all or some of the trays by structured packing. The choice between a packed column and a tray-type column is based mainly on economics when factors of contacting efficiency, loadability, and pressure drop must be considered.

*Cross flow plates* are the most common type of plate contactors used in tray distillation column. In a cross-flow plate, the liquid flows across and down from the plate and the vapour flows up through the plate.

The flowing liquid is transferred from plate to plate through vertical channels called downcomers. A pool of liquid is retained on the plate by an outlet weir. Other types of plates, such as bubble-cap and valve trays, are also used.[7-10]

For a packed column, the principle requirements for the packing is that it should (1) provide a large surface area between the gas and liquid; (2) promote uniform liquid distribution on the packing surface; (3) promote uniform vapour gas flow across the column cross-section and (4) have an open structure to give a low resistance to gas flow. Many diverse types and shapes of packing have been developed to satisfy these requirements and details of the most recent ones can be found on the web sites of column internals vendors, *e.g.* ref. 9 or 10.

### 5.5.2   Design Range

When designing a distillation column, the main degrees of freedom are:

   (i) The height of the column section
   (ii) The diameter of the column section

The height of the column section is determined mainly by the number of stages or plates required which in turn depends on how easy or difficult the mixture is to separate, *i.e.* the extent of separation. If the boiling points of the components to be separated are close, then a large number of stages are required leading to a very tall column. The column height is limited, however, by factors such a wind conditions, *etc.*

The diameter of the column section is determined mainly by the amount of material flowing in the column, which in turn depends on the reflux ratio, *i.e.* how much liquid is returned from the top of the column, and the boilup, *i.e.* how high the heat input is at the bottom.

### 5.5.3   Operating Range

When operating a distillation column, the main *degrees of freedom* are:

   (i) The flow rate, composition, temperature, pressure and phase condition of the feed.
   (ii) The column pressure.
   (iii) The degree of subcooling, if any, in the condenser.

(iv) The reflux ratio, *i.e.* the ratio between the amount of distillate returned to the column as liquid and that taken off as product $(R = L/D)$.

(v) The vapour rate given by the heat input to the reboiler.

(vi) The amount of product taken off at the bottom or at the top.

Of these, the feed mixture may or may not vary, but is generally taken as given. The column pressure and the degree of subcooling are normally fairly constant. The main operational variables are the reflux ratio $R$ and the heat input to the reboiler $Q_R$ and once these are set, the amount of product withdrawal at the bottom or at the top will also be given by the product specifications. An optimum exists for the reflux ratio in terms of operating costs, and normally a number of ratios are tested, and the economics of each scenario is investigated, before a decision is reached.

Note that the reflux ratio and the boilup rate cannot be chosen independently as there is a strong interaction between them. The upper limit to the vapour flow is set by the condition of *flooding*. At flooding there is a sharp drop in plate efficiency and increase in pressure drop. Flooding is caused either by the excessive carryover of liquid to the next plate by entrainment, or by liquid backing up in the downcomer. The lower limit of the vapour flow is set by the condition of *weeping*. Weeping occurs when the vapour flow is insufficient to maintain a level of liquid on the plate. *Coning* occurs at low liquid rates, and is the term given to the condition where the vapour pushes the liquid back from the holes and jets upwards with poor liquid contact.[2,3]

## 5.5.4 Design Calculations

The basis for any design calculations for distillation is the thermodynamic data, *i.e.* the vapour–liquid equilibrium characteristics of the mixture which is to be separated. As mentioned earlier, it is the *extent of separation* which determines whether or not a separation is feasible and how easy or difficult it will be to separate the mixture. Vapour–liquid equilibrium (VLE) data can be found either tabulated in textbooks,[7] given by equations or correlations such as Antoine's equation[1,2,4,5] or calculated using commercial programmes. For hand-calculations, VLE data is needed to generate the McCabe–Thiele diagram. If designing using commercial software, most programmes will have VLE databases built in. It is important to note that accurate thermodynamic data is vital for design. If the data is wrong, the design will be wrong, regardless of how accurate the calculations are. Spending time getting reliable thermodynamic data is, therefore, a good investment and is highly recommended.

The equilibrium stage calculations discussed earlier can easily be applied to distillation in order to find the required operating and design parameters. Consider the following problem specification.

### 5.5.4.1 Problem

A mixture of methanol and water containing 40 mol per cent of methanol is to be separated to give a product of at least 90 mol per cent of methanol at the top, and a bottom product with no more than 10 mol per cent of methanol. The feed flow rate is 100 kmol h$^{-1}$ and the feed is heated so that it enters the column at its boiling point. The vapour leaving the column is condensed, but not sub-cooled, and provides reflux and product. Since all the vapour from the column is condensed, the composition of the vapour from the top plate must equal that of the top product as well as that returned as reflux.

It is proposed to operate the unit with a reflux ratio of 3 kmol kmol$^{-1}$ product (reflux ratio $R = L/D$, where $L$ is the flow returned to the column and $D$ is the distillate product). It is required to find the top and bottom product flow rates, as well as the liquid and vapour flow rates above and below the feed point.

### 5.5.4.2 Solution

A total material balance over the whole column, where F, D and B are the feed, top product and bottom product flow rates, respectively, yields:

$$F = D + B$$

The component balance for the most volatile component (methanol), where $x_F$, $x_D$ and $x_B$ are the feed, top product and bottom product mole fractions (of methanol), respectively, gives:

$$Fx_F = Dx_D + Bx_B$$

Combining the equations:

$$B = F(x_F - x_D)/(x_B - x_D) = 100(0.4 - 0.9)/(0.1 - 0.9) = 62.5 \text{ kmol h}^{-1}$$

and

$$D = F - B = 100 - 62.5 = 37.5 \text{ kmol h}^{-1}$$

A total mass balance over the top of the column as given by the dotted area in Figure 5.8 gives:

$$V_n = L_{n+1} + D$$

The reflux ratio is given by $R = L_{n+1}/D = 3$.

Combining the top total material balance with the equation for the reflux ratio, gives an expression for the internal liquid flow L above the feed stage (assuming $L_n = L_{n+1} = L$):

$$V = L + D = RD + D = (R+1)D$$

Hence the internal vapour and liquid flow rates *above* the feed plate are:

$$V = 150.0 \text{ kmol h}^{-1} \quad \text{and} \quad L = 112.5 \text{ kmol h}^{-1}$$

Since the feed is liquid at its boiling point, it will run down the column together with the liquid from the top part of the column:

$$L' = L + F$$

Hence the internal liquid flow rate *below* the feed plate is:

$$L' = 212.5 \text{ kmol h}^{-1}$$

The vapour flow is unchanged (as the feed is all liquid):

$$V' = 150.0 \text{ kmol h}^{-1} (=V)$$

Given the heat of vapourisation of the mixture, the required heat input $Q_R$ (in kW) can be found from the vapour flow rate $V$. In the calculations above, the reflux ratio was given. As mentioned earlier, different values are typically considered, and the calculations repeated, to determine the reflux ratio which gives the most economical column operation and design.

### 5.5.5  Distillation Column Height

The height of a distillation column depends on the feed conditions, the product purity specifications and the extent of separation through the vapour–liquid equilibrium relationship, but also on the type of tray or packing used in the column as this affects the *rate* of separation. Column vendors will normally provide information on tray or packing efficiencies.[9,10]

Using design software or McCabe–Thiele's method, the number of stages required for the separation outlined above is five plates. As equilibrium has been assumed when performing the calculations, an efficiency factor needs to be applied, which can range from 0.3 to 1 depending on the type of trays or packing chosen. With an efficiency of 0.5, 10 actual stages would be required in the column section.

The diameter of the column is determined mainly by the vapour flow rate and correlations for this can be found in textbooks. If using commercial programmes, suitable correlations for different trays/packings will normally be incorporated.

## 5.6  Separation by Absorption

The removal of one of more selected components from a mixture of gases by absorption into a suitable solvent (Mass Separating Agent, MSA) is the second major operation of chemical engineering after distillation. Absorption is based on interface mass transfer controlled largely by rates of diffusion. It is worth noting that absorption followed by a chemical reaction in the liquid phase is often used to get more removal of a solute from a gas mixture.

In considering the design of equipment to achieve gas absorption, the main requirement is that the gas is brought into intimate contact with the liquid, and the effectiveness of the equipment will largely be determined by the success with which it promotes contact between the two phases. The general form of equipment is similar to that described earlier for distillation and both tray and packed columns are generally used for large installations.[9,10] The method of operation, however, is not the same. In absorption, the feed is a gas and is introduced at the bottom of the column, and the solvent (MSA) is fed to the top as a liquid. The solvent with the absorbed gas leaves at the bottom as liquid and the unabsorbed components leave as gas from the top (see Figure 5.9).

The essential difference between distillation and absorption is that in the former, the vapour has to be introduced in each stage by partial vapourisation of the liquid, which is therefore at its boiling point, whereas in absorption the liquid is well below its boiling point. In distillation, there is diffusion of molecules in both directions, so that for an ideal system, equimolar counter-diffusion exists. In absorption, gas molecules are diffusing into the liquid and the movement in the reverse direction is negligible. In general, the ratio of the liquid flow rate to the gas flow rate is considerably greater in absorption than in distillation, with the result that the layout of the columns is different in the two cases. Furthermore, with the higher liquid rates in absorption, packed columns are much more commonly used.

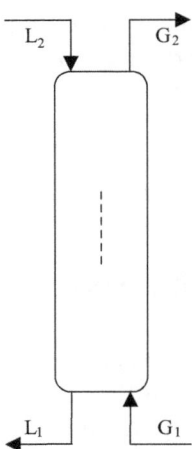

**Figure 5.9**   General absorption column.

## 5.6.1   Design Range

When designing an absorption column, the main degrees of freedom are:

(1) The choice of solvent.
(2) The height of the column section.
(3) The diameter of the column section.

One of the main degrees of freedom when designing absorption processes is the choice of solvent. The main criteria are that: (1) the component which is to be removed absorbs well, and fast, into the solvent and (2) the absorption of the other components present in the vapour is negligible. The choice, therefore, depends on the thermodynamic properties of the components in the feed and will not be considered here.

The same issues apply to the height and diameter of an absorption column as for a distillation column; the more difficult the separation, the taller the column and the more liquid and vapour flowing inside the column, the wider the column diameter.

## 5.6.2   Operating Range

When operating an absorption column, the main *degrees of freedom* are:

(1) The flow rate, composition, temperature and pressure of the gas feed.

(2) The column pressure.
(3) The amount of solvent introduced at the top.

The conditions of the gas feed may or may not vary, but is normally taken as given. The column pressure is normally fairly constant. The main operational variable is the amount of MSA required – the more solvent that is used, the more gas will be absorbed. However, the more liquid is present in the column, the wider the column needs to be, which will make the column more expensive. As for distillation, a trade-off therefore exists between operation and design.

## 5.6.3  Design Calculations

Most absorption operations are carried out in counter-current flow contactors in which the gas phase is introduced at the bottom of the column and the liquid solvent is introduced at the top of the column.

The material balances for an absorption column is normally set up per unit cross-sectional area, *i.e.*

$$G_1 + L_2 = G_2 + L_1$$

where $G_1$ and $G_2$ are the vapour flow rate per unit cross-sectional area into the column at the bottom and out of the column at the top, respectively; and $L_1$ and $L_2$ are the liquid flow rate per unit cross-sectional area out of the column at the bottom and into the column at the top, respectively. (In other words, index 1 denotes the bottom of the column and index 2 the top of the column. This may vary from textbook to textbook, so always check which is which before using an equation.)

Note that the material balances for an absorption column are normally written on a *solute-free* basis (the solute is the component being absorbed). In other words, we give the flow rates in terms of the components which are *not* being absorbed. This makes the calculations easier as the solute-free flow rates of both gas and liquid in or out of the column are constant.

The solute-free liquid and gas flow rates are given by:

$$L = L_i(1 - x_{A,i}) \quad \text{where } i = 1 \text{ or } 2, \textit{ i.e. } \text{bottom or top of column;}$$

$$G = G_i(1 - y_{A,i}) \quad \text{where } i = 1 \text{ or } 2, \textit{ i.e. } \text{bottom or top of column;}$$

where $L$ and $G$ are the solute-free liquid and gas flow rate in or out of the column, respectively; and $x_{Ai}$ and $y_{Ai}$ are the liquid and gas mole fraction of component A (the solute which is absorbed) in or out of the column, respectively.

When setting up the component material balance, we express the amounts of each component in terms of mole *ratios* $X_A$ and $Y_A$ instead of mole *fractions* $x_A$ and $y_A$:

$$X_A = \frac{x_A}{1 - x_A} = \frac{\text{mole fraction of component A in the liquid}}{\text{mole fraction of non-A components in the liquid}}$$

$$Y_A = \frac{y_A}{1 - y_A} = \frac{\text{mole fraction of component A in the vapour}}{\text{mole fraction of non-A components in the vapour}}$$

The component material balance then becomes:

$$GY_{A1} + LX_{A2} = GY_{A2} + LX_{A1}$$

### 5.6.3.1  Problem

A hydrocarbon gas stream is to be purified by continuous counter-current contact with a liquid organic solvent in an absorption column. The inlet gas contains 1.5% by volume of toxic DMSO of which 95% is to be removed. The gas flows at a rate of $G = 0.1$ kmol s$^{-1}$ on a DMSO-free basis. Calculate

   (i) the mol fractions of vapour $y_{Ai}$ and the corresponding mol ratios of gas $Y_{Ai}$
  (ii) the total gas flow rates (DMSO *and* hydrocarbon), $G_i$
 (iii) the component A vapour flow rates (DMSO only), $G_{Ai}$

at the bottom $(i=1)$ and at the top $(i=2)$ of the column.

### 5.6.3.2  Solution

The conditions of the inlet vapour feed are given. The mole *fraction* of A at the inlet is therefore:

$$y_{Ai} = 1.5 \text{ vol\%} = 0.015$$

The mole *ratio* of A is:

$$Y_{A1} = \frac{y_{A1}}{1 - y_{A1}} = \frac{0.015}{1 - 0.015} = 0.0152$$

The total vapour flow rate (hydrocarbon *and* DMSO) into the column at the base is:

$$G_1 = G/(1 - y_{A1}) = 0.1 \text{ kmol s}^{-1}/(1 - 0.015) = 0.10155 \text{ kmol s}^{-1}$$

The component A gas flow rate (DMSO only) into the column at the base is:

$$G_{A1} = G_1 y_{A1} = 0.10155 \text{ kmol s}^{-1} \, 0.015 = 0.00152 \text{ kmol s}^{-1}$$

For the top of the column, it is specified that 95% of the DMSO must be removed, *i.e.* only 0.5% remains:

$$G_{A2} = G_{A1} \, 0.05 = 0.00152 \text{ kmol s}^{-1} \, 0.05 = 7.6 \times 10^{-5} \text{ kmol s}^{-1}$$

The total gas flow rate (hydrocarbon *and* DMSO) out of the column is then:

$$G_2 = G + G_{A2} = 0.1 \text{ kmol s}^{-1} + 7.6 \times 10^{-5} \text{ kmol s}^{-1} = 0.100076 \text{ kmol s}^{-1}$$

The mole fraction of component A at the top of the column is then:

$$y_{A2} = G_{A2}/G_2 = 7.6 \times 10^{-5} \text{ kmol s}^{-1}/0.100076 \text{ kmol s}^{-1} = 7.6 \times 10^{-4}$$

and the mole ratio of A is:

$$Y_{A2} = \frac{y_{A2}}{1 - y_{A2}} = \frac{7.6 \times 10^{-4}}{1 - 7.6 \times 10^{-4}} = 7.6 \times 10^{-4}.$$

### 5.6.4   Amount of Solvent

As previously mentioned, one of the main degrees of freedom in absorption operations is the amount of solvent which is required to achieve the required absorption. This amount depends on how well the undesired gas component (the solute) is absorbed into the solvent. The vapour–liquid equilibrium relationship for absorption of a gas component into a liquid solvent is expressed as:

$$Y^* = f(X)$$

where $Y^*$ and $X$ are the mole ratio of the absorbing component in the gas and liquid phases at equilibrium, respectively. For some separations, the equilibrium relationship is linear, which greatly simplifies the calculations. For non-linear relationships, graphical calculation methods exists.[2,4,5]

As for distillation, the thermodynamic data form the basis for the design for absorption and the absorption equilibrium data can be found either tabulated, from correlations or from commercial software. Again, good data is vital to ensure good design.

We know that the column will not operate at equilibrium, but to avoid having to use complex calculation methods based on diffusion, we find the solvent flow rate which *would have been* required to achieve the separation had the column operated at equilibrium, and then assume that the actual flow rate will be a fraction larger than this, typically 1.4–1.5 times larger.

### 5.6.4.1 Problem

Consider the same separation as previously, where a hydrocarbon gas stream is to be purified by continuous counter-current contact with a liquid organic solvent in an absorption column. The inlet gas contains 1.5% by volume of toxic DMSO of which 95% is to be removed. The gas flows at a rate of $G = 0.1$ kmol s$^{-1}$ on a DMSO-free basis.

The organic liquid solvent initially contains 0.001 mole fraction DMSO. The required solvent flow rate is to be 1.5 times the minimum.

The gas–liquid equilibrium relationship is given by:

$$Y^* = 0.2X$$

where $Y^*$ is the mole fraction of DMSO in the gas phase and $X$ is the mole ratio of DMSO in the liquid phase.

Calculate

(1) the required solvent flow rate $L$
(2) the actual mole fraction of the solvent at the bottom of the column $x_{A1}$.

### 5.6.4.2 Solution

The mole ratio of the solvent entering the column at the top is:

$$X_{A2} = \frac{x_{A2}}{1 - x_{A2}} = \frac{0.001}{1 - 0.001} = 0.001001$$

Had the column operated at equilibrium, then the maximum amount of solute would have been removed and we would have used the minimum amount of liquid. Given the gas–liquid equilibrium relationship:

$$Y^* = 0.2X$$

the maximum mole ratio of solute in the solvent, $X_{A1,max}$, is given by:

$$Y_{A1} = Y_{A1*} = 0.2X_{A1,max}$$

therefore (with $Y_{A1}$ found earlier):

$$X_{A1,max} = Y_{A1}/0.2 = 1.52 \times 10^{-2}/0.2 = 7.6 \times 10^{-2}$$

From an overall mass balance between the top and bottom of the column:

$GY_{A1} + L_{min}X_{A2} = GY_{A2} + L_{min}X_{A1,max}$

$G(Y_{A1} - Y_{A2}) = L_{min}(X_{A1,max} - X_{A2})$

$L_{min} = 0.1(1.52 \times 10^{-2} - 7.6 \times 10^{-4})/(7.6 \times 10^{-2} - 1.001 \times 10^{-3})$

$L_{min} = 1.9 \times 10^{-2} \text{ kmol s}^{-1}$

The actual liquid solvent flow rate which is required is then given by:

$$L = 1.5\,L_{min} = 2.9 \times 10^{-2} \text{ kmol s}^{-1}$$

The *actual* liquid mole ratio $X_{A1}$ can then be found from an overall mass balance between the top and bottom of the column:

$GY_{A1} + LX_{A2} = GY_{A2} + LX_{A1}$

$G(Y_{A1} - Y_{A2}) = L(X_{A1} - X_{A2})$

$X_{A1} = X_{A2} + G/L(Y_{A1} - Y_{A2})$

$X_{A1} = 1.001 \times 10^{-3} + 0.1/(2.9 \times 10^{-2})(1.52 \times 10^{-2} - 7.6 \times 10^{-4})$

$X_{A1} = 0.051.$

## 5.6.5 Absorption Column Height

The height of an absorption column depends on the feed conditions, the product purity specifications, the solvent used and the extent of separation through the absorption equilibrium relationship, but also on the *rate* of separation. If the rate of mass transfer of the gaseous component from the gas phase into the liquid phase is slow, then the column needs to be longer to ensure that the required amount is removed. The rate of mass transfer depends on the *mass-transfer coefficient*, normally denoted $k_G$ or $k_L$. The value of the mass-transfer coefficient depends on the components in the gas feed and on the solvent used and is often determined experimentally. The type of packing used in the column will also have an impact on the column height as for distillation.

The diameter of the column is determined mainly by the vapour flow rate and correlations for this can be found in textbooks. If using commercial programmes, suitable correlations for different trays/packings will normally be incorporated.

## 5.7   Further Reading

This chapter has mainly been based on textbooks generally used in undergraduate courses on mass transfer and separation processes.[1–6] Most of these textbooks contain sections on both basic calculations as well as some design considerations. Coulson *et al.*[3] is devoted entirely to design. Perry,[7] in addition to basic theory, also includes thermodynamic data for many common mixtures which may be used for preliminary calculations. Most of the more recent textbooks, such as Seader & Henley,[1] also consider less common separation methods, such as membrane separations, ion exchange and chromatography, in some detail.

Stichlmair and Fair[11] and Rose[8] are textbooks devoted entirely to design and operation of distillation columns, Ruthven[12] considers absorption and Astarita *et al.*[13] absorption with chemical reaction. Ho[14] is a handbook for membrane separations and Guiochon[15] considers chromatographic separations.

## 5.8   Relevant Material not Considered in this Chapter

### 5.8.1   Combined Reaction and Separation

The possibility of combining a reaction unit and a separation unit into one reactive separation unit has not been considered in this chapter. However, some of the textbooks, Coulson *et al.*[2] and Seader and Henley,[1] give a brief discussion of this type of equipment for distillation and absorption, respectively. A lot of research has also been devoted to this type of operation in recent years and it is generally believed that it will become more widely used in the future.

### 5.8.2   Combined Separation Units

The possibility of combining two different separation units into one, hybrid, process has not been considered in this chapter. Hybrid processes are quite novel and have only very recently been considered by industry and have, therefore, so far not made it into the standard textbooks. A hybrid process has the combined benefits of both of the component units and the benefits should theoretically outweigh the disadvantages. An example is a hybrid of a distillation column and a pervaporation unit for azeotropic separation, where the distillation unit alone is limited by the azeotropic point. Again, a lot of research is currently devoted to this type of operation and it is generally believed that it will become more widely used in the future.

### 5.8.3 Batch Operation

In this chapter, continuous operation has been implicitly assumed. However, batch operation is very often used in both the fine chemical and pharmaceutical industries. Analysing and designing batch equipment is generally different from the equivalent continuous units, in that the usual assumption of steady-state operation does not apply. However, the main characteristics of the processes are still the same, such as suitability of plates or packing, trade-offs between reflux ratio and boilup rate in a distillation column, *etc.* Unfortunately, most textbooks, including Perry and Green,[7] contain very little information on batch equipment or batch design.

# References

1. J. D. Seader and E. J. Henley, *Separation Process Principles*, Wiley, New York, 1998.
2. J. M. Coulson, J. F. Richardson, J. R. Backhurst and J. H. Harker, *Coulson & Richardson's Chemical Engineering*, Pergamon Press, Oxford, 4th edn (revised), 1993, vol. 2 (Particle Technology and Separation Processes).
3. J. M. Coulson, J. F. Richardson and R. K. Sinnott, *Coulson & Richardson's Chemical Engineering*, Pergamon Press, Oxford, 1st edn (revised), 1986, vol. 6 (Design).
4. W. L. McCabe, J. C. Smith and P. Harriott, *Unit Operations of Chemical Engineering*, McGraw-Hill, Singapore, 5th edn, 1993.
5. R. E. Treybal, *Mass-Transfer Operations*, McGraw-Hill, New York, 3rd edn, 1981.
6. J. R. Welty, C. E. Wicks and R. E. Wilson, *Fundamentals of Momentum, Heat and Mass Transfer*, Wiley, 3rd edn, 1984.
7. R. H. Perry and D. W. Green, *Perry's Chemical Engineers' Handbook*, McGraw-Hill, New York, 7th edn, 1997.
8. L. M. Rose, *Distillation Design in Practice*, Elsevier, New York, 1985.
9. Sulzer Chemstech at www.sulzer.com.
10. Koch-Glitsch at www.koch-glitsch.com.
11. J. G. Stichlmair and J. R. Fair, *Distillation – Principles and Practice*, Wiley, New York, 1998.

12. D. M. Ruthven, *Principles of Adsorption and Adsorption Processes*, Wiley, New York, 1984.
13. G. Astarita, D. W. Savage and A. Bisio, *Gas Treating With Chemical Solvents*, Wiley, New York, 1983.
14. *Membrane Handbook*, ed. W. Ho and K. Sirkar, van Nostrand Reinhold, New York, 1st edn, 1992.
15. G. Guiochon, *Fundamentals of Preparative and Nonlinear Chromatography*, Academic Press, Boston, 1994.

# 6 Scale-up in Chemical Engineering

Tim Elson

## 6.1 Introduction and Objectives

In only a few cases can theoretical analysis alone provide full description of chemical engineering process units. Normally it is necessary to turn to experimental work, often upon small- or pilot-scale models, to complete such a study. Even if a complete quantitative theory is available, experimental results are still necessary to verify it, since theories are invariably based on assumptions that may not be completely satisfied in the real systems.

Although it is desirable to conduct the experimental work in the system for which the results are required, this is not always easy. The system of interest, or *prototype*, may be hazardous or expensive to build and run, while the fluids involved may be corrosive or toxic. In this case scale *models* are used, which overcome the above problems and allow extensive experimentation. In the majority of cases the model is smaller in size than the prototype. In a number of cases, however, the model and the prototype may be of the same size but the fluids involved are different.

Examples of the use of models for the design of large-scale systems include the measurement of pressure drop and heat transfer in model heat exchangers, the mixing and rate of reaction in a bench-top batch reactor and the prediction of pressure drops in pipelines.

The results from the model will have to be applied to the prototype. Since the model is usually smaller than the prototype, this procedure

Concepts of Chemical Engineering for Chemists: 2nd Edition
Edited by Stefaan Simons
© The Royal Society of Chemistry 2017
Published by the Royal Society of Chemistry, www.rsc.org

is called *scale-up*. Scale-up is only possible when both the model and the prototype are *physically similar*.[1-3] Similar comments apply to scale-down, where the model is larger than the prototype.

Engineering data are often presented in dimensionless form. Applying this data to a particular problem in hand is a form of scale-up, the literature data or correlation being the model here. The dimensionless form is obtained by *dimensional analysis*.

*The objectives of this chapter are*

- to show how dimensionless variables (groups) are obtained from primary variables;
- to show some of the advantages and limitations of dimensional analysis; and
- to show how to obtain and use engineering data in dimensionless form.

## 6.2  Units and Fundamental Dimensions

Chemical engineers and scientists will know that the data they use are expressed in a great variety of different units. The length of a rod may be variously described as 12 in, 1 ft, 0.3048 m, 304.8 mm, *etc.* These lengths are all equivalent. Inch, foot, metre and millimetre define the size of the *unit* and 12, 1, 0.3048 and 304.8 define the number of the units in each system. There are many useful references in the literature that discuss conversion between units and tabulate conversion factors.[1,4,5]

Although the units for each of the measurements above are different, they share the same *fundamental dimension*, length **L**. Each of these measurements has the dimensions of length.

Velocity is the distance travelled per unit time and so has the dimensions of length, **L**, over time, **T**, or $\mathbf{LT}^{-1}$. The quantity of material, given by its mass, has the dimensions of mass, **M**.

Mass, **M**, length, **L** and time, **T**, are called *fundamental dimensions*. With heat transfer problems, temperature, **Θ**, and heat, **H**, are introduced as fundamental dimensions (see Section 6.7.4).

Astarita[6] presents a brief history of the concept of dimensions and dimensional analysis from Euclid to the present day.

## 6.3  Physical Similarity

Two systems are said to be *physically similar* in respect to certain specified physical quantities when the ratio of corresponding magnitudes of these quantities between the two systems is the same everywhere.

Table 6.1 lists the types of physical similarity, and the associated similar quantities, commonly used in chemical engineering.

We will discuss two of these types of physical similarities in the following sections.

### 6.3.1 Geometric Similarity

*Geometric similarity* is the most obvious requirement in a model system designed to correspond to a given prototype system. It is the similarity of shape. *Two systems are geometrically similar when the ratio of any length in one system to the corresponding length in another system is everywhere the same.*

Figure 6.1 illustrates two geometrically similar objects. Here $P$ and $P'$ are known as *corresponding points*; $X$ and $X'$, $Y$ and $Y'$ are known as *corresponding lengths*. The ratio of corresponding lengths is known as the *scale factor*:

$$\text{Scale factor} = \frac{X}{X'} = \frac{Y}{Y'} \tag{6.1}$$

**Table 6.1** Types of physical similarity.

| Type | Physical quantity | Example system |
|---|---|---|
| Geometric | Lengths | Stirred tank |
| Kinematic | Lengths + time intervals, or velocities | Planetarium, tidal models |
| Dynamic | Forces | Flow models, wind tunnel |
| Thermal | Temperature differences | Pilot-plant heat exchanger |
| Chemical | Concentration differences | Bench scale reactor |

(a) model      (b) prototype

**Figure 6.1** Geometric similarity.

Perfect geometric similarity is not always easy to achieve and difficulties can arise due to

- scaling of surface roughness or finish,
- scaling of surface tension (*e.g.* scale-down of a river will result in a thin water layer where surface tension effects will be more important than in the river).

## 6.3.2  Dynamic Similarity

Dynamic similarity is similarity of forces. *A model and prototype are dynamically similar when all forces acting at corresponding points, on fluid elements or corresponding boundaries, form a constant ratio between model and prototype.*

Forces of the same kind, *e.g.* gravitation, centripetal, viscous, *etc.*, acting at *corresponding points* at *corresponding times* are *corresponding forces*.

Dynamic similarity is especially important in fluid flow systems. A fluid's flow pattern is determined by the forces acting on the fluid elements. The net force acting on a fluid element gives the acceleration of that element over the next time interval and, hence, determines its motion. If the net forces on corresponding fluid elements at corresponding times are similar, then their motion will be similar.

Many forces may be involved, *e.g.*

- viscous forces – due to the fluid's viscosity;
- inertial forces – due to its density and velocity;
- pressure forces – due to pressure difference acting over an area of fluid;
- body forces, *e.g.* gravitational – due to earth's gravitational field;
- surface tension forces – due to the presence of interfaces;
- boundary forces, *e.g.* forces imparted by a moving boundary, a rotating impeller.

## 6.3.3  Dimensionless Groups

Considering all the different types of physical quantities involved in a system, it is obvious that complete physical similarity is very difficult to attain and to ensure. For this reason, *dimensionless groups* are used for scale-up purposes.

If the model and prototype illustrated in Figure 6.1 are geometrically similar, then the ratios of the lengths, $X/Y$ and $X'/Y'$, are equal. Now, $X/Y$ [=] (length/length), *i.e.* a ratio of lengths and is therefore dimensionless, a dimensionless group. [=] stands for "*... has the* *dimensions of...*". Thus, "$X$ has the dimensions of length" may be abbreviated to "$X$ [=] length" or "$X$ [=] $\mathbf{L}$".

Dimensionless groups allow us to perform scale-up, and we will see a number of examples of this in later sections of this chapter.

## 6.4 Dimensional Homogeneity

Take, for illustrative purposes, the equation relating the distance $s$, covered by an object in time $t$, travelling at an initial velocity $u$, when subject to a constant acceleration $a$:

$$s = ut + \frac{1}{2}at^2 \qquad (6.2)$$

Each term in eqn (6.2), $s$, $ut$ and $\frac{1}{2}at^2$, has the same dimensions,

$s$ [=] length, $\mathbf{L}$;

$ut$ [=] velocity $\times$ time [=] length/time $\times$ time [=] $\mathbf{LT}^{-1} \times \mathbf{T}$ [=] $\mathbf{L}$;

and $\frac{1}{2}at^2$ [=] acceleration $\times$ time$^2$ [=] length/time$^2$ $\times$ time$^2$ [=] $\mathbf{LT}^{-2} \times \mathbf{T}^2$ [=] $\mathbf{L}$.

Thus eqn (6.2) is said to be dimensionally homogeneous or dimensionally consistent.

*Any equation relating physical quantities must be dimensionally homogeneous or dimensionally consistent.* It is not permissible to add a length to a velocity because they are quantities of different types. Even Einstein's $E = mc^2$ is dimensionally homogeneous.

Note that in some equations, the values of constants are a function of the system of units used. These constants have dimensions and are known as *dimensional constants*. For example, many flowmeters measure volumetric flow rates, $Q_v$, by measuring heads of fluids, $\Delta h$. Typically, $Q_v \propto \sqrt{\Delta h}$ or $Q_v = c\sqrt{\Delta h}$. Here, as $Q_v [\neq] \sqrt{\Delta h}$, $c$ is a dimensional constant. Dimensionally, $\mathbf{L}^3\mathbf{T}^{-1} [=] c\sqrt{\mathbf{L}}$ or $c[=] \mathbf{L}^{2.5}\mathbf{T}^{-1}$, and the value of $c$ will depend upon the units of both $Q_v$ and $\Delta h$.

## 6.5 Dimensional Analysis and Dimensionless Groups

Chemical process systems can be described by a number of physical parameters. In the majority of cases, however, these systems are very complex and the number of parameters needed to describe them very large. Dimensional analysis is used to reduce the number of parameters needed to describe a system without destroying the generality of the relationship.[1,2]

## 6.5.1 Example: Dimensional Analysis

The distance that an accelerating object covers in a certain time is given by eqn (6.2), $s = ut + \dfrac{1}{2}at^2$. So, four variables, $s$, $u$, $a$ and $t$, are needed to describe this system. If we are interested in the distance travelled, we can write

$$s = f(u, t, a) \tag{6.3}$$

Figure 6.2 represents eqn (6.3) graphically, showing curves of $s$ *versus* $t$ for various values of $a$ at one particular value of $u$.

To represent the effects of all the possible values of $a$, we require an infinite number of curves on this figure as $a$ can take an infinite number of values. Now, this is just for one value of $u$, and $u$ may itself take an infinite number of values. Therefore to represent $s$ for all the possible values of $t$, $a$ and $u$, we need an infinite number of figures (all possible $u$ values), each with an infinite number of curves (all possible $a$ values). This, clearly, is not very convenient.

We saw earlier that eqn (6.2), $s = ut + \dfrac{1}{2}at^2$, was dimensionally homogeneous; each term has the same dimensions. Ratios of the terms in this equation will therefore be dimensionless. So we can make the original expression dimensionless by, for example, dividing each term by $ut$, giving

$$\frac{s}{ut} = 1 + \frac{1}{2}\frac{at}{u} \quad \text{or} \quad \frac{s}{ut} = f\left(\frac{at}{u}\right) \tag{6.4}$$

where $s/ut$ and $at/u$ are *dimensionless groups*. Plotting $s/ut$ *versus* $at/u$, we get Figure 6.3.

Here we have reduced the number of variables required to describe this system from four to two and the number of curves needed to one

**Figure 6.2**   Distance–time plots.

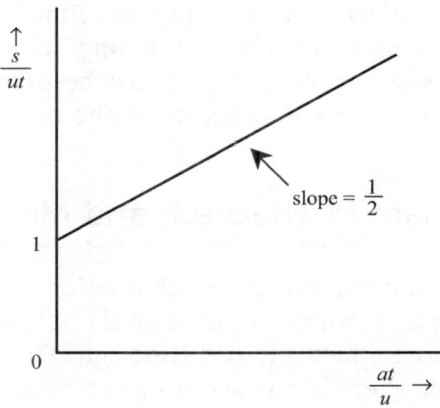

**Figure 6.3**   Dimensionless distance–time plots.

(Figure 6.3). The new variables are the two dimensionless groups in eqn (6.4). This form is useful when we want to see how distance travelled, $s$, varies with acceleration, $a$. Each variable of interest only occurs once and in separate groups.

### 6.5.1.1 Alternate Form

If instead we were interested in distance travelled $s$, as a function of initial velocity $u$, we would wish $s$ and $u$ to appear only once each and in separate groups. So, taking $s = ut + \dfrac{1}{2}at^2$ and dividing each term by $at^2$ gives

$$\frac{s}{at^2} = \frac{u}{at} + \frac{1}{2} \quad \text{or} \quad \frac{s}{at^2} = f\!\left(\frac{u}{at}\right) \tag{6.5}$$

where $s/at^2$ and $u/at$ are new dimensionless groups.

### 6.5.1.2 Points to Note

- With dimensional analysis an equation can take several equally valid forms, *e.g.* $s/ut = f(at/u)$ or $s/at^2 = f(u/at)$.
- Although four dimensionless groups have been found here, there are only two independent dimensionless groups. The extra, alternate, groups are just products of the powers of the original groups, *i.e.* $s/at^2 = (s/ut) \times (at/u)^{-1}$ and $u/at = (at/u)^{-1}$.
- Dimensional analysis gives fewer independent dimensionless groups (two in this case) than the number of variables (four in this case).

- Dimensional analysis does not give the functional relationship. It does not say what form the function $f$ takes. The functional relationship between the groups must be determined by other theory, experiment or combination of the two.

## 6.6   Buckingham Pi Theorem and Method

In the previous example, the physical relation that described the system was known and dimensional analysis helped to represent experimental data more effectively and concisely. Dimensional analysis is more useful when there is no governing equation for the system. In this case, the dimensionless groups are formed by following a more general procedure known as the *Buckingham method.* The objective of this method is to assemble all the parameters describing the problem into a smaller number of dimensionless groups ($\Pi$s). The relation connecting the different groups can be obtained experimentally or, in some cases, from mathematical analysis.

### 6.6.1   Buckingham Pi Theorem

The Buckingham method is based on the *Buckingham Pi Theorem,* which states

> "If a system is characterised by $m$ variables, which are completely described by $n$ fundamental dimensions (**M, L, T,** *etc.*), then these $m$ variables can be grouped into $m - n$ independent dimensionless groups $\Pi$s which describe the same situation."

Thus, if a system is described by $m$ variables $Q_1, Q_2, \ldots, Q_m$, then $Q_1 = f(Q_2, Q_3, \ldots Q_m)$ or

$$f(Q_1, Q_2, Q_3, \ldots, Q_m) = 0 \qquad (6.6)$$

and these $m$ variables possess $n$ fundamental dimensions, mass **M**, length **L** and time **T**. With dimensional analysis, the system can be written as

$$f = (\Pi_1, \Pi_2, \ldots, \Pi_{m-n}) = 0 \qquad (6.7)$$

where, $\Pi_1, \Pi_2, \ldots, \Pi_{m-n}$ are dimensionless groups.

The number of variables needed to describe the system has been reduced from $m$ to $m - n$.

## 6.6.2  Buckingham's Method – Procedure

(1) List all $m$ variables describing the problem and their fundamental dimensions.

(2) Find the number of fundamental dimensions, $n$, that appear and the number, $m - n$, of the dimensionless groups $\Pi$s that can be formed.

(3) Select $n$ variables, the *recurring set.* How?
- Recurring sets must, among them, include all $n$ fundamental dimensions.
- Selected variables should not have the same dimensions or the same dimensions but to a different power.
- Experience.
- Exclude the dependent variable and those variables whose effect one tries to isolate.
- If a different recurring set is chosen, the resulting dimensionless groups will be different, although these new groups will be products of the powers of the original groups.

(4) The dimensionless groups, $\Pi$s, are found one at a time. Take the product of one of the remaining variables to a known power (usually 1) and recurring variables to unknown powers. Determine the exponents of the recurring variables, using the principle of dimensional homogeneity, so that the group becomes dimensionless.

(5) Repeat the procedure in the previous step (4) $m - n$ times to define all the $m - n$ dimensionless groups.

## 6.6.3  Example: Drag Force on a Sphere

### 6.6.3.1  Problem Statement

A sphere, illustrated in Figure 6.4, has a diameter of $d$ and is held inside a duct where fluid flows with a relative velocity $u_\infty$. The fluid has density $\rho$ and viscosity $\mu$. Owing to the relative motion between the

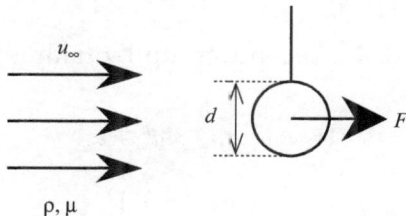

**Figure 6.4**  Drag on a sphere.

fluid and the sphere, a force $F$, called drag force, is exerted on the sphere. Which are the dimensionless groups that describe this system?

### 6.6.3.2  Buckingham Method

(1) List all variables $m$, describing the problem and their fundamental dimensions:

| *Variable* | $F$ | $=f$ | $(u_\infty$ | $d$ | $\mu$ | $\rho)$ |
|---|---|---|---|---|---|---|
| *Dimensions* | $\mathbf{MLT}^{-2}$ | | $\mathbf{LT}^{-1}$ | $\mathbf{L}$ | $\mathbf{ML}^{-1}\mathbf{T}^{-1}$ | $\mathbf{ML}^{-3}$ |

(2) Find the number of fundamental dimensions, $n$, that appear and the number, $m - n$, of the dimensionless groups, $\Pi$s, that can be formed:

$m = 5$ ($F$, $u_\infty$, $d$, $\mu$ and $\rho$), $n = 3$ ($\mathbf{M}$, $\mathbf{L}$ and $\mathbf{T}$); therefore, $\Pi s = m - n = 5 - 3 = 2$

(3) Select $n$ variables, the *recurring set:*

Select $\rho$, $d$ and $u_\infty$, dimensions $[\mathbf{ML}^{-3}]$, $[\mathbf{L}]$ and $[\mathbf{LT}^{-1}]$, respectively

- Recurring set includes all $n$ fundamental dimensions, $\mathbf{M}$, $\mathbf{L}$ and $\mathbf{T}$.
- Set excludes $F$ (the dependent variable) and $\mu$.
- Selected variables do not have the same dimensions.

(4) Find the first dimensionless group:

Take the product of one of the remaining variables ($F$ the dependent variable) to a known power (1) and recurring variables to unknown powers:

$$\Pi_1 = F^1 \times \rho^{a_1} \times d^{b_1} \times u_\infty^{c_1}$$

$$[=](\mathbf{MLT}^{-2}) \times (\mathbf{ML}^{-3})^{a_1} \times (\mathbf{L})^{b_1} \times (\mathbf{LT}^{-1})^{c_1}$$

Equating the powers of each dimension in turn and solving simultaneously for the unknown powers:

$$\left. \begin{array}{ll} \mathbf{M}: 0 = 1 + a_1 & a_1 = -1 \\ \mathbf{L}: \ 0 = 1 - 3a_1 + b_1 + c_1 & b_1 = -2 \\ \mathbf{T}: \ 0 = -2 - c_1 & c_1 = -2 \end{array} \right\} \quad \text{and so} \quad \Pi_1 = \frac{F}{\rho d^2 u_\infty^2}$$

(5) Repeat to find the second group by taking the one remaining variable ($\mu$):

$$\Pi_2 = \mu^1 \times \rho^{a_2} \times d^{b_2} \times u_\infty^{c_2}$$

$$\left. \begin{array}{ll} \mathbf{M}: 0 = 1 + a_2 & a_2 = -1 \\ \mathbf{L}: 0 = -1 - 3a_2 + b_2 + c_2 & b_2 = -1 \\ \mathbf{T}: 0 = -2 - c_2 & c_2 = -2 \end{array} \right\} \quad \text{and so} \quad \Pi_2 = \frac{F}{\rho d u_\infty}$$

Now,

$$f(\Pi_1, \Pi_2) = 0 \quad \text{or} \quad f\left(\frac{F}{\rho d^2 u_\infty^2}, \frac{\mu}{\rho d u_\infty}\right) = 0$$

Thus,

$$\frac{F}{\rho d^2 u_\infty^2} = f\left(\frac{\mu}{\rho d u_\infty}\right) \tag{6.8}$$

Or, in conventional form,

$$C_{\mathrm{D}} = \frac{8F}{\pi d^2 \rho u_\infty^2} = f\left(\frac{\rho u_\infty d}{\mu}\right) = f(\mathrm{Re}) \tag{6.9}$$

$C_{\mathrm{D}}$ is known as the drag coefficient. Drag coefficients are normally expressed in the form $C_{\mathrm{D}} = (F/A) \Big/ \left(\frac{1}{2}\rho u_\infty^2\right)$, where $A$ is the projected or cross-sectional area of the particle in the direction of flow. Hence the particular form shown in eqn (6.9). Re is the particle Reynolds number, $\mathrm{Re} = (\rho u_\infty d/\mu)$, and is the reciprocal of the second dimensionless group found above.

Most dimensionless groups can be thought of as ratios:

- Drag coefficient:

$$C_{\mathrm{D}} = \frac{8F}{\mu d^2 \rho u_\infty^2} \equiv \frac{\text{drag forces on particle}}{\text{fluid inertial forces}}$$

- Particle Reynolds number:

$$\mathrm{Re} = \frac{\rho u_\infty d}{\mu} \equiv \frac{\text{fluid inertial forces}}{\text{fluid viscous forces}}$$

The Reynolds number characterises the type of flow (see Chapter 3, "*Concepts of Fluid Flow*", for more information).

### 6.6.3.3 Experimental Results

Figure 6.5 shows the variation of $C_{\mathrm{D}}$ with Re for a sphere. The curve shown is valid for spheres of any diameter. The form of this chart and the related flow regimes around the sphere are discussed in Chapter 7: "*An Introduction to Particle Systems*" and in many texts.[1,3]

## 6.7 Use of Dimensionless Groups in Scale-up

Since the prototype (P) and the model (M) should be physically similar systems, they should be governed by the same equations. These

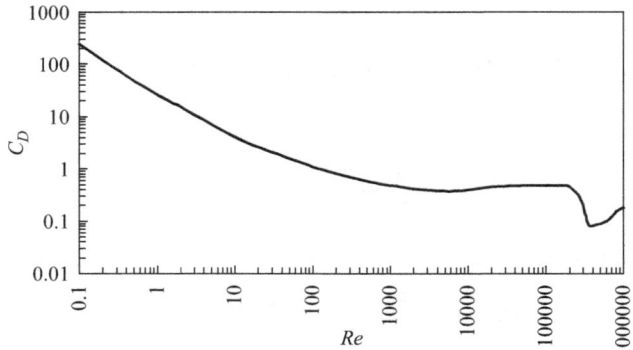

**Figure 6.5**   Drag coefficients for a sphere.

equations can be expressed in terms of dimensionless groups in eqn (6.7):

$$f(\Pi_1, \Pi_2, \ldots, \Pi_{m-n}) = 0 \tag{6.7}$$

For the sake of argument, let $m - n = 3$ and so $f(\Pi_1, \Pi_2, \Pi_3) = 0$ or $\Pi_1 = f(\Pi_2, \Pi_3)$. So, if $\Pi_{2M} = \Pi_{2P}$ and $\Pi_{3M} = \Pi_{3P}$, then $\Pi_{1M}$ must equal $\Pi_{1P}$, even though we do not know the form of the function $f$.

Thus, if we arrange for the values of all but one of the dimensionless groups ($\Pi_2$ and $\Pi_3$, above) to be equal between model and prototype, then the remaining dimensionless group ($\Pi_1$ above) must be equal between model and prototype.

The following sections show examples of scale-up using dimensionless groups.

## 6.7.1   Drag Force on a Particle

Dimensional analysis (see drag force on a sphere example) yields $C_D = f(\text{Re})$. If we ensure that the value of Re is the same from prototype to model, then we will have dynamic similarity, and therefore $C_D$ must have the same value from model to prototype.

### 6.7.1.1   Problem

Calculate the drag force exerted on a spherical particle, diameter 25 mm, as it moves through water at 2 m s$^{-1}$.

### 6.7.1.2   Solution

It is often convenient to tabulate the calculations, showing clearly the values of the variables for prototype and model in separate columns. The solution to this problem is given in Table 6.2.

**Table 6.2** Solution, drag force on a particle problem.

| Variable | Prototype | Model – graphical data |
|---|---|---|
| $U$ | $u_P = 2$ m s$^{-1}$ | |
| $D$ | $d_P = 0.025$ m | |
| $\rho$ | $\rho_P = 1000$ kg m$^{-3}$ | |
| $\mu$ | $\mu_P = 0.001$ Pa s | |
| $Re = \dfrac{\rho \bar{u} d}{\mu}$ | $Re_P = \dfrac{1000 \times 2 \times 0.025}{0.001} = 50\,000$ | $Re_M = 50\,000$ |

If we have dynamic similarity, *i.e.* $Re = $ constant, then $C_D$ must be constant between model (M) and prototype (P)

$$C_D = \frac{8F}{\pi \rho d^2 u^2}$$

Literature data:
from Figure 6.5,
at $Re_M = 50\,000$,
$C_{D_M} = 0.48$

$$C_{D_P} = \frac{8F_P}{\pi \rho_P d_P^2 u_P^2}$$

$$0.48 = \frac{8F_P}{\pi \times 1000 \times 0.025^2 \times 2^2}$$

| $F$ | Answer: $F_P = 0.47$ N | |

### 6.7.1.3 Achievements

Calculation of drag force on prototype based upon graphical experimental model data.

## 6.7.2 Pressure Drop in a Pipe

The frictional pressure drop per unit length of smooth pipe, $\Delta p_f / l$, is expected to be a function of the pipe diameter $d$, mean flow velocity $\bar{u}$ across the cross-section and fluid density $\rho$ and viscosity $\mu$ (neglects end effects, *i.e.* $l \gg d$). Thus,

$$\frac{\Delta p_f}{l} = f(d, \bar{u}, \rho, \mu) \tag{6.10}$$

These five variables (considering $\Delta pf/l$ as a single variable), $m = 5$, may be expressed in terms of $n = 3$ fundamental dimensions, *i.e.* mass, length and time. Using Buckingham's $\Pi$ theorem, these five variables may be rearranged into $m - n = 5 - 3 = 2$ new dimensionless variables:

$$c_f = \frac{\Delta p_f}{l} \frac{d}{2\rho u^2} = f\left(\frac{\rho \bar{u} d}{\mu}\right) = f(Re) \tag{6.11}$$

where $c_f$ is the "Fanning" friction factor and Re here is the pipe flow Reynolds number.

- Fanning friction factor:

$$c_f = \frac{\Delta p_f}{l} \frac{d}{2\rho u^2} \equiv \frac{\text{frictional forces}}{\text{inertial forces}}$$

- Pipe flow Reynolds number:

$$\text{Re} = \frac{\rho \bar{u} d}{\mu} \equiv \frac{\text{inertial forces}}{\text{viscous forces}}$$

Note for different flow situations, the Reynolds number may take different forms. See Chapter 3, *"Concepts of Fluid Flow"* for a discussion of the role of Reynolds number.

### 6.7.2.1  *Problem*

Methanol at 50 °C flows at a mean velocity of 2 m s$^{-1}$ through a smooth pipe of 25 mm internal diameter and 12 m length. Calculate the frictional pressure drop.

### 6.7.2.2  *Solution*

The solution to this problem is given in Table 6.3.

### 6.7.2.3  *Achievements*

Calculation of pipe frictional pressure drop based upon correlated experimental model data.

## 6.7.3  Modelling Flow Around a Body Immersed in a Fluid

### 6.7.3.1  *Problem*

In this example, we consider the flow around a body. Air, at atmospheric pressure, flows at 20 m s$^{-1}$ across a bank of heat exchanger tubes. A 1/10th-scale model is built. At what velocity must air flow over the model bank of tubes to achieve dynamic similarity?

### 6.7.3.2  *Solution*

From dynamic similarity,

$$\text{Re}_M = \text{Re}_P \quad \text{or} \quad \left(\frac{\rho u l}{\mu}\right)_M = \left(\frac{\rho u l}{\mu}\right)_P \tag{6.12}$$

**Table 6.3** Solution, pressure drop in a pipe problem.

| Variable | Prototype | Model-literature correlation |
|---|---|---|
| $\bar{u}$ | $\bar{u}_P = 2 \text{ m s}^{-1}$ | |
| $D$ | $d_P = 0.025 \text{ m}$ | |
| $\rho$ | $\rho_P = 790 \text{ kg m}^{-3}$ (from charts[1]) | |
| $\mu$ | $\mu_P = 0.00055 \text{ Pa s}$ (from charts[1]) | |
| $Re = \dfrac{\rho \bar{u} d}{\mu}$ | $Re_P = \dfrac{790 \times 2 \times 0.025}{0.00055} = 71\,800$ | $Re_M = 71\,800$ |

If we have dynamic similarity, *i.e.* $Re = $ constant, then $c_f$ must be constant between model and prototype

$$c_f = \frac{\Delta p_f d}{2 \rho u^2 l}$$

Literature data: for smooth pipes at moderate Reynolds numbers (see Chapter 3, "Concepts of Fluid Flow"): $c_f = 0.079$ $Re^{-0.25}$ giving $c_{f_M} = 4.8 \times 10^{-3}$

$$c_{f_P} = \frac{\Delta p_{f_P} d_P}{2 \rho_P u_P^2 l_P}$$

$$4.8 \times 10^{-3} = \frac{\Delta p_{f_P} \times 0.025}{2 \times 790 \times 2^2 \times 12}$$

$\Delta p_f$ — Answer: $\Delta p_{f_P} = 14\,600 \text{ Pa} = 14.6 \text{ kPa}$

---

where $l$ is the characteristic length scale of the body, giving

$$u_M = u_P \left(\frac{l_P}{l_M}\right) \left(\frac{\rho_P}{\rho_M}\right) \left(\frac{\mu_M}{\mu_P}\right) \tag{6.13}$$

Using the same fluid, air gives $u_M = u_P(l_P/l_M) = 20(1/0.1) = 200 \text{ m s}^{-1}$.

A $u_M$ value of 200 m s$^{-1}$ is too high. It is approximately 60% of the speed of sound and compressibility effects will be significant, and so we cannot use air.

Let us keep a value of 20 m s$^{-1}$ for $u_M$ and look for a different fluid with suitable properties. From eqn (6.12),

$$\left(\frac{\mu_M}{\rho_M}\right) = \left(\frac{u_M}{u_P}\right) \left(\frac{l_M}{l_P}\right) \left(\frac{\mu_P}{\rho_P}\right) = \left(\frac{20}{20}\right) \left(\frac{0.1}{1}\right) \left(\frac{\mu_P}{\rho_P}\right) = 0.1 \left(\frac{\mu_P}{\rho_P}\right)$$

Thus, we require a fluid with a kinematic viscosity ($\mu/\rho$) one tenth that of air at atmospheric pressure for our experiments. Water at moderate temperatures should be suitable. See Chapter 3, "*Concepts of Fluid Flow*" for information on kinematic viscosity.

### 6.7.3.3 Achievements

- Found that use of the same fluid for model and prototype was not feasible.
- Determined the fluid physical properties required for the model fluid to achieve dynamic similarity between prototype and model.

## 6.7.4 Heat Transfer

A novel shell-and-tube heat exchanger is being designed. It is decided to build a small-scale model to determine the tube-side process fluid heat transfer characteristics of this type of heat exchanger.

The rate of heat transfer per unit area of heat exchanger (heat flux), $q$, will be a function of the temperature driving force $\Delta T$, tube diameter $d$, the mean fluid flow velocity $\bar{u}$, fluid flow properties – density $\rho$ and viscosity $\mu$ – and fluid thermal properties – specific heat capacity $c_p$ and thermal conductivity $k$.

$$q = f(\Delta T, d, \bar{u}, \rho, \mu, c_p, k) \tag{6.14}$$

$m = 8$ and $n = 5(\mathbf{M}, \mathbf{L}, \mathbf{T}, \mathbf{H}, \mathbf{\Theta})$, therefore $m - n = 3$.

*Note* that temperature may be expressed in terms of kinetic energy of molecules. Heat, a form of energy, may be expressed in terms of $\mathbf{M}$, $\mathbf{L}$ and $\mathbf{T}$, $[\mathbf{M} \, \mathbf{L}^2 \, \mathbf{T}^{-2}]$. However, in most heat transfer problems, heat is conserved and is not transformed into other forms of energy. Here we consider heat and temperature as new fundamental dimensions, $\mathbf{H}$ and $\mathbf{\Theta}$. This has the advantage of increasing the number of fundamental dimensions, thus reducing the number of dimensionless groups required to describe the problem.

Dimensional analysis yields

$$\text{Nu} = f(\text{Re}, \text{Pr}) \tag{6.15}$$

- Nusselt number:

$$\text{Nu} = \frac{q}{\Delta T} \frac{d}{k} = \frac{hd}{k} \equiv \frac{\text{convective heat transfer rate}}{\text{conductive heat transfer rate}}$$

- Tube flow Reynolds number:

$$\text{Re} = \frac{\rho \bar{u} d}{\mu} \equiv \frac{\text{inertial forces}}{\text{viscous forces}}$$

- Prandtl number

$$\text{Pr} = \frac{c_p \mu}{k} \equiv \frac{\text{momentum diffusivity}}{\text{thermal diffusivity}}$$

In the Nusselt number, the term $(q/\Delta T)$, the rate of heat transfer per unit area of heat exchanger per unit temperature driving force, is known as the heat transfer coefficient and is given the symbol $h$. The heat transfer coefficient is used to characterise heat transfer rates. Heat transfer processes are described in more detail in Chapter 4, "*An Introduction to Heat Transfer*".

### 6.7.4.1 Problem

(a) A geometrically similar 1/10th-scale model of a prototype heat exchanger has been built and tests are to be performed upon it using the same process fluid, operating at similar temperatures as in the prototype. The prototype is designed to operate at a process fluid flow rate of 300 m³ h⁻¹. What flow rate of fluid through the model will give similar flow conditions to those in the prototype?

### 6.7.4.2 Solution

Table 6.4 tabulates the solution to part (a).

### 6.7.4.3 Alternative Representation

Dynamic similarity, $Re_M = Re_P$, or $(\rho \bar{u} d/\mu)_M = (\rho \bar{u} d/\mu)_M$. Since the physical properties are the same in model and prototype, the above

**Table 6.4** Solution to heat transfer problem, part (a).

| Variable | Prototype | Model |
|---|---|---|
| $Q$ | $Q_P = 300$ m³ h⁻¹ | $Q_M =$ unknown |
| $d$ | $d_P$ | $d_M = 0.1 d_P$ 1/10th scale |
| $\bar{u} = \dfrac{Q}{csa} \propto \dfrac{Q}{d^2}$ | $\bar{u}_P \propto \dfrac{Q_P}{d_P^2}$ | $\bar{u}_M \propto \dfrac{Q_M}{d_M^2} = \dfrac{Q_M}{(0.1 d_P)^2}$ |
| $\rho$ | $\rho_P$ | $\rho_P$ (same fluid) |
| $\mu$ | $\mu_P$ | $\mu_P$ (same fluid) |
| $c_p$ | $c_{pP}$ | $c_{pP}$ (same fluid) |
| $k$ | $k_P$ | $k_P$ (same fluid) |
| $Re = \dfrac{\rho \bar{u} d}{\mu} \propto \dfrac{\rho(Q/d^2)d}{\mu}$ | $Re_P \propto \dfrac{\rho_P(Q_P/d_P^2)d_P}{\mu_P}$ | $Re_M \propto \dfrac{\rho_P(Q_M/0.1 d_P)^2 0.1 d_P}{\mu_P}$ |
| | | Dynamic similarity, $Re_P = Re_M$, so |
| | | $\dfrac{\rho_P(Q_P/d_P^2)d_P}{\mu_P} = \dfrac{\rho_P[Q_M/(0.1 d_P)^2 0.1 d_P]}{\mu_P}$ |
| $Q$ | | $\dfrac{Q_M}{0.1} = Q_P = 300$ |
| | | Answer: $Q_M = 30$ m³ h⁻¹ |

equation becomes $\bar{u}_M d_M = \bar{u}_P d_P$, and as $d_M = 0.1 d_P$, so $\bar{u}_M = (\bar{u}_P d_P / d_M) = (\bar{u}_P d_P / 0.1 d_P) = 10 \bar{u}_P$:

$$Q_M = \bar{u}_M \frac{\pi d_M^2}{4} = 10 \bar{u}_P \frac{\pi (0.1 d_P)^2}{4} = \frac{1}{10} \bar{u}_P \frac{\pi d_P^2}{4}$$

$$= \frac{1}{10} Q_P = \frac{300}{10} = 30 \, \text{m}^3 \, \text{h}^{-1}$$

### 6.7.4.4   Problem Continued

(b) At the model flow rate obtained above, the tube-side heat transfer coefficient for the model was found to be 250 W m$^{-2}$ K$^{-1}$. What will be the heat transfer coefficient for the prototype?

### 6.7.4.5   Solution

Table 6.5 tabulates the solution to part (b).

### 6.7.4.6   Alternative Representation

As $\text{Re}_M = \text{Re}_P$ (dynamic similarity achieved with $Q_M = 30 \, \text{m}^3 \, \text{h}^{-1}$) and as $\text{Pr}_M = \text{Pr}_P$ (same fluid), then $\text{Nu}_M = \text{Nu}_P$, or $(hd/k)_M = (hd/k)_P$. Thus, as we are using the same fluid in model and prototype, $h_M d_M = h_P d_P$:

$$h_P = \frac{h_M d_M}{d_P} = \frac{h_M d_M}{10 d_M} = \frac{h_M}{10} = \frac{250}{10} = 25 \, \text{W m}^2 \, \text{K}^{-1}$$

### 6.7.4.7   Achievements

- Determined the conditions under which to operate a geo-metrically similar model so that physical similarity is obtained with the prototype.
- Determined the heat transfer coefficient for the prototype based upon the measured heat transfer coefficient of the model.

## 6.7.5   Mass Transfer

Analysis of mass transfer systems is complicated by the range of units in which gas and liquid concentrations may be defined, *e.g.* mass concentration, molar concentration, mass fractions, mole fractions, partial pressures, *etc.* This leads to a wide range of units for the mass transfer coefficient.

**Table 6.5**  Solution to heat transfer problem, part (b).

| Variable | Prototype | Model |
|---|---|---|
| $h$ | $h_P$ = unknown | $h_M$ = 250 W m$^{-2}$ K$^{-1}$ |
| $\mathrm{Pr} = \dfrac{c_p \mu}{k}$ | $\mathrm{Pr}_P$ | $\mathrm{Pr}_M = \mathrm{Pr}_P$ (same fluid) |
| $\mathrm{Nu} = \dfrac{hd}{k}$ | $\mathrm{Nu}_P = \dfrac{h_P d_P}{k_P}$ | $\mathrm{Nu}_M = \dfrac{250 \times 0.1 d_P}{k_P}$ |

Nu = $f$(Re, Pr), as Re$_P$ = Re$_M$ and Pr$_P$ = Pr$_M$, then Nu$_P$ must equal Nu$_M$

By Nusselt number similarity, $\dfrac{h_P d_P}{k_P} = \dfrac{250 \times 0.1 d_P}{k_P}$

Answer: $h_p$ = 25 W m$^{-2}$ K$^{-1}$

Let us consider, as an example, the mass transfer of component "A" from the inside surface of a pipe to the bulk of a liquid flowing through the pipe. The liquid-phase mass transfer coefficient for this case, $k_L$, is defined by

$$N_A = k_L (c_{A_S} - c_{A_L})$$

where $N_A$ is the molar flux of A [kmol A m$^{-2}$ s$^{-1}$] and $c_A$ is the molar concentration of A [kmol A m$^{-3}$] at the liquid surface (S) and in the bulk of the liquid (L). In this case, $k_L$ has the units of m s$^{-1}$.

The mass transfer coefficient is expected to be a function of

- fluid flow properties, density, $\rho$, and viscosity, $\mu$
- a characteristic length, the pipe diameter, $d$
- a characteristic velocity, mean velocity $\bar{u}$
- diffusivity of A in B, $D_{AB}$

giving $k_L = f(\rho, \mu, d, \bar{u}, D_{AB})$.

Dimensional analysis gives the following dimensionless groups:

$$\frac{k_L d}{D_{AB}} = f\left( \frac{\rho \bar{u} d}{\mu}, \frac{\mu}{\rho D_{AB}} \right)$$

or

$$\mathrm{Sh} = f(\mathrm{Re}, \mathrm{Sc})$$

- Schmidt number:

$$\mathrm{Sc} = \frac{\mu/\rho}{D_{AB}} \equiv \frac{\text{momentum diffusivity}}{\text{mass diffusivity}}$$

- Sherwood number:

$$\text{Sh} = \frac{d/D_{\text{AB}}}{1/k_{\text{L}}} \equiv \frac{\text{diffusive resistance to mass transfer}}{\text{convective resistance to mass transfer}}$$

### 6.7.5.1  Example Correlation

Pipe flow, liquids:[7]

$$\text{Sh} = 0.023 \ \text{Re}^{0.83} \ \text{Sc}^{0.33}$$

$$2100 < \text{Re} < 35\,000; \quad 0.6 < \text{Sc} < 3000$$

(6.16)

### 6.7.5.2  Problem

(a) A long pipe, of 25 mm internal diameter, has a 25 mm long section of its wall made of solid naphthalene. Estimate the rate of mass transfer from the naphthalene if liquid benzene at 7 °C flows through the pipe at a mean velocity of 1 m s$^{-1}$.

Benzene properties:[8] density, $\rho = 890 \ \text{kg m}^{-3}$; viscosity, $\mu = 0.00076 \ \text{Pa s}$. Naphthalene solubility in benzene:[9] $c_{\text{C}_{10}\text{H}_{8\text{S}}} = 1.28 \ \text{kmol m}^{-3}$. Naphthalene diffusivity in benzene:[9] $D_{\text{C}_{10}\text{H}_8} - \text{C}_6\text{H}_6 = 1.19 \times 10^{-9} \ \text{m}^2 \ \text{s}^{-1}$.

#### 6.7.5.2.1  Dimensionless Groups

Reynolds number:

$$\text{Re} = \frac{\rho \bar{u} d}{\mu} = \frac{890 \times 1 \times 0.025}{0.00076} = 29\,276$$

Schmidt number:

$$\text{Sc} = \frac{\mu}{\rho D_{\text{C}_{10}\text{H}_8 - \text{C}_6\text{H}_6}} = \frac{0.00076}{890 \times 1.19 \times 10^{-9}} = 717.6$$

Sherwood number, from eqn (6.16) noting that both Re and Sc are within the specified ranges:

$$\text{Sh} = 0.023 \ \text{Re}^{0.83} \ \text{Sc}^{0.33} = 0.023 \times 29\,276^{0.83} \times 717.6^{0.33} = 1026.5$$

#### 6.7.5.2.2  Mass Transfer Coefficient

$$k_{\text{L}} = \text{Sh} \times \frac{D_{\text{C}_{10}\text{H}_8 - \text{C}_6\text{H}_6}}{d} = 1026.5 \times \frac{1.19 \times 10^{-9}}{0.025}$$

$$= 4.89 \times 10^{-5} \ \text{m s}^{-1}$$

### 6.7.5.2.3 Mass Transfer Rate

The molar flux, $N_A = k_L \left( c_{C_{10}H_{8_S}} - c_{C_{1C}H_{8_L}} \right)$, assuming the concentration of naphthalene in the bulk of the liquid phase is negligible, *i.e.* $c_{C_{10}H_{8_L}} = 0$, becomes

$$N_A = k_L \times c_{C_{10}H_{8_S}} = 4.89 \times 10^{-5} \, \text{m s}^{-1} \times 1.28 \, \text{kmol m}^{-3}$$

$$= 6.26 \times 10^{-5} \, \text{kmol m}^{-2} \, \text{s}^{-1}.$$

$$\text{Mass flux} = 6.26 \times 10^{-5} \, \text{kmol m}^{-2} \, \text{s}^{-1} \times 128.174 \, \text{kg kmol}^{-1}$$

$$= 0.0080 \, \text{kg m}^{-2} \, \text{s}^{-1}.$$

$$\text{Mass transfer rate} = \text{flux} \times \text{area} = 0.0080 \, \text{kg m}^{-2} \, \text{s}^{-1} \times \pi 0.025^2 \, \text{m}^2$$

$$= 1.58 \times 10^{-5} \, \text{kg s}^{-1}.$$

### 6.7.5.2.4 Problems with Different Units for *k*

If we are dealing with gas-phase mass transfer, $N_A = k_G(p_{A_i} - p_{A_G})$, where $N_A$ is again the molar flux of A [kmol A m$^{-2}$ s$^{-1}$] and $p_A$ is the partial pressure of A in the gas phase (*e.g.* Pa) at the gas–solid interface (i) and in the bulk of the gas (G). In this case, the gas-phase mass transfer coefficient $k_G$, has the units kmol m$^{-2}$ s$^{-1}$ Pa$^{-1}$, and the Sherwood number becomes Sh $= (k_G R T d / D_{AB})$, where R has the units J kmol$^{-1}$ K$^{-1}$.

### 6.7.5.3 *Example Correlation*

Pipe flow, gases:[7]

$$\text{Sh} = 0.023 \, \text{Re}^{0.83} \text{Sc}^{0.44}$$

$$2000 < \text{Re} < 35\,000; \quad 0.6 < \text{Sc} < 2.5. \tag{6.17}$$

### 6.7.5.4 *Problem Continued*

(b) Estimate the rate of mass transfer from the naphthalene if, instead of the liquid benzene in (a) above, air at 40 °C and 1 bar flows through the pipe with a mean velocity of 10 m s$^{-1}$.

Air properties:[8] density, $\rho = 1.11$ kg m$^{-3}$; viscosity, $\mu = 0.000019$ Pa s
Naphthalene vapour pressure:[9] $p_{C_{10}H_{8_i}} = 44.72$ Pa
Naphthalene diffusivity in air:[9] $D_{C_{10}H_8-\text{Air}} = 9.02 \times 10^{-6}$ m$^2$ s$^{-1}$

#### 6.7.5.4.1   Dimensionless Groups

Reynolds number:

$$\text{Re} = \frac{\rho \bar{u} d}{\mu} = \frac{1.11 \times 10 \times 0.025}{0.000019} = 14\,656$$

Schmidt number:

$$\text{Sc} = \frac{\mu}{\rho D_{C_{10}H_8-\text{Air}}} = \frac{0.000019}{1.11 \times 9.02 \times 10^{-6}} = 1.89$$

Sherwood No. from eqn (6.17), noting that both Re and Sc are within the specified ranges:

$$\text{Sh} = 0.023\,\text{Re}^{0.83}\,\text{Sc}^{0.44} = 0.023 \times 14\,656^{0.83} \times 1.89^{0.44} = 87.4$$

#### 6.7.5.4.2   Mass Transfer Coefficient

$$k_L = \text{Sh} \times \frac{D_{C_{10}H_8-\text{Air}}}{RTd} = 87.4 \times \frac{9.02 \times 10^{-6}}{8314 \times 313 \times 0.025}$$

$$= 1.21 \times 10^{-8}\,\text{kmol}\,\text{m}^{-2}\,\text{s}^{-1}\,\text{Pa}^{-1}$$

#### 6.7.5.4.3   Mass Transfer Rate

The molar flux, $N_A = k_G(p_{C_{10}H_{8i}} - p_{C_{10}H_{8G}})$, assuming the concentration of naphthalene in the bulk of the gas phase is negligible, *i.e.* $p_{C_{10}H_{8G}} = 0$, becomes

$$N_A = k_G \times p_{C_{10}H_{8i}}$$

$$= 1.21 \times 10^{-8}\,\text{kmol}\,\text{m}^{-2}\,\text{s}^{-1}\,\text{Pa}^{-1} \times 44.72\,\text{Pa}$$

$$= 5.41 \times 10^{-7}\,\text{kmol}\,\text{m}^{-2}\,\text{s}^{-1}.$$

$$\text{Mass flux} = 5.41 \times 10^{-7}\,\text{kmol}\,\text{m}^{-2}\,\text{s}^{-1} \times 128.174\,\text{kg}\,\text{kmol}^{-1}$$

$$= 6.9 \times 10^{-5}\,\text{kg}\,\text{m}^{-2}\,\text{s}^{-1}.$$

$$\text{Mass transfer rate} = \text{flux} \times \text{area} = 6.9 \times 10^{-5}\,\text{kg}\,\text{m}^{-2}\,\text{s}^{-1} \times \pi 0.025^2\,\text{m}^2$$

$$= 1.36 \times 10^{-7}\,\text{kg}\,\text{s}^{-1}.$$

### 6.7.5.5   Achievements

Estimated the rate of mass transfer from a pipe wall, in (a) liquid and (b) gas flows.

These two cases illustrate the use of dimensionless correlations for mass transfer and some of the problems with units in mass transfer.

Correlations for mass transfer coefficients for a wide range of geometries and flow conditions are available in the literature.[7]

## 6.7.6 Correlation of Experimental Data – Formation of Gas Bubbles at an Orifice

### 6.7.6.1 Problem

Benzing[10] reports measurements of the diameter of bubbles $D$, formed in a liquid at an orifice and how they varied with the orifice diameter $d$, and the physical properties of the liquid. An extract from the data presented is summarised in Table 6.6.

Carry out a dimensional analysis and correlate this data in a form that is convenient to use if it is required to predict the bubble diameter for a given system. Represent the effects of viscosity and surface tension so that they each appear in only one group.

### 6.7.6.2 Solution

Now, $D=f(d, \mu, \rho, \sigma, g)$; $m=6, n=3$ (**M, L, T**); therefore, $m-n=3$. The term $g$ is included as gravity (buoyancy) affects the detachment of the bubbles at the orifice. Liquid viscosity, $\mu$, and surface tension, $\sigma$, are thought to be significant in the formation of bubbles at orifices, so the recurring set is chosen as $d$, $\rho$ and $g$; leaving $D$ – the independent variable – $\mu$ and $\sigma$ are to appear in separate groups. Dimensional analysis gives the following:

$$\frac{D}{d} = f\left(\frac{\mu^2}{d^3 \rho^2 g}, \frac{\sigma}{d^2 \rho g}\right)$$

**Table 6.6** Experimental data of bubble formation at an orifice.[10]

| System | Liquid viscosity $\mu$ (P) | Liquid density $\rho$ (g cm$^{-1}$) | Surface tension $\sigma$ (dynes cm$^{-1}$) | Bubble diameter, $D$ (cm) | |
| | | | | Orifce diameter, $d=0.293$ cm | Orifice diameter, $d=0.438$ cm |
| --- | --- | --- | --- | --- | --- |
| Water–air | 0.0081 | 0.996 | 70.2 | 0.512 | 0.612 |
| Water–H$_2$ | 0.0081 | 0.996 | 70.2 | 0.516 | 0.630 |
| 7.5% Ethanol solution–air | 0.0107 | 0.982 | 51.4 | 0.480 | 0.583 |
| 45% Sugar solution–air | 0.52 | 1.197 | 58.1 | 0.486 | 0.585 |
| Oil–air | 0.57 | 0.920 | 36.5 | 0.411 | 0.491 |

We have now reduced the number of variables from 6 to 3.

Trying a correlation of the form

$$\frac{D}{d} = a\left(\frac{\mu^2}{d^3 \rho^2 g}\right)^b \left(\frac{\sigma}{d^2 \rho g}\right)^c$$

least-squares analysis of the data in Table 6.6 yields $a = 1.83$, $b = -0.002$ and $c = 0.276$. The small value of $b$ indicates that the effect of liquid viscosity on bubble formation is small or negligible. Removing the $(\mu^2/d^3 \rho^2 g)$ group from the analysis yields

$$\frac{D}{d} = 1.86\left(\frac{\sigma}{d^2 \rho g}\right)^{0.281}$$

with negligible change in the variance. Figure 6.6 compares the experimental values of D/d (circles) with this final correlation (solid line).

### 6.7.6.3  Achievements

- Reduced the number of variables for the analysis of some experimental data.
- Produced a correlation of the experimental data in dimensionless form.

*Note* that Rowe[11] presents a cautionary tale of the dangers of blind usage of the results of least squares correlation of engineering data. He simulated experimental data by selecting values from a table of random numbers and obtained some seemingly reasonable

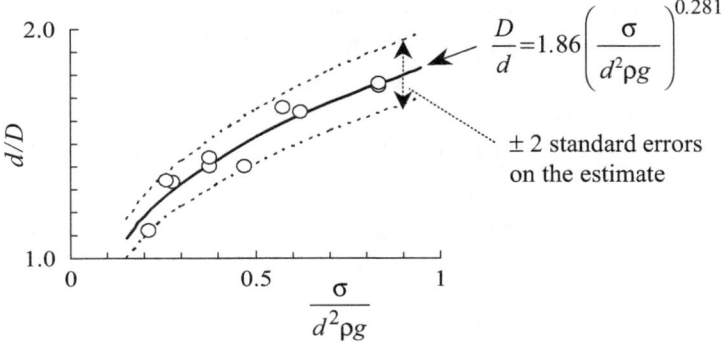

**Figure 6.6**  Correlation of experimental data.

correlations between them. The principal reasons for the apparent reasonableness were two-fold:

- the use of log–log plots (often used to present dimensionless data), which tend to concentrate data and make the correlation look better; and
- the inclusion of the same primary variable (in separate dimension-less groups) on each axis resulting in a plot of that variable against itself.

## 6.7.7 Application of Scale-up in Stirred Vessels

During the mixing of a single-phase liquid in a stirred tank, the important parameters are the rate of power input, $P$, the impeller stirring speed, $N$, gravitational acceleration, $g$, the fluid density, $\rho$ and viscosity, $\mu$, the impeller diameter, $D$ and the other geometric characteristic lengths of the vessel (*i.e.* baffle width and depth, impeller shape, *etc.*).[5,12]

$$P = f(D, N, \rho, \mu, g, \text{geometry}) \tag{6.18}$$

Assuming geometric similarity, dimensional analysis yields the following dimensionless groups:

$$\text{Po} = f(\text{Re}_\text{I}, \text{Fr}) \tag{6.19}$$

- Power number:

$$\text{Po} = \frac{P}{\rho N^3 D^5} \equiv \frac{\text{drag forces on impeller}}{\text{inertial forces}}$$

- Impeller Reynolds number:

$$\text{Re}_\text{I} = \frac{\rho N D^2}{\mu} \equiv \frac{\text{inertial forces}}{\text{viscous forces}}$$

- Froude number:

$$\text{Fr} = \frac{N^2 D}{g} \equiv \frac{\text{inertial forces}}{\text{gravitational forces}}$$

The Froude number is associated with the formation of a vortex on the liquid free surface around the impeller shaft. At low impeller speeds ($\text{Re}_\text{I} < 300$) or when baffles are present in the vessel, a vortex is not formed and the Fr can be omitted from the above relation.

The formation of a vortex is best avoided, since it limits power input and can result in poor mixing.

$$\text{Po} = f(\text{Re}_I \text{ only}) \tag{6.20}$$

See Chapter 3, "Concepts of Fluid Flow" for experimental details of Po *versus* $\text{Re}_I$ relationships.

### 6.7.7.1  Scale-up at Constant Re_I

For a given fluid, $\text{Re}_I \propto ND^2 = \text{constant}$, *i.e.* $N \propto D^{-2}$. Scale-up at constant $\text{Re}_I$ implies constant Po, yielding, for a given fluid, $P \propto N^3 D^5$. Therefore, scale-up at constant $\text{Re}_I$ gives $P \propto N^3 D^5 \propto (D^{-2})^3 \times D^5 \propto D^{-1}$. Power requirement decreases as scale increases! This is not a suitable scale-up criterion.

### 6.7.7.2  Scale-up Based on Constant Power Per Unit Volume

In this approach, the power curve is used for scale-up, similar to single-phase systems. If we assume that mixing takes place in the turbulent region, $\text{Po} = (P/\rho N^3 D^5) = \text{constant}$. For a given fluid, with Po constant, $P \propto N^3 D^5$ and $(P/V) \propto (N^3 D^5/D^3) \propto N^3 D^2$. According to this criterion, constant power per unit volume implies that $N^3 D^2$ should be constant between model and prototype.

   Table 6.7 lists these and a number of other scale-up criteria for stirred tanks. Here it is assumed that the fluid is the same on scale up, *i.e.* density and viscosity are both constant, and that the vessel is operating in the fully turbulent, constant power number regime.

## 6.7.8  Incompatibility of Some Equations

It is decided to model a full-scale prototype, unbaffled, stirred vessel with a one-tenth scale model. The liquid in the prototype has a kinematic viscosity, $v$, of $10^{-7}$ m$^2$ s$^{-1}$. As we have seen above, power number is a function of both Reynolds number and Froude number for unbaffled vessels. To ensure power number similarity, we need to ensure both Reynolds number and Froude number are similar from prototype to model.

   Froude number similarity, as $g$ is constant, $N_M^2 D_M = N_M^2 D_P$, Thus,

$$\frac{N_M}{N_P} = \sqrt{\frac{D_P}{D_M}} = \sqrt{\frac{10}{1}} = \sqrt{10}$$

**Table 6.7** Some scale-up criteria for geometrically similar stirred vessels in the fully turbulent regime.[12]

| Scale-up criteria | Keep constant | $P \propto$ | Comment |
|---|---|---|---|
| Constant Reynolds number | $ND^2$ | $D^{-1}$ | Unrealistic scale-up criterion, larger scales require less power! |
| Constant power per unit volume: implies similar turbulence micro-scales | $N^3D^2$ | $N^3D^5$ | May be difficult to achieve due to large power requirements |
| Constant impeller tip speed: (often used to limit shear damage to sensitive fluids) | $ND$ | $D^2$ | Industrial rule-of-thumb: keep tip speed below 10 m s$^{-1}$ for shear sensitive products |
| Constant average rate of shear | $N$ | $D^5$ | May be impossible to achieve due to large power requirements |
| Constant mixing time Chemical reaction: (constant residence time distribution and constant mean residence time)[13] | | | |
| Constant heat transfer coefficient $h$: Nu $\propto$ Re$^{2/3}$ | $N^{2/3}D^{1/3}$ | $D^{3.5}$ | May be difficult to achieve due to large power requirements |
| Solids suspension: speed to "just suspend" solids[14] | $ND^{0.85}$ | $D^{2.45}$ | One of many correlations for particle suspension |

Reynolds number similarity:

$$(ND^2/v)_M = (ND^2/v)_P$$

rearranging

$$v_M = v_P \frac{N_M}{N_P} \left(\frac{D_M}{D_P}\right)^2 = 1 \times 10^{-7} \times \sqrt{10} \left(\frac{1}{10}\right)^2$$

$$= 3 \times 10^{-9} \text{ m}^2 \text{ s}^{-1}$$

Thus we require a model liquid with a kinematic viscosity of $3 \times 10^{-9}$ m$^2$ s$^{-1}$. There is no such liquid.[9]

### 6.7.8.1 Notes

In many situations, it may be impossible to maintain similarity between all dimensionless groups. Therefore, one should try to maintain similarity with the most important dimensionless group and take into account the effect(s) of the other group(s) in some other way.

## 6.8 Conclusions

- Dimensional analysis is a powerful tool.
- Dimensional analysis reduces the number of variables to be analysed in a problem. $Q_1$, $Q_2$, $Q_3$, $\ldots$, $Q_m$ is reduced to $\Pi_1$, $\Pi_2$, $\Pi_3$, $\ldots$, $\Pi_{m-n}$
- Dimensional analysis tells us that the new, dimensionless variables are related in some way, $\Pi_1 = f(\Pi_2, \Pi_3, \ldots, \Pi_{m-n})$.
- Dimensional analysis does not tell us the form of the function $f$.
- Selection of variables $Q_1$, $Q_2$, $Q_3$, $\ldots$, $Q_m$?
  - experience
  - too few, data will not correlate
  - too many, redundant dimensionless groups will appear
- Multiple forms of $\Pi$s. Selection of different recurring sets within the rules will give different dimensionless groups, although these new groups will be products of the powers of the original groups. Any number of dimensionless groups can be formed in this way.
- Similarity of dimensionless groups may be used for scale-up of physically similar systems.
- It may not be feasible to maintain similarity between all dimensionless groups on scale-up.

## Nomenclature

| Symbol | Description | Dimensions | SI unit |
|---|---|---|---|
| ***Variables*** | | | |
| $a$ | Acceleration | $L\,T^{-2}$ | $m\,s^{-2}$ |
| $a, b, c$ | Constants | — | — |
| $A$ | Area, cross-sectional area | $L^2$ | $m^2$ |

| Symbol | Description | Dimensions | SI unit |
|--------|-------------|------------|---------|
| A, B | Component A, B | — | — |
| $c$ | Molar concentration | $\mathbf{M\,L^{-3}}$ | kmol m$^{-3}$ |
| $C_p$ | Specific heat capacity | $\mathbf{H\,M^{-1}\,\Theta^{-1}}$ | J kg$^{-1}$ K$^{-1}$ |
| $d$ | Diameter, pipe diameter | $\mathbf{L}$ | m |
| $D$ | Impeller diameter | $\mathbf{L}$ | m |
| $D_{AB}$ | Diffusivity of component A in component B | $\mathbf{L^2\,T^{-1}}$ | m$^2$ s$^{-1}$ |
| $f$ | Function | n/a | n/a |
| $F$ | Force | $\mathbf{M\,L\,T^{-2}}$ | N |
| $g$ | Gravitational acceleration | $\mathbf{L\,T^{-2}}$ | m s$^{-2}$ |
| $h$ | Heat transfer coefficient | $\mathbf{H\,L^{-2}\,T^{-1}\,\Theta^{-1}}$ | Wm$^{-2}$ K$^{-1}$ |
| $k$ | Thermal conductivity | $\mathbf{H\,L^{-1}\,T^{-1}\,\Theta^{-1}}$ | Wm$^{-1}$ K$^{-1}$ |
| $k_G$ | Gas-phase mass transfer coefficient | $\mathbf{L^2\,T^{-1}}$ | m$^2$ s$^{-1}$ |
| $k_L$ | Liquid-phase mass transfer coefficient | Various | Various |
| $l$ | Characteristic length, pipe length | $\mathbf{L}$ | m |
| $m$ | Number of variables | — | — |
| $n$ | Number of fundamental dimensions | — | — |
| $N$ | Rate of rotation | $\mathbf{T^{-1}}$ | Hz |
|  | Molar flux | $\mathbf{M\,L^{-2}\,T^{-1}}$ | kmol m$^{-2}$ s$^{-1}$ |
| $p$ | Pressure, partial pressure | $\mathbf{M\,L^{-1}\,T^{-2}}$ | Pa |
| $P$ | Power | $\mathbf{M\,L^{-2}\,T^{-3}}$ | W |
| $q$ | Heat flux | $\mathbf{H\,L^{-2}\,T^{-1}}$ | W m$^{-2}$ |
| $Q$ | Quantity | n/a | n/a |
| $Q_v$ | Volumetric flow rate | $\mathbf{L^3\,T^{-1}}$ | m$^3$ s$^{-1}$ |
| $R$ | Universal gas constant | $\mathbf{L^2\,T^{-2}\,\Theta^{-1}}$ | J kmol$^{-1}$ K$^{-1}$ |
| $s$ | Distance moved | $\mathbf{L}$ | m |
| $t$ | Time | $\mathbf{T}$ | s |
| $T$ | Absolute temperature | $\mathbf{\Theta}$ | K |
| $u$ | Initial velocity | $\mathbf{L\,T^{-1}}$ | m s$^{-1}$ |
| $u_\infty$ | Velocity relative to infinity or undisturbed flow | $\mathbf{L\,T^{-1}}$ | m s$^{-1}$ |

| Symbol | Description | Dimensions | SI unit |
|--------|-------------|------------|---------|
| $\bar{u}$ | Average velocity across a cross-section | $\mathbf{L\ T^{-1}}$ | m s$^{-1}$ |
| $V$ | Volume | $\mathbf{L^3}$ | m$^3$ |
| $X$ | Length | $\mathbf{L}$ | m |
| $Y$ | Length | $\mathbf{L}$ | m |
| $\beta$ | Coefficient of thermal expansion | $\mathbf{\Theta^{-1}}$ | K$^{-1}$ |
| $\mu$ | Dynamic viscosity | $\mathbf{M\ L^{-1}\ T^{-1}}$ | Pa s, N s m$^{-2}$ |
| $\Pi$ | Dimensionless group | — | — |
| $\rho$ | Fluid density | $\mathbf{M\ L^{-3}}$ | kg m$^{-3}$ |
| $\sigma$ | Surface tension | $\mathbf{M\ T^{-2}}$ | N m$^{-1}$ |

**Subscripts**

| | |
|---|---|
| f | Frictional |
| G | Gas, gas-phase |
| I | Impeller |
| i | Interface |
| L | Liquid, liquid-phase |
| M | Model |
| P | Prototype |
| s | Sphere |
| S | Surface |

**Symbols**

| | |
|---|---|
| $\Delta$ | "The change in ..." |
| [=] | "Has the dimensions of ..." |

**Fundamental Dimensions**

| | |
|---|---|
| M | Mass |
| L | Length |
| T | Time |
| $\Theta$ | Temperature |
| H | Heat |

# Dimensionless Groups

Table 6.8 lists some common dimensionless groups used in chemical engineering. More complete lists of dimensionless groups are available in the literature.[15–17]

**Table 6.8** Some common dimensionless groups used in chemical engineering.

| Symbol | Name | Formula | Ratio | Usage |
|---|---|---|---|---|
| $C_D$ | Drag coefficient | $\dfrac{F}{A\frac{1}{2}\rho u^2}$ | $\dfrac{\text{drag force on body}}{\text{inertial forces}}$ | Drag forces on bodies |
| $c_f$ | Friction factor | $\dfrac{\Delta p_f d}{2l\rho u^2}$ | $\dfrac{\text{frictional forces}}{\text{inertial forces}}$ | Fluid friction in pipes, etc. |
| Fl | Flow | $\dfrac{Qv}{ND^3}$ | $\dfrac{\text{volumetric flow rate}}{\text{rate of swept volume}}$ | Flow created by impellers, fans and pumps |
| Fr | Froude | $\dfrac{u^2}{lg}$ | $\dfrac{\text{inertial forces}}{\text{gravitational forces}}$ | Free surfaces |
| Gr | Grashof | $\dfrac{l^3\rho^2\beta g\Delta T}{\mu^2}$ | $\dfrac{\text{buoyancy} \times \text{inertial forces}}{\text{viscous forces}^2}$ | Natural convection heat transfer |
| Le | Lewis | $\dfrac{Sc}{Pr}$ | $\dfrac{\text{thermal diffusivity}}{\text{mass diffusivity}}$ | Simultaneous heat and mass transfer |
| Nu | Nusselt | $\dfrac{hl}{k}$ | $\dfrac{\text{convective heat transfer}}{\text{conductive heat transfer}}$ | Heat transfer and flow |
| Pe | Peclet (heat) <br> Peclet (mass) | Re Pr <br> Re Sc | | Heat transfer and flow <br> Mass transfer and flow |
| Po | Power | $\dfrac{Po}{\rho N^3 D^5}$ | $\dfrac{\text{drag force on impeller}}{\text{inertial forces}}$ | Power consumed by impellers, fans and pumps |
| Pr | Prandtl | $\dfrac{c_p\mu}{k}$ | $\dfrac{\text{momentum diffusivity}}{\text{thermal diffusivity}}$ | Heat transfer and flow |
| Re | Reynolds | $\dfrac{\rho ul}{\mu}$ | $\dfrac{\text{inertial forces}}{\text{viscous forces}}$ | Fluid flow |
| Sc | Schmidt | $\dfrac{\mu}{\rho D_{AB}}$ | $\dfrac{\text{momentum diffusivity}}{\text{mass diffusivity}}$ | Mass transfer and flow |
| Sh | Sherwood | $\dfrac{k_l l}{D_{AB}}$ | $\dfrac{\text{diffusive resistance to mass transfer}}{\text{convective resistance to mass transfer}}$ | Mass transfer and flow |
| St | Stanton | $\dfrac{Nu}{RePr}$ | | Heat transfer and flow |
| We | Weber | $\dfrac{\rho u^2 l}{\sigma}$ | $\dfrac{\text{inertial forces}}{\text{surface tension forces}}$ | Flow with interfaces |

# References

1. J. M. Coulson and J. F. Richardson, in *Chemical Engineering*, ed. J. R. Backhurst and J. H. Harker, Elsevier Butterworth-Heinemann, Oxford, 5th edn, 1999, ch. 1, vol 1.
2. B. S. Massey, in *Mechanics of Fluids*, ed. J. Ward-Smith, Taylor & Francis, Abingdon, Oxford, 8th edn, 2006, ch. 5.
3. J. R. Welty, C. E. Wicks, R. E. Wilson and G. Rorrer, *Fundamentals of Momentum, Heat and Mass Transfer*, Wiley, New York, 4th edn, 2001, ch. 11.
4. J. W. Mullin, *AIChE J.*, 1972, **18**(1), 222.
5. R. H. Perry and D. Green, *Perry's Chemical Engineers' Handbook*, McGraw Hill, New York, 7th edn, 1997, Section 1.
6. G. Astarita, *Chem. Eng. Sci.*, 1997, **54**(24), 4681.
7. R. H. Perry and D. Green, *Perry's Chemical Engineers' Handbook*, McGraw Hill, New York, 7th edn, 1997, Section 5.
8. R. H. Perry and D. Green, *Perry's Chemical Engineers' Handbook*, McGraw Hill, New York, 7th edn, 1997, Section 2.
9. DETHERM, *Thermophysical Properties Database*, Vn. 2.1.0.0, DECHEMA e.V., 2006.
10. R. J. Benzing and J. E. Myers, *Ind. Eng. Chem.*, 1955, **47**, 2087.
11. P. N. Rowe, *Trans. Inst. Chem. Eng.*, 1963, **41**, CE 70.
12. N. Harnby, M. F. Edwards and A. W. Nienow, *Mixing in the Process Industries*, Butterworth-Heinemann, Oxford, 2nd edn, 1997.
13. A. W. Nienow, *Chem. Eng. Sci.*, 1974, **29**, 1043.
14. T. N. Zwietering, *Chem. Eng. Sci.*, 1958, **8**, 244.
15. R. H. Perry and D. Green, *Perry's Chemical Engineers' Handbook*, McGraw Hill, New York, 7th edn, 1997, Section 6.
16. D. F. Boucher and G. E. Alves, *Chem. Eng. Prog.*, 1959, **55**(9), 55.
17. J. P. Catchpole and G. Fulford, *Ind. Eng. Chem.*, 1966, **58**(3), 46.

# 7 An Introduction to Particle Systems

Paola Lettieri

## 7.1 Introduction

Particles and processes involving particles are of enormous importance in the chemical and allied industries. It has been estimated that particulate products, which include pharmaceuticals, detergents, agrochemicals, pigments and plastics, generate $1000M annually for the US economy and Du Pont estimates that 60% of its manufactured products rely heavily on particle science and technology. Despite this, particle technology is a relatively new subject and although the science is advancing rapidly there is still much reliance on empiricism.

The behaviour of powders is often quite different from the behaviour of liquids and gases. Engineers and scientists are used to dealing with liquids and gases whose properties can be readily measured, tabulated and even calculated. With particle systems the picture is quite different. The flow properties of certain powders may depend not only on the particle size, size distribution and shape, but also on surface properties, on the humidity of the atmosphere and the state of compaction of the powder. These variables are not easy to characterise and so their influence on the flow properties is difficult to predict. In the case of particle systems it is almost always necessary to perform appropriate measurements on the actual powder in question rather than to rely on tabulated data. The measurements made are generally measurements of bulk properties, such as bulk density and shear stress.

Concepts of Chemical Engineering for Chemists: 2nd Edition
Edited by Stefaan Simons
© The Royal Society of Chemistry 2017
Published by the Royal Society of Chemistry, www.rsc.org

Given the wide range of industrial applications that involve particle systems, the influence of the operating conditions such as temperature, pressure, velocities, reactor design and any other special conditions, such as the presence of liquid in the reactor, can also significantly affect the fluid–particle contacting efficiency and the interaction between particles, thus affecting the overall flow properties of the powders.

In light of the complexity of such systems, research in particle technology has been devoted to the development of both experimental techniques and theoretical methodologies to determine the particle properties, the fluid–particle and particle–particle interactions. Substantial mathematical effort is currently being directed towards the development of computational fluid dynamics (CFD) modelling as a new tool to support engineering design and research in multiphase systems. The recent development of mathematical modelling of particle systems behaviour, together with the increased computing power, will enable us to link fundamental particle properties directly to powder behaviour and predict the interaction between particles and gaseous or liquid fluids.

The aim of this chapter is to provide a brief introductory guide to the:

(i) Characterisation of solid materials (size analysis, shape and density);
(ii) Interaction between particles and a fluid;
(iii) Fluidisation, in relation to fluidised beds (as an example of a common particle process).

A number of relevant worked examples are also presented, to provide the reader with the direct application of the theory presented.

## 7.2 Characterisation of Solid Materials

The starting point of any study of particle systems is the characterisation of particles in terms of their size, shape and density. Figure 7.1 shows the range of sizes of some commonly encountered particulate materials compared with other quantities, such as the wavelengths of electromagnetic radiation and the sizes of molecules.

### 7.2.1 Size and Size Distribution

Only a minority of systems of industrial interest contain powders with a uniform particle size, *i.e. monodisperse*. Most systems generally

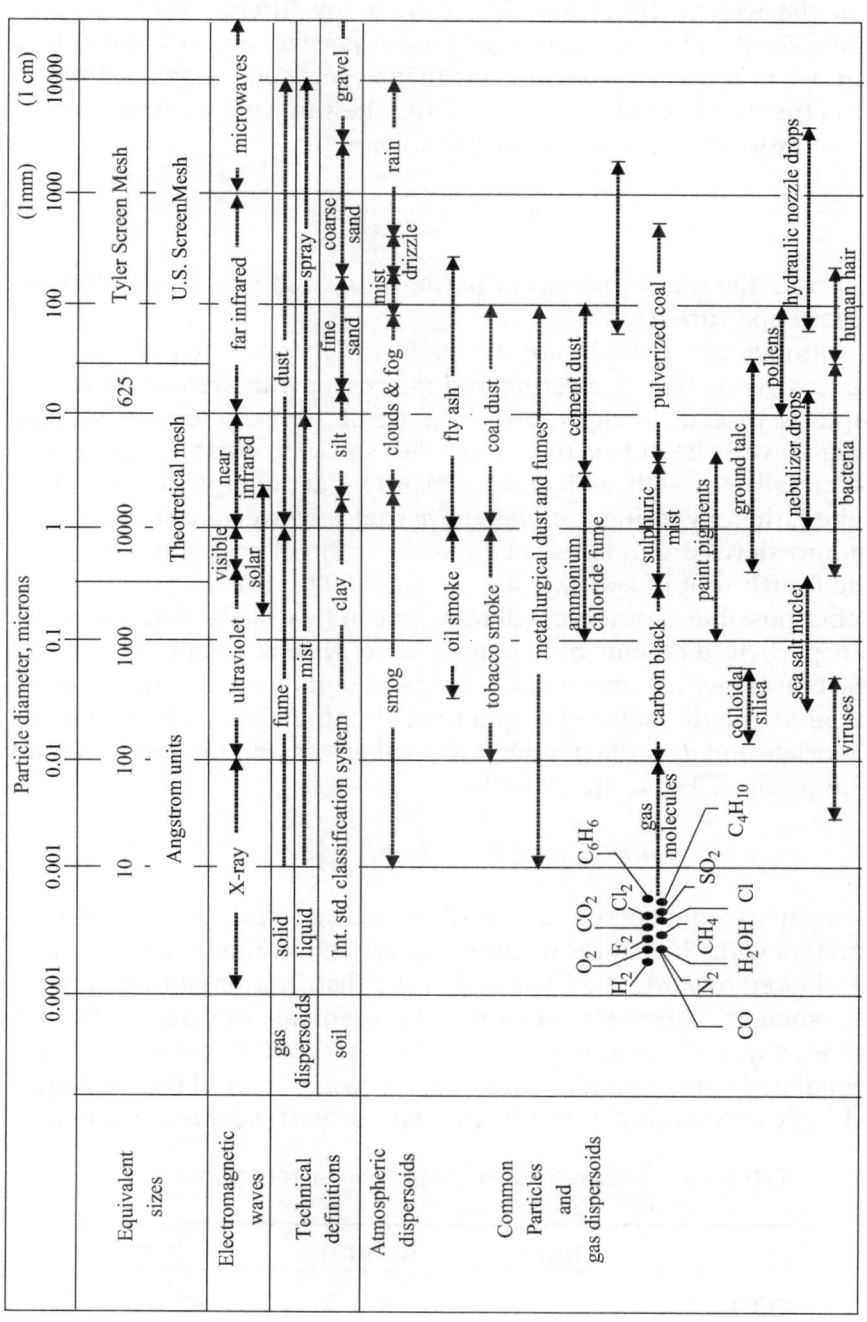

**Figure 7.1** Particle size ranges (Adapted from ref. 5).

show a distribution of sizes (*polydisperse*) and it is then necessary to define the average dimension. There are many different definitions of particle size,[1,2] the most commonly used, particularly in fluidisation, is the so called *volume-surface* mean or the *Sauter mean* diameter. This is the diameter of a particle having the same external surface-to-volume ratio as a sphere and is given by:

$$d_{VS} = \frac{1}{\sum x_i / d_{pi}} \quad (7.1)$$

where $x_i$ is the mass fraction of particles in each size range given by the sieve aperture $d_{pi}$.

A number of methods are available for measuring particle size, such as sieving, light scattering and microscopy. In sieve analysis the sample is placed on the top of a stack of sieves whose mesh size decreases with height of the stack. The stack is vibrated and as the powder falls through it is separated into fractions, which are then weighed, thereby giving the mass of particles between each sieve size. Standard sieves are in sizes such that the ratio of adjacent sieve sizes is the fourth root of two (*e.g.* 45, 53, 63, 75, 90 μm).

Other possible geometrical diameters can be used to determine the mean particle diameter of a polydisperse system. Examples are the surface average, $d_s$, and volume average diameters, $d_v$; where $d_s$ is defined as the diameter of a sphere having the same surface area as the particle and $d_v$ is the diameter of a sphere having the same volume as the particle. These are given by:

$$d_s = \sqrt{\sum x_i d_{pi}^2}; \quad d_v = \sqrt[3]{\sum x_i d_{pi}^3} \quad (7.2)$$

A comparison between the surface average and volume average diameters with the Sauter mean diameter shows that the latter is always skewed towards the finer end of the distribution and, therefore, is the smaller equivalent mean particle diameter in value. Table 7.1 provides a quantitative comparison of the mean diameters $d_{vs}$, $d_s$ and $d_v$ calculated using eqn (7.1) and (7.2) for two industrial fluid catalytic cracking catalysts (FCC1 and FCC2)[3] whose particle size distribution

**Table 7.1** Geometrical mean particle diameters for two FCC catalysts.

|        | $d_{vs}$ (μm) | $d_s$ (μm) | $d_v$ (μm) |
|--------|---------------|------------|------------|
| FCC1   | 71            | 91         | 102        |
| FCC2   | 57            | 104        | 124        |

is shown in Figure 7.2. See worked example E1 (Section 7.5.1) for the determination of the mean particle diameter and size distribution of the FCC1 catalyst sample.

The relative diameter spread $\sigma_d/d_{50}$ can then be used to compare the width of size distribution of particles having different mean diameter. Plotting the cumulative size fraction *versus* particle size, the diameter spread is defined as:

$$\sigma_d = \frac{d_{84\%} - d_{16\%}}{2} \tag{7.3}$$

The median size diameter, $d_{50}$, is the size corresponding to the 50% value on the graph of cumulative size fraction *versus* particle size, see example E1. The values of $\sigma_d/d_{50}$ are used to give an indication of the width of the size distribution, as summarised in Table 7.2.[4]

**Figure 7.2** Particle size distributions of two industrial FCC catalysts.[3]

**Table 7.2** Width of size distribution based on relative spread.

| $\sigma_d/d_{50}$ | Type of distribution |
|---|---|
| 0 | Very narrow |
| 0.03 | Narrow |
| 0.17 | Fairly narrow |
| 0.25 | Fairly wide |
| 0.33 | Fairly wide |
| 0.41 | Wide |
| 0.48 | Wide |
| 0.6 | Very wide |
| 0.7 | Very wide |
| >0.8 | Extremely wide |

## 7.2.2  Particle Shape

This is a fundamental property affecting powder packing, bulk density, porosity, permeability, flowability, attrition and the inter-action with fluids. It is generally considered that shape can be de-scribed in terms of two characteristics, form and proportions. Form refers to the degree to which a particle approaches some standard such as a sphere, cube or tetrahedron, while proportion distinguishes one spheroid from another of the same class. A simple form of shape factor is the sphericity, $\phi$ which is defined as the ratio of surface area of a sphere having the same volume as the particle to the surface area of the particle. Thus:

$$\phi = \frac{\text{surface area of equivalent volume sphere}}{\text{surface area of the particle}} = \left(\frac{d_v}{d_s}\right)^2 \qquad (7.4)$$

and $0 < \phi < 1$. $d_v$ and $d_s$ are as defined in the previous section.

The sphericity $\phi$ can be calculated exactly for such geometrical shapes as cuboids, rings and manufactured shapes, see Table 7.3. However, most particles are irregular and there is no simple generally accepted method for measuring their sphericity. Values for some common solids have been published (see ref. 4). However, these

**Table 7.3**  Sphericity for some regular solids.

| Shape | Relative proportions | Sphericity $\phi$ |
|---|---|---|
| Spheroid | 1 : 1 : 2 | 0.93 |
| | 1 : 2 : 2 | 0.92 |
| | 1 : 1 : 4 | 0.78 |
| | 1 : 4 : 4 | 0.70 |
| Ellipsoid | 1 : 2 : 4 | 0.79 |
| Cylinder | Height = diameter | 0.87 |
| | Height = 2 × diameter | 0.83 |
| | Height = 4 × diameter | 0.73 |
| | Height = 1/2 × diameter | 0.83 |
| | Height = 1/4 × diameter | 0.69 |
| Rectangular parallelepiped | 1 : 1 : 1 | 0.81 |
| | 1 : 1 : 2 | 0.77 |
| | 1 : 2 : 2 | 0.77 |
| | 1 : 1 : 4 | 0.68 |
| | 1 : 4 : 4 | 0.64 |
| | 1 : 2 : 4 | 0.68 |
| Rectangular tetrahedron | — | 0.67 |
| Rectangular octahedron | — | 0.83 |

**Table 7.4** Sphericity of some common solids.

|  | Sphericity $\phi$ |
| --- | --- |
| Crushed coal | 0.75 |
| Crushed sandstone | 0.8–0.9 |
| Round sand | 0.92–0.98 |
| Crushed glass | 0.65 |
| Mica flakes | 0.28 |
| Sillimanite | 0.75 |
| Common salt | 0.84 |

**Figure 7.3** SEM pictures of two FCC catalyst particles.[3]

should be regarded as estimates only. Table 7.4 shows that $\phi$ is between 1 and 0.64 for most materials.

Other methods are available for quantifying shape factors and these are described in detail in ref. 2 and 5. Using a Scanning Electron Microscope (SEM), for example, the shape and surface characteristics of particles can be observed. Photographs of two fluid catalytic cracking catalysts particles are reported as typical examples,[3] see Figure 7.3. The sphericity of the FCC2 catalyst particle is more accentuated and therefore a higher value of the shape factor is expected compared to FCC1. Using the Ergun Equation (see Section 7.3.2, eqn (7.25)), a value of the particle's sphericity can be evaluated, which gives a sphericity $\phi_{\text{FCC1}} = 0.70$ for FCC1 and a value $\phi_{\text{FCC2}} = 0.92$ for FCC2.

### 7.2.3 Particle Density

The density of a particle immersed in a fluid is the *particle density*, defined as the mass of the particle divided by its hydrodynamic volume, $V_{\text{p}}$.[2] This is the volume "seen" by the fluid in its fluid dynamic

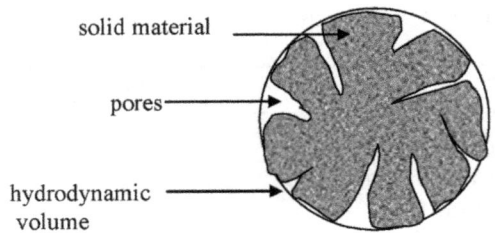

solid material

pores

hydrodynamic
volume

**Figure 7.4**   Hydrodynamic volume of a particle (Adapted from ref. 2).

interaction with the particle and includes the volume of all open and closed pores, see Figure 7.4.

$$\rho_p = \frac{\text{mass of particle}}{\text{volume the particle would displace if its surface were non-porous}} = \frac{M}{V_p}$$

(7.5)

For non-porous solids the particle density is equal to the true, skeletal, or absolute density, $\rho_{ABS}$, which can be measured using either a specific gravity bottle or air pycnometer:

$$\rho_p = \frac{\text{mass of particle}}{\text{volume of solids material making up the particle}}$$

(7.6)

For porous materials $\rho_p < \rho_{ABS}$ and cannot be measured with such methods. A mercury porosimeter can be used to measure the density of coarse porous solids but is not reliable for fine materials, since the mercury cannot penetrate the voids between small particles. In this case, helium is used to obtain a more accurate value of the particle density. Methods to measure the particle density of porous solids can be found in ref. 2 and 5.

In the case of a bed of particles immersed in a fluid, the bulk density of the bed $\rho_b$, which includes the voids between the particles, is defined as follows:

$$\rho_b = (1 - \varepsilon)\rho_b$$

(7.7)

where $\varepsilon$ is the bed voidage, *i.e.* the volume fraction occupied by the fluid.

## 7.3   Interaction Between Particles and Fluids

The fluid–particle and particle–particle interactions dominate the fluid dynamics of particle systems and are of major importance in predicting the behaviour of complex operations, such as fluidisation, pneumatic transport and flow of slurries. In this section, the simple case study of a single particle immersed in a fluid is the starting point for the subsequent examination of the behaviour of fluidisation processes, which involve many particles in up-flowing fluids. This extends what is presented in Chapter 3, in which the concepts of Fluid Flow are introduced and parameters such as drag coefficient and Reynolds Number are defined.

## 7.3.1  Single Particles

We consider the motion of a single spherical particle immersed in a stationary fluid and falling at a constant velocity and we examine the forces exerted on the particle to estimate the steady velocity of the particle relative to the fluid, *the particle terminal fall velocity, $u_t$*. It is important to emphasise at this point that the equilibrium conditions experienced by the particle falling at $u_t$ are equivalent to that of a motionless particle suspended in an upwardly flowing fluid with velocity $u_t$, as represented in Figure 7.5.

A stationary particle suspended in a fluid experiences a *buoyancy force $F_b$*, evaluated from Archimedes' principle as the weight of fluid displaced, $\rho_f g v$, where $\rho_f$ is the fluid density, $g$ is the acceleration due to gravity and $v$ is the volume of the particle. If the particle begins to move the fluid will exert an additional force, the *drag force*, made up of two components; the skin friction drag, which is a direct result of the shear stress at the surface due to fluid viscosity, and the form drag due to differences of pressure over the surface of the particle.

The drag force on a single particle falling through a fluid is a function of a dimensionless drag coefficient, $C_D$, the projected area of the particle and the inertia of the fluid.[6] For a sphere:

$$F_d = C_D \frac{\pi d_p^2}{4} \frac{\rho_f u^2}{2} \tag{7.8}$$

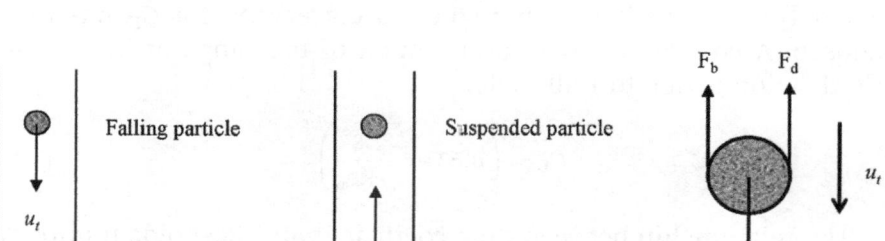

**Figure 7.5**  Fluid–particle interaction.

From a balance of forces on a stationary particle, as represented in Figure 7.5:

$$\text{Gravitation} - \text{Buoyancy} - \text{Drag} = \text{acceleration} = 0$$

Hence:

$$C_D \frac{\pi}{8} d_p^2 \rho_f u^2 = \frac{\pi d_p^2}{6}(\rho_p - \rho_f)g \tag{7.9}$$

or

$$C_D Re_t^2 = \frac{4}{3} Ar \tag{7.10}$$

where

$$Re_t = \frac{\rho_f u_t d_p}{\mu}; \quad Ar = \frac{d_p^3 \rho_f (\rho_p - \rho_f)g}{\mu^2} \tag{7.11}$$

From eqn (7.10), $u_t$ can be written as follows:

$$u_t = \left[ \frac{4 d_p (\rho_p - \rho_f)g}{3 \rho_f C_D} \right]^{1/2} \tag{7.12}$$

Eqn (7.12) is implicit, as $u_t$ is a function of $C_D$.

At low Reynolds numbers ($Re_p < 1$) Stokes' law applies and:

$$F_d = C_D \frac{\pi d_p^2}{4} \frac{\rho_f u^2}{2} = 3\pi d_p \mu u \tag{7.13}$$

with

$$C_D = \frac{24}{Re_p} \quad \text{and} \quad Re_p = \frac{\rho_f u d_p}{\mu} \tag{7.14}$$

In the region where Newton's law applies ($750 < Re_p < 3.4 \times 10^5$), $C_D$ is almost constant at 0.445.

For the intermediate flow regime, a correlation for $C_D$ has to be chosen. A correlation that covers much of the range of interest for fluidisation is due to Dallavalle:[7]

$$C_D = \left( 0.63 + \frac{4.8}{\sqrt{Re_p}} \right)^2 \tag{7.15}$$

The relationship between drag coefficient and Reynolds number is shown in Figure 7.6. Changes in drag coefficient correspond to changes in the flow pattern around a sphere, as also represented in Figure 7.6. At low Reynolds numbers, up to $Re_p = 1$, the flow pattern

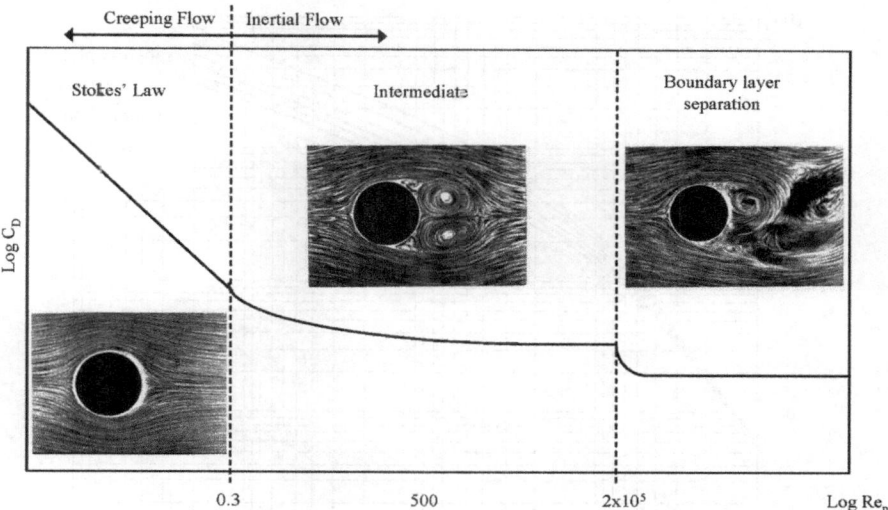

**Figure 7.6** Drag coefficient of a sphere and flow pattern development.

past the particle is symmetrical (*i.e.* the flow does not separate); in the intermediate regime, with values of $1 < Re_p < 500$, the flow will tend to separate, giving rise to the formation of the so-called Karman vortices. A further increase in the Reynolds number will result in fully separated flow. The higher the Reynolds number, the lower the drag coefficient, *i.e.* the resistance applied to the particle to flow.

When the flow regime is not known, the *analytical calculation* of $u_t$ from eqn (7.12) requires choosing an equation for $C_D$, such as, for example, eqn (7.15), and solving a "non-trivial" system of two non-linear equations. To avoid the analytical solution, an *iterative procedure* can be applied by guessing an initial value of $C_D$ or $Re_p$:

- Step 1: Guess an initial value of $C_D$ (or $Re_p$).
- Step 2: Calculate value of $u_t$ from eqn (7.12).
- Step 3: Calculate Reynolds number from eqn (7.11).
- Step 4: Check initial guess for $C_D$ by computing it from knowledge of $Re_p$ and using, for example, eqn (7.15). If $C_D$ is equal to the initial guess, then the $u_t$ value calculated at step 2 is correct, otherwise go to step 5.
- Step 5: Choose a new value of $C_D$ and go back to step 2.

Alternatively, a *graphical solution* can be applied to determine $u_t$ from an empirical plot based on the two dimensionless groups reported in eqn (7.16) and (7.17), which are depicted in Figure 7.7.[8] Worked

**Figure 7.7**  Empirical diagram for the calculation of the particle terminal fall velocity (Adapted from ref. 8).

example E2 (Section 7.5.2) shows the application of this procedure to the calculation of $u_t$.

$$d_p^* = d_p \left[ \frac{\rho_f(\rho_p - \rho_f)g}{\mu^2} \right]^{1/3} = Ar^{1/3} \qquad (7.16)$$

$$u_t^* = u_t \left[ \frac{\rho_f^2}{\mu(\rho_p - \rho_f)g} \right]^{1/3} = \frac{Re_t}{Ar^{1/3}} \qquad (7.17)$$

## 7.3.2  Flow Through Packed Beds

We now examine the case in which the particles are stationary and in direct contact with their neighbours, which is known as a packed bed.

In a packed bed, the fluid–particle interaction force is insufficient to support the weight of the particles. Hence, the fluid that percolates through the particles loses energy due to frictional dissipation. This results in a loss of pressure that is greater than can be accounted for by the progressive increase in gravitational potential energy. We now determine this additional energy loss, which is also the energy required to overcome the weight of the particles in a fluidised bed.

The total drop in fluid pressure across a length $L$ of bed is given by:

$$\Delta p = \Delta P + \rho_f g L \tag{7.18}$$

where $\rho_f g L$ represents the hydrostatic contribution and $\Delta P$ is the portion lost due to the frictional interaction between the fluid and the particles. It represents the energy lost by the fluid and dissipated as heat.

In laminar flow, the flow through a packed bed of solid particles has been analysed in terms of the fluid-flow through parallel straight tubes, in light of the similarity in the expression of the pressure drop through a porous media as given by Darcy[9] and that in cylindrical tubes proposed by Hagen–Poiseuille.[6]

When a fluid passes through a bed of porous material the pressure drop per unit length of bed is given by Darcy's law as:

$$\frac{\Delta P}{L} = k \mu U \tag{7.19}$$

where $U$ is the volumetric flux (volumetric flow rate per unit cross section of the empty tube), $\mu$ is the fluid viscosity and $k$ is the permeability of the medium. The Hagen–Poiseuille equation for laminar flow through cylindrical tubes is:

$$\frac{\Delta P}{L} = \frac{32 \mu U}{D^2} \tag{7.20}$$

The application of the analogy involves simply replacing the fluid flux $U$ and the bed diameter $D$ in the Hagen–Poiseuille equation with equivalent terms relating to flow through porous media in order to take into account that a fraction of the cross-sectional area of the bed is occupied by particles. Following this reasoning, the expression for the pressure drop through a bed of spherical particles for laminar flow can be obtained:

$$\frac{\Delta P}{L} = \frac{72(1 - \varepsilon)^2}{\varepsilon^3} \frac{\mu U}{d_p^2} \tag{7.21}$$

where the flux $U$ in the Hagen–Poiseuille equation has been taken as indicating the superficial velocity of fluid relative to the tube wall (volume flow of gas/cross section of the bed). Using some empirical evidence, the constant in eqn (7.21) has been subsequently increased from 72 to 150 and eqn (7.21) with the new constant value has become known as the *Blake and Kozeny* equation (see ref. 10 for the full theoretical development). Eqn (7.21) applies to the flow region where the pressure drop is due solely to viscous losses. For beds of particles larger than about 150 μm in diameter, inertial forces become important. For the turbulent flow regime the pressure drop in a straight tube can be written as:

$$\Delta P = 4f \frac{1}{2D} \rho_f L U^2 \tag{7.22}$$

where $f$ is the friction factor. Substituting the fluid flux $U$ and the diameter $D$ in eqn (7.22), we obtain:

$$\Delta P = \frac{3f(1-\varepsilon)}{\varepsilon^3} \frac{\rho_f L U^2}{d_p} \tag{7.23}$$

It was found experimentally that $3f = 1.75$. With this value, eqn (7.23) becomes known as the *Burke–Plummer* equation:

$$\Delta P = \frac{1.75(1-\varepsilon)}{\varepsilon^3} \frac{\rho_f L U^2}{d_p} \tag{7.24}$$

Experiments have shown that eqn (7.21) and (7.24) can be combined to cover for any flow condition, giving what is known as the Ergun equation:[11]

$$\frac{\Delta P}{L} = \frac{150(1-\varepsilon)^2}{\varepsilon^3} \frac{\mu U}{d_p^2} + \frac{1.75(1-\varepsilon)\rho_f U^2}{\varepsilon^3 d_p} \tag{7.25}$$

where the first term is the viscous term and the second is the inertial term.

In general, the Ergun equation for flow through a packed bed of non-spherical and polydisperse-sized particles becomes:

$$\frac{\Delta P}{L} = \frac{150(1-\varepsilon)^2}{\varepsilon^3} \frac{\mu U}{(\phi d_{sv})^2} + \frac{1.75(1-\varepsilon)}{\varepsilon^3} \frac{\rho_f U^2}{\phi d_{sv}} \tag{7.26}$$

The Ergun equation may be rearranged to form dimensionless groups, as graphically represented in Figure 7.8. Using eqn (7.25) we obtain:

$$\left(\frac{\Delta P}{L}\right)\left(\frac{d_p}{\rho_f U^2}\right)\left(\frac{\varepsilon^3}{1-\varepsilon}\right) = 150 \frac{\mu(1-\varepsilon)}{d_p \rho_f U} + 1.75 \tag{7.27}$$

It is worth noting the similarity between Figure 7.8 and the friction factor for tube flow represented in Figure 7.6. This highlights the physical meaning of pressure drop across the bed as energy dissipation due to friction between the particles.

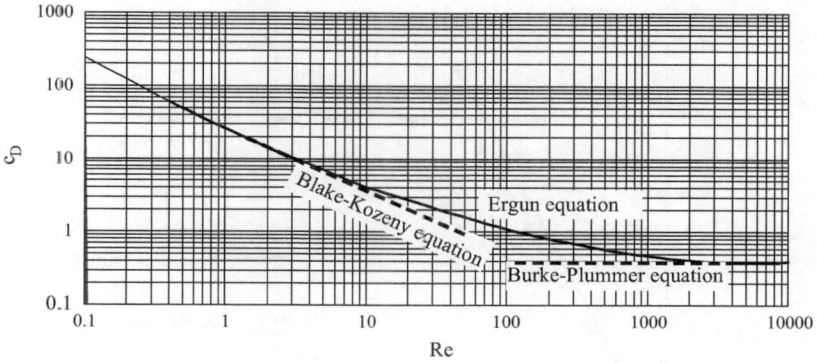

**Figure 7.8** The Ergun equation (Adapted from ref. 11).

## 7.4 Fluidised Beds

As the velocity of the fluidising fluid (gas or liquid) is increased through a bed of solid particles, there comes a point where the drag force exerted by the fluid on the particles, which is proportional to the global pressure drop across the bed, is balanced by the buoyant weight of the suspension. At this point, the particles are lifted by the fluid, the separation between them increases and the bed becomes fluidised. Thus:

$$\Delta P = \frac{\text{weight of particles and fluid} - \text{upthrust}}{\text{bed cross-sectional area}} = \frac{Mg}{A} \qquad (7.28)$$

where $M$ is the mass of particles in the bed and $A$ is the cross sectional area of the bed.

For a bed of particles of density $\rho_p$, fluidised by a fluid of density $\rho_f$ to form a bed of depth $L$ and voidage $\varepsilon$ in a vessel of cross-sectional area $A$, eqn (7.28) becomes:

$$\Delta P = \frac{[AL(1 - \varepsilon)\rho_p g - AL\varepsilon\rho_f g] - AL\rho_f g}{A} \qquad (7.29)$$

from which:

$$\frac{\Delta P}{L} = (\rho_p - \rho_z)(1 - \varepsilon)g \qquad (7.30)$$

An important property of fluidised beds follows immediately from this simple relation; as the bed expands, the product $(1 - \varepsilon)\,L$, which represents the total volume of particles per unit cross section, remains unchanged. As the fluid flux is increased, $L$ increases and $(1 - \varepsilon)$ decreases so as to maintain their product at a constant value.

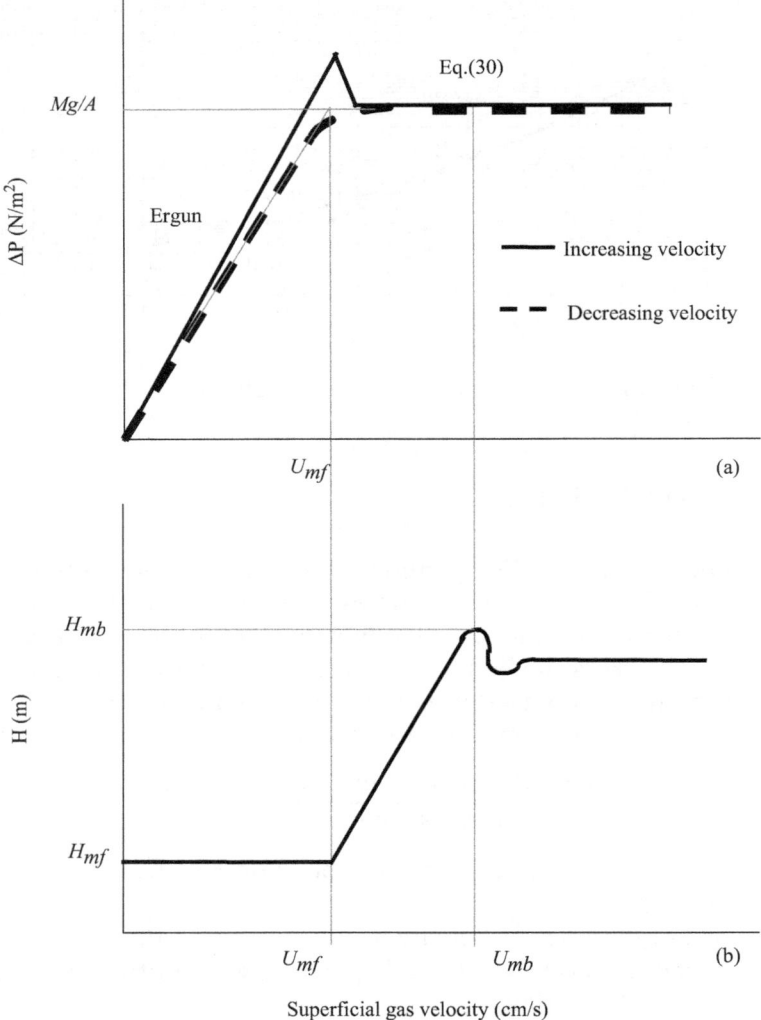

**Figure 7.9**   Pressure drop and bed expansion profiles for gas–solid systems of fine particles.

As a consequence, the pressure drop through the fluidised bed remains constant for further increases in fluid velocity, as shown in Figure 7.9.

## 7.4.1   Minimum Fluidisation Velocity

The fluid velocity at which the particles become suspended is the minimum fluidisation velocity, $u_{mf}$. For spherical particles, the velocity at minimum fluidisation can be predicted by combining eqn (7.25) and (7.30), given that the values of the bed voidage at minimum fluidisation, $\varepsilon_{mf}$, are known *a priori*:

$$(\rho_p - \rho_f)g = \frac{150(1 - \varepsilon_{mf})}{\varepsilon_{mf}^3}\frac{\mu u_{mf}}{d_p^2} + \frac{1.75}{\varepsilon_{mf}^3}\frac{\rho_f u_{mf}^2}{d_p} \tag{7.31}$$

which can also be written as:

$$Ar = \frac{150(1 - \varepsilon_{mf})}{\varepsilon_{mf}^3}Re_{mf} + \frac{1.75}{\varepsilon_{mf}^3}Re_{mf}^2 \tag{7.32}$$

From this, the minimum fluidisation velocity can be obtained:

$$u_{mf} = \frac{\varepsilon_{mf}^3(\rho_p - \rho_f)d_p^2 g}{150(1 - \varepsilon_{mf})\mu} \tag{7.33}$$

Eqn (7.33) is strictly applicable only to conditions of viscous flows where the pressure drop is solely due to viscous energy losses.

If eqn (7.31) is expressed as a function of the sphericity factor, $\phi$, as shown in eqn (7.26), then a value of $\phi$ can be back-calculated by knowing $u_{mf}$ and $\varepsilon_{mf}$, as done for the FCC catalysts shown in Figure 7.3.

If $\varepsilon_{mf}$ is not known, than we can use the equation proposed by Wen–Yu,[12] which expresses $u_{mf}$ solely as a function of the physical characteristics of the fluid and the particle. Wen–Yu used experimental evidence to reformulate eqn (7.32) as follows:

$$Ar = 1650\,Re_{mf} + 24.5\,Re_{mf}^2 \tag{7.34}$$

This formulation assumes values of the voidage at minimum fluidisation of around 0.38.

For fine spherical particles (*below about 100 μm*) $u_{mf}$ is obtained from the viscous term of eqn (7.34)

$$u_{mf} = \frac{d_p^2(\rho_p - \rho_f)g}{1650\mu} \tag{7.35}$$

For larger particles, it becomes:

$$u_{mf}^2 = \frac{d_p(\rho_p - \rho_f)g}{24.5\rho_f} \tag{7.36}$$

The sensitivity of $u_{mf}$ on the bed voidage is shown in the worked example E3 (Section 7.5.3), in which the calculation of $u_{mf}$ using the Ergun and Wen–Yu equations is compared.

Most fluidisation processes are operated at high temperatures and pressures. It is important, therefore, to be able to predict changes in fluidisation with the operating conditions. Using eqn (7.35) and (7.36), the effect of temperature and pressure can be determined. With increasing temperature, *gas viscosity increases* while *gas density decreases*. For small particles, the fluid–particle interaction is dominated by the viscous effects. Eqn (7.35) shows that $u_{mf}$ varies with $1/\mu$, and $u_{mf}$ should therefore decrease with temperature. For large particles, the inertial effects dominate; eqn (7.36) predicts that $u_{mf}$ will vary with $(1/\rho_f)^{0.5}$, $u_{mf}$ should therefore increase with temperature.

With increasing pressure, gas viscosity is essentially independent of pressure, while gas density increases. For small particles, $u_{mf}$ remains almost constant. For large particles, $u_{mf}$ decreases with pressure.

## 7.4.2 Types of Fluidisation Regimes

At fluid velocities higher than at minimum fluidisation, different types of fluidisation behaviour are possible:

(i) uniform expansion;
(ii) bubbling or slugging;
(iii) turbulent;
(iv) lean fluidisation.

In liquid–solid systems, and for gas–solid systems of small particles, increasing the fluidising velocity beyond $u_{mf}$ gives rise to a uniform expansion of the bed. The homogeneous expansion of a gas–solid system is generally described using the Richardson–Zaki equation:[13]

$$u = u_t \varepsilon^n \tag{7.37}$$

Eqn (7.37) describes the relationship between fluidising velocity and voidage as a function of the particle terminal fall velocity and of a parameter $n$ which depends on the particle Reynolds number according to the following relations:[10]

$$
\begin{array}{ll}
n = 4.8 & Re_t < 0.2 \\
n = 4.6\ Re_t^{-0.03} & 0.2 < Re_t < 1 \\
n = 4.6\ Re_t^{-0.1} & 1 < Re_t < 500 \\
n = 2.4\ Re_t & > 500
\end{array}
\tag{7.38}
$$

In gas–solid systems, as the gas velocity is further increased, gas bubbles form within the bed to give a bubbling fluidised bed. The velocity at which the first bubble forms is called the minimum bubbling velocity, $u_{mb}$. An empirical correlation typically used to predict $u_{mb}$ at ambient conditions has been given by Abrahamsen and Geldart:[14]

$$u_{mb} = 2.07 \, \exp(0.716 \, F_{45}) \frac{d_p \, \rho_g^{0.06}}{\mu^{0.347}} \qquad (7.39)$$

where $F_{45}$ is the mass fraction of particles smaller than 45 μm.

The pressure drop and bed expansion profiles shown in Figure 7.9 represent the different stages from fixed to bubbling fluidisation for a typical gas–solid system of fine particles. Knowing the bed height, the average bed voidage can be obtained:

$$\varepsilon = 1 - \frac{M}{\rho_p HA} \qquad (7.40)$$

An increase in gas velocity beyond minimum bubbling leads to more vigorous bubbling with larger bubbles that may eventually become as large as the vessel diameter, giving a *slugging bed*. This may occur typically with high aspect ratio ($H/D$) beds for Geldart Group B particles. Turbulent fluidisation occurs for even higher velocities. Here the two-phase structure disappears and bed takes on a foam-like appearance. Still higher gas velocities result in particles being transported out of the bed, until only a low concentration of rapidly moving particles remains in contact with the gas, as in *pneumatic conveying systems*.[2] These types of systems are used for transporting particulate solids in a conveying fluid, where the solids are typically at least three times denser than the transporting fluid. A typical example is the transport of particles in a pipeline using gas. The distinction between pneumatic conveying systems and *hydraulic conveying systems* lies in the ratio between the solid and fluid density, with the solid density being generally of the order of magnitude of the fluid density in the latter. In pneumatic conveying systems, *two types* of flows are generally encountered. "Dilute phase" is characterised by low particle concentrations (less than 1% by volume) and relatively high velocities (greater than 20 m s$^{-1}$), which results in high attrition of the particles. Dilute phase conveying is carried out in systems in which the solids are fed to the pipe from hoppers at a controlled flow rate and cyclone separators are used to separate the solids from the gas stream at the end of the line. "Dense phase" conveying systems are characterised instead by high particle concentrations (greater than 30% by volume) and low velocities (1–5 m s$^{-1}$). In this case the solids are not entirely suspended in the gas stream; particle attrition is reduced due to the low solid velocities, reducing in turn pipeline

| Fixed bed | Minimum fluidization | Particulate | Bubbling | Slugging | Lean |
|---|---|---|---|---|---|

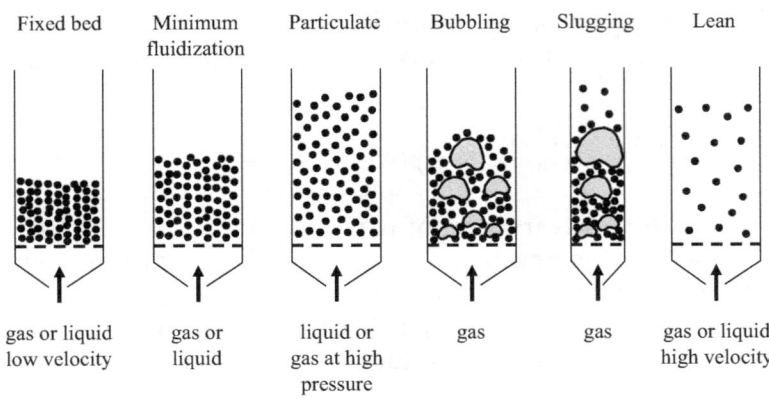

| gas or liquid low velocity | gas or liquid | liquid or gas at high pressure | gas | gas | gas or liquid high velocity |
|---|---|---|---|---|---|

**Figure 7.10**   Types of fluid–particle contacting.

erosion and product degradation. Both types of flows are used in horizontal and vertical transport systems. Figure 7.10 shows various types of fluid–particle contacting, from fixed bed, to bubbling fluidisation, slugging and pneumatic transport.

The minimum fluidisation velocity represents the lower limit of the range of operative conditions at which a fluid-bed process can be operated, while the particle terminal fall velocity represents the upper limit beyond which the particles will start leaving or elutriating from the bed. To avoid or reduce carryover of particles from a fluidised bed, gas velocity has to be kept between $u_{mf}$ and $u_t$. For polydisperse systems, in calculating $u_{mf}$, the mean diameter $d_{vs}$ is used for the particle size distribution present in the bed. In calculating $u_t$, the smallest size of solids present in appreciable quantity in the bed is used.

Using the dimensionless parameters introduced with eqn (7.16) and (7.17) (substituting $u_t^*$ and $u_t$ with $u^*$ and $u$), for a given particle system and a given fluidising velocity, the fluidisation regime can be determined using the diagram reported in Figure 7.11.[15] Similarly, the operating velocity needed to attain a certain fluidisation regime can also be determined from Figure 7.12, see worked example E4 (Section 7.5.4).

## 7.4.3   Classification of Powders

In an attempt to classify the different fluidisation behaviours, Geldart[16] proposed an empirical classification, which divides fluidisation behaviour according to the mean particle size and gas and particle densities. Boundaries between these groups were proposed in the form of a dimensional plot of $(\rho_p - \rho_g)$ *versus* $d_p$, as shown in

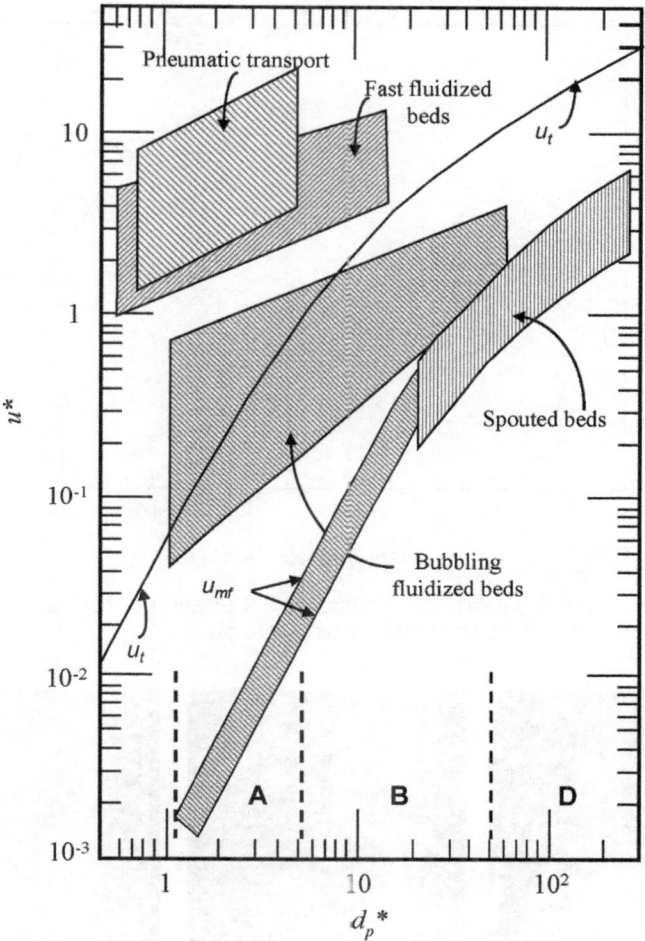

**Figure 7.11** Empirical diagram for the prediction of the fluidisation regime (Adapted from ref. 15).

Figure 7.12. Geldart proposed four typical fluidisation regimes for materials fluidised with air at ambient conditions, which he indicated as A, B, C and D.

*Group C* materials are usually cohesive and very difficult to fluidise. The behaviour of this group of powders is strongly influenced by the interparticle forces, which are greater than those exerted by the fluidising gas. As a result, the particles are unable to achieve the separation they require to be totally supported by drag and buoyancy forces and true fluidisation does not occur, with the pressure drop across the bed being lower than the theoretical value given by

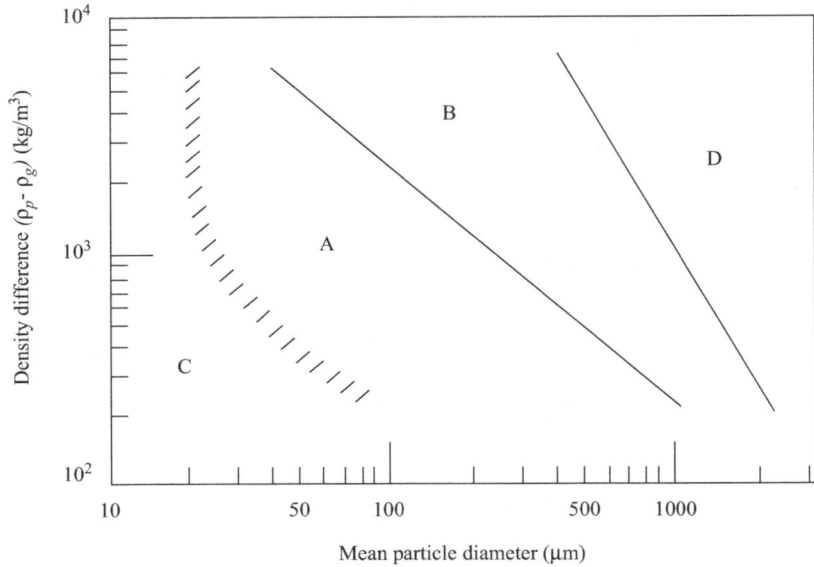

**Figure 7.12**  Classification of fluidisation behaviour for air at ambient conditions (Adapted from ref. 16).

**Figure 7.13**  Fluidisation behaviour of a group C powder: an attempt to fluidise a group C powder produces channels or a discrete plug.

eqn (7.30). Bubbles, as such, do not appear; instead the gas flow forms channels through the powder. An attempt to fluidise a Group C powder is shown in Figure 7.13. Fluidisation of Group C materials can sometimes be improved using mechanical stirring or vibrations, which break up the stable channels. In the case of plastic materials, fluidisation may be promoted by adding some sub-micron particles, which modify the contact geometry, thereby reducing the interparticle forces.

*Group A* powders are those which exhibit a stable region of non-bubbling expansion between $u_{mf}$ and $u_{mb}$, as shown in Figure 7.10. Geldart[4] distinguished these powders as those for which $u_{mb}/u_{mf} > 1$. At gas velocities above $u_{mb}$ bubbles begin to appear, which constantly split and coalesce, and a maximum stable bubble size is achieved. The flow of bubbles produces high solids and gas back-mixing, which makes the powders circulate easily, giving good bed-to-surface heat transfer.

*Group B* powders are characterised by having $u_{mb} = u_{mf}$. Bubbles rise faster than the interstitial gas velocity, coalescence is the dominant phenomenon and there is no evidence of a maximum bubble size, as defined for Group A materials. Bubble size increases with increasing fluidising gas velocity, see Figure 7.14. The interparticle forces are considered to be negligible for these powders.

*Group D* powders are large particles that are distinguished by their ability to produce deep spouting beds. The distinction between this and the previous groups concerns also the rise velocity of the bubbles, which is, in general, less than the interstitial gas velocity. The gas velocity in the dense phase is high and solids mixing is relatively poor, consequently, back-mixing of the dense phase is small. Segregation by size is likely when the size distribution is broad, even at high gas velocities. The flow regime around these particles may be turbulent, causing particle attrition and, therefore, elutriation of the fines produced.

## 7.4.4 Interparticle Forces, Measurement of Cohesiveness and Flowability of Solid Particles at Ambient Conditions

When trying to describe the fluidisation of different materials, the nature of the forces acting between adjacent particles becomes of major importance in order to predict their contribution to the fluid-bed behaviour. As described in the previous section, fluidisation of Geldart Group C materials is dominated by interparticle forces, whose effects become negligible in the fluidisation behaviour of Group B powders. The role of the interparticle forces on the fluidisation behaviour of Group A powders is, on the other hand, far from being

**Figure 7.14** Fluidisation behaviour of a Group B powder: effect of gas flow on bubbles in a two-dimensional fluidised bed, (i) lower gas velocity, (ii) higher gas velocity. A max. stable bubble size is never achieved.

unequivocally understood, mainly due to the difficulty in recognising the nature of the forces involved and, therefore, of quantifying their effect on the fluidisation behaviour.

Particle–particle contacting can be the result of different mechanisms of adhesion, which in turn can influence, in different ways, the fluidisation behaviour of solid materials, see ref. 17. An extensive review on the subject is also reported in Israelachvili.[18] Table 7.5 lists various mechanisms of adhesion with and without material bridges.

It would be desirable to have simple tests capable of characterising the fluidisation behaviour or "flowability" of particulate materials on the basis of their *bulk* properties. To this end, Carr[19] developed a system to characterise bulk solids with respect to flowability. Table 7.6 summarises the properties which are determined. In Carr's method a numerical value is assigned to the results of each of these tests, and is summed to produce a relative flowability index for that particular bulk material. Given the extensive use of these empirical techniques in academia and industry, a brief review on the subject is reported here. Nevertheless, it should be emphasised that these techniques allow measurements of the flowability or cohesion of materials solely in their stationary or compressed status and at ambient conditions. A direct relationship between these measurements and the fluidisation behaviour can therefore prove very difficult, in particular when operating conditions cause the properties of the materials to change.

**Table 7.5** Mechanisms of adhesion.

| Without material bridges | With material bridges |
| --- | --- |
| Van der Waals forces | Capillary forces |
| Electrostatic forces | Solid bridges |
| Magnetic forces | Sintering |
| Hydrogen bonding | |

**Table 7.6** Properties of bulk solids to determine flowability with the Carr method.

| | |
| --- | --- |
| Bulk density loose, $\rho_{BDL}$ | Angle of repose |
| Bulk density packed, $\rho_{BDP}$ | Angle of fall |
| Hausner ratio | Angle of difference |
| Compressibility | Angle of spatula |
| Cohesion | Angle of internal friction |

### 7.4.4.1    Loose and Packed Bulk Density

The loose bulk density is determined by gently pouring a sample of powder into a container through a screen, and is measured before settlement takes place. The packed or tapped bulk density is determined after settling and deaeration of the powder has occurred due to tapping of the sample. Details on the techniques are reported by Geldart and Wong.[20]

### 7.4.4.2    Hausner Ratio and Powder Compressibility

The ratio between the loose and packed bulk density, $\rho_{BDL}$ and $\rho_{BDP}$ respectively, is known as the Hausner Ratio ($HR$) and is used as an indication of the cohesiveness of the materials, see ref. 20. In addition to the $HR$, the powder compressibility is also used to define cohesiveness. This is expressed as $100(\rho_{bdp} - \rho_{bdl})/\rho_{bdp}$.

### 7.4.4.3    Cohesion

This test is used for very fine powders (below 70 mm). Material is passed through three vibrating sieves in series. The material left on each sieve is weighed and a cohesion index is determined from the relative amounts retained. Carr[19] defined cohesion as the apparent surface force acting on the surface of powders, which are composed of millions of atoms. The number of points of contact within the powder mass determines the effect of this force. Thus, cohesiveness increases with decreasing particle size, since the number of contact points increases as the particle size decreases.

### 7.4.4.4    Angle of Repose, Angle of Fall and Angle of Difference

*The angle of repose* is defined as the angle between a line of repose of loose material and a horizontal plane. This is determined by pouring the powder into a conical pile from a funnel and screen assembly. Materials with good flowability are characterised by low angles of repose. The *angle of fall* is determined by dropping a small weight on the platform on which a loose or poured angle of repose has been formed. The fall causes a decrease of the angle of repose forming a new angle identified as the angle of fall. The more free-flowing the powder, the lower the angle of fall. *The angle of difference* is determined by the difference between the angle of repose and the angle of fall. The greater this angle the better the flow.

## 7.4.4.5   Angle of Spatula and Angle of Internal Friction

*The angle of spatula* is a quick measurement of the angle of internal friction. This is the angle, measured from the horizontal, that a material assumes on a flat spatula that has been inserted into a pile and then withdrawn vertically. A free-flowing material will have formed one angle of repose on the spatula's blade. A cohesive material will have formed several angles of repose on the blade, the average of these is then taken. The higher the angle of spatula of a material, the less its flowability. *The angle of internal friction* has been defined as the angle at which the dynamic equilibrium between the moving particles of a material and its bulk solid is achieved. This is of particular interest for flows in hoppers and bins. The measurement of this angle is described below.

### 7.4.4.5.1   Measurements of Angles of Friction

Shear cells are generally used to determine the properties which influence powder flow and handling.[21] Some of these properties are called "failure properties". These include the effective angle of internal friction, the angle of wall friction, the failure (or flow) function, the cohesion and ultimate tensile strength. The requirement for getting powders to flow is that their strength is less than the load put on them, *i.e.* they must fail, hence the name "failure properties". From soil mechanics, a solid is characterised by a yield locus that defines the limiting shear strength under any normal stress. Plotting shear stress $\tau$ and normal stress $\sigma_n$, the yield locus for a Coulomb material intersects the $\tau$ axis at a value of $\tau$ which is defined as cohesion $C$ at an angle $\phi_i$, which is defined as the angle of internal friction of that material.

In a shear cell, such as the Jenike shear cell,[22] the test powder is consolidated in a shallow cylindrical chamber which is split horizontally. The lower half of the cell is fixed and a shear force is applied to the upper moveable part at a constant low rate (see example E5, Section 7.5.5, where experiments were obtained at a constant rate of 0.03 rpm, corresponding to a linear velocity of 1 mm min$^{-1}$). Shearing can be carried out for each of a series of normal loads on pre-consolidated samples, so that at the end of the test the relationships between the shear stress and normal stress at various bulk densities are obtained.

### 7.4.4.5.2   Construction of the Static Internal Yield Locus

Once the shear cell is loaded with the powder specimen, the pre-consolidation load $N_c$ is applied on the cell lid and the pre-shearing phase starts. When a steady-state value is attained for the measured shear stress, the pre-shearing is stopped. Then the shearing phase starts under a normal load $N_1$ lower than $N_c$ and a peak value of the shear stress is reached corresponding to the material failure. The sequence of pre-shearing and shearing phases is then repeated, keeping the same value of the normal load for each pre-shearing phase, $N_c$, and using decreasing normal loads for each shearing phase, *i.e.* $N_1 > N_2 > N_3 > \ldots > N_N$.

Following the theory of Jenike,[23] the solid normal stress acting in the shear plane can, in general, be calculated by simply dividing the external normal load by the shear surface area, which is assumed to correspond to the cross-sectional area of the cell.

$$\sigma_i = \frac{N_i}{A_{\text{cell}}} \tag{7.41}$$

Figure 7.15 reports a typical shear stress chart and the derived yield locus corresponding to the applied consolidation stress. The yield locus is the interpolation of the experimental points $(\sigma, \tau)$ corresponding to failure. It is worth pointing out that different yield loci are obtained when applying different consolidation stresses $\sigma_c$. The choice of the normal stresses used in the shearing phases and the number of phases used in each experiment, on the other hand, are arbitrary, as all the $(\sigma, \tau)$ points obtained in the shearing phases will belong to the same yield locus, provided that the same consolidation stress is applied in the pre-shearing phases. The values of the static angle of internal friction $\phi$ and the cohesion $C$ were worked out as the slope and the intercept with the $\tau$ axis of the yield locus, respectively, as shown in Figure 7.15. The circle passing through $\sigma_c$ and tangent to the yield locus is called the Mohr Circle. The two intercepts of this Mohr circle and the $\sigma$ axis are the major principal stresses $\sigma_1$ and $\sigma_2$

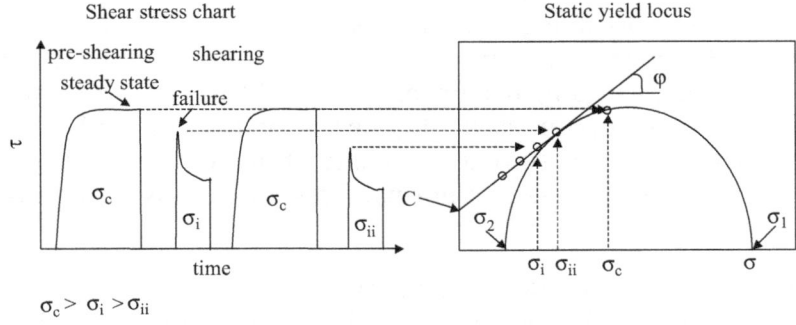

**Figure 7.15** Construction of the static yield locus.

**Figure 7.16** Yield loci of (a) free-flowing materials; (b) cohesive materials.

corresponding to the consolidation applied. By repeating the whole experiment with different consolidation loads $N_c$, a family of incipient yield loci can be obtained. As a result, different angles of internal friction and values of the cohesion can be obtained as a function of the major principal stress $\sigma_1$.

Jenike carried out many experimental measurements on free-flowing and cohesive materials. He found that the yield locus of a dry material would be a straight line passing through the origin, as shown in Figure 7.16(a). The term "Cohesionless" was therefore used to describe materials which have a negligible shear strength under zero normal load $(\sigma_n = 0)$. On the other hand, Jenike found that the yield loci of cohesive materials differ significantly from a straight line and have a nonzero intercept, indicated by $C$. Moreover, the position of the locus for a cohesive powder is a strong function of the interstitial voidage of the material. Figure 7.16(b) shows the typical yield locus for cohesive materials.

### 7.4.4.5.3 Construction of the Dynamic Internal Yield Locus

The dynamic yield locus represents the steady state deformation, as opposed to the static yield locus which represents the incipient failure. The dynamic yield locus is constructed by plotting on a $(\sigma, \tau)$ plane the principal Mohr circles obtained for various consolidation stresses. The dynamic yield locus will be the curve or straight line tangent to all circles, as shown in Figure 7.17. The dynamic angle of internal friction $\delta$ and cohesion $C_\delta$ are independent of the consolidation stress. $\delta$ and $C_\delta$ are obtained as the slope and the intercept at $\sigma = 0$ of the dynamic yield locus of the powder.

An approximation of the dynamic yield locus is given by the effective yield locus, provided that $C_\delta = 0$. The effective yield locus, shown

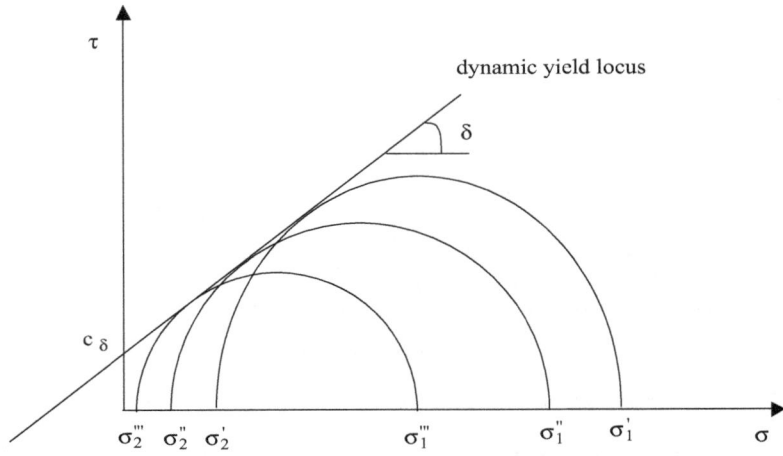

**Figure 7.17**  Dynamic yield locus.

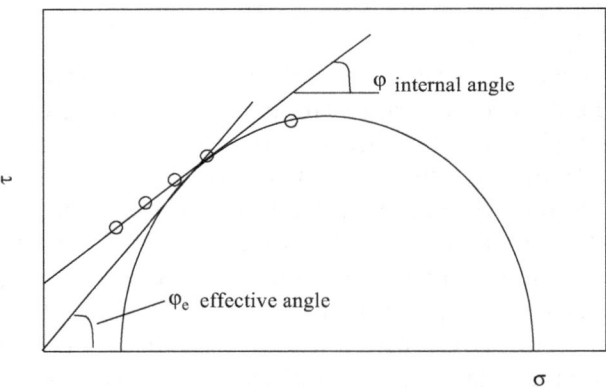

**Figure 7.18**  Effective angle of internal friction.

in Figure 7.17, is the straight line drawn through the origin and tangentially to the Mohr's circle represented in Figure 7.15. Since it passes through the origin, it represents the angle of internal friction that the material would have if it were cohesionless. This angle is denoted by the effective angle of internal friction of the material, $\phi_e$ (Figure 7.18). The difference between the effective and the internal angle of friction can be an indication of the degree of cohesiveness of the powder. An application of the above procedure is presented in the worked example E5, which illustrates the results obtained from experiments carried out on two different alumina materials and ballotini.[24]

## 7.4.5 Some Industrial Fluid-bed Applications

The fluidised bed is only one of the many reactors employed in industry for gas–solid reactions, as reported by Kunii and Levenspiel.[25] Whenever a chemical reaction employing a particulate solid as a reactant or as a catalyst requires reliable temperature control, a fluidised bed reactor is often the choice for ensuring nearly isothermal conditions by a suitable selection of the operating conditions. The use of gas–solid fluidised beds spans across many industries, ranging from the pharmaceutical, refining and petroleum sectors to power-generation and can be classified depending on whether the physical mechanisms of heat and mass transfer are the governing factors in the process or whether chemical reactions are taking place in the fluid-bed. Examples are given in Table 7.7.

Examples of chemical reaction processe that employ a fluid-bed include catalytic cracking of petroleum, metallurgical processes, titanium refining, where the fluidised bed reactor is used for extracting titanium from naturally occurring ore, coal gasification, combustion and incineration, where the recent emphasis is on using

**Table 7.7** Some industrial applications of fluidised beds.

| Physical mechanism | Application |
|---|---|
| Heat and/or mass transfer between gas/particle | Solids drying |
| | Absorption of solvents |
| | Food freezing |
| Heat and mass transfer between particle/ particle or particle/surface | Coating of pharmaceutical tablets |
| | Granulation |
| | Mixing of solids |
| | Plastic coating of surfaces |
| Heat transfer between bed/surface | Heat treatment of glasses, rubber, textile fibres |

| Chemical mechanism | Application |
|---|---|
| Gas/gas reaction in which solid acts as catalyst | Oil cracking |
| | Manufacture of: |
| | Acrylonitrile |
| | Polyethylene |
| Gas/gas reaction in which solids are transformed | Coal combustion |
| | Gasification |
| | Incineration of waste |
| | Catalyst regeneration |

**Table 7.8**   Operating conditions employed in fluidised bed processes.

| Process | Example/Products | Process conditions |
|---|---|---|
| Drying of solids | Inorganic materials | 60–110 °C; 1 atm |
|  | Pharmaceuticals | 60 °C; 1 atm |
| Calcination | Limestone | 770 °C; 1 atm |
|  | Alumina | 550–600 °C, 1 atm |
| Granulation | Soap powders | 5 °C; 1 atm |
|  | Food; Pharmaceuticals | 20–40 °C; 2–3 atm |
| Coating | Food; Pharmaceuticals | 20–80 °C; 2–3 atm |
| Roasting | Food industry products | 200 °C; 1 atm |
|  | Sulfide ores ($FeS_2$) | 650–1100 °C; 1 atm |
| Synthesis Reactions | Phthalic anhydride | 340–380 °C; 2.7 atm |
|  | Acrylonitrile | 400–500 °C; 1.5–3 atm |
|  | Ethylene dichloride | 260–310 °C; 1–10 atm |
|  | Maleic anhydride | 400–500 °C; 4 atm |
|  | Polyethylene (low/ high density) | 75–120 °C; 15–30 atm |
| Thermal cracking | Ethylene; propylene | 750 °C; 1 atm |
| *Some other applications* |  |  |
| Combustion and incineration of waste solid |  | 800–900 °C; 1–10 atm |
| Gasification of coal and coke/ solid waste |  | 800 °C; 1–10 atm |
| Biofluidisation, cultivation of micro-organism |  | 30 °C; 1 atm |
| Semiconductor industries |  | 300–1100 °C; 1 atm |

fluidised bed reactors for the treatment of clean biomass and urban waste in order to produce a clean syngas from which electric energy can subsequently be obtained. Applications where mass and heat transfer are the key mechanisms include processes such as granulation, drying and mixing.

It has been shown earlier that the fluidisation properties of a powder in air at ambient conditions may be predicted by establishing in which Geldart group it lies. Table 7.8 shows that most fluidised bed industrial processes are operated at temperatures and pressures well above ambient. It is important to note that at operating temperatures and pressures above ambient a powder may appear in a different group from that which it occupies at ambient conditions.[26] This is due to the effect of gas properties on the grouping and may have serious implications as far as the operation of the particle system is concerned.

Reliable design of commercial scale particle systems requires a good understanding of the highly complex flow phenomena and also requires detailed knowledge of how the hydrodynamics are affected by both reactor design and plant scale-up. Factors such as powder characteristics (*i.e.* density, surface properties, fines content), operating conditions (*i.e.* temperature, pressure, velocity) and reactor design (*i.e.* cooling coils, dip-legs, grid), plus any special conditions, such as the presence of liquids and other additives in the reactor, can significantly affect the gas–solid contacting efficiency. In light of this and of its wide range of industrial applications, particle science and technology remains a complex and challenging area for scientific research.

## 7.5 Worked Examples

### 7.5.1 Example E1

Table 7.9 shows the results of the sieve analysis obtained for a sample of a FCC1 powder. The values reported in the first two columns of the table are the standard diameters of the sieve apertures. From these the mean diameter $d_p$ is obtained for each two adjacent sieve sizes and the values are reported in the third column. From the mass fraction of powder in each sieve (values in the fourth column) the weight percentage is obtained and reported in the fifth column. Thus, the sum of the mass fraction over the mean diameter allows the calculation of the *volume-surface* mean particle diameter of the distribution using eqn (7.1):

$$d_{VS} = \frac{1}{\sum x_i/d_p} \tag{7.1}$$

Plotting the weight percentage *versus* the mean particle diameter $d_p$ allows the *size distribution* of the powder to be obtained, as shown in Figure 7.2. The relative diameter spread $\sigma/d_{50}$ is then calculated to obtain an indication of the width of the size distribution of the sample under analysis. The diameter spread, $\sigma_d$, is given by:

$$\sigma_d = \frac{d_{84\%} - d_{16\%}}{2} \tag{7.3}$$

and the median size diameter, $d_{50}$, are determined by plotting the cumulative size fraction, reported in Table 7.9, *versus* particle size, see

**Table 7.9** Worked example E1: calculation of mean particle diameter.

| Min particle diameter | Max particle diameter | Mean particle diameter | Mass fraction | % Mass fraction in sieve | Mass fraction/ mean diameter | Cumulative mass fraction |
|---|---|---|---|---|---|---|
| 0 | 38 | 19 | 0.015 | 1.5 | 7.89E-04 | 1.5 |
| 38 | 45 | 41.5 | 0.027 | 2.7 | 6.51E-04 | 4.2 |
| 45 | 53 | 49 | 0.086 | 8.6 | 1.76E-03 | 12.8 |
| 53 | 63 | 58 | 0.157 | 15.7 | 2.71E-03 | 28.5 |
| 63 | 75 | 69 | 0.204 | 20.4 | 2.96E-03 | 48.9 |
| 75 | 90 | 82.5 | 0.201 | 19.8 | 2.44E-03 | 69.0 |
| 90 | 106 | 98 | 0.132 | 13.2 | 1.35E-03 | 82.2 |
| 106 | 125 | 115.5 | 0.085 | 8.5 | 7.36E-04 | 90.7 |
| 125 | 150 | 137.5 | 0.050 | 5.0 | 3.64E-04 | 95.7 |
| 150 | 180 | 165 | 0.025 | 2.5 | 1.52E-04 | 98.2 |
| 180 | 212 | 196 | 0.011 | 1.1 | 5.61E-05 | 99.3 |
| 212 | 250 | 231 | 0.004 | 0.4 | 1.73E-05 | 99.7 |
| 250 | 300 | 275 | 0.001 | 0.1 | 3.64E-06 | 99.8 |
| 300 | 355 | 327.5 | 0.001 | 0.1 | 3.05E-06 | 99.9 |
| 355 | 425 | 390 | 0.001 | 0.1 | 2.56E-06 | 100 |
| 425 | 500 | 462.5 | 0.001 | 0.1 | 2.16E-06 | 100 |
| 500 | 600 | 550 | 0.000 | 0.0 | 0.00E-00 | 100 |
| Total | | | | 100.0 | 1.40E-02 | |

Mean particle diameter (μm)          71.54

Mean particle diameter at 16%  84% of cumulative mass fraction

                                  51          102

Diameter spread          26
  (μm)

Median particle          69
  diameter (μm)
(at 50% of the
  cumulative mass
  fraction)

Particle size          0.37
  distribution

also Figure 7.19. This gives a value for $\sigma_d/d_{50} = 0.37$, indicating a wide particle size distribution according to Table 7.2.

## 7.5.2 Example E2

A non-spherical FCC catalyst particle falls in a column of nitrogen at ambient conditions and attains a terminal fall velocity $u_t$. The density and viscosity of the nitrogen are $\rho_f = 1.2 \times 10^{-3}$ g cm$^{-3}$ and

**Figure 7.19**   Worked example E1: cumulative percentage under size *versus* particle size.

$\mu = 1.8 \times 10^{-4}$ g cm$^{-1}$ s$^{-1}$ respectively. Calculate the terminal velocity of the particle given its physical characteristics, $d_p = 70$ μm, sphericity $\phi = 0.7$ and $\rho_p = 1.40$ g cm$^{-3}$, using a graphical solution method based on eqn (7.16) and (7.17).

*Solution:*
First, calculate $d_p^*$ from eqn (7.16):

$$d_p^* = d_p \left[ \frac{\rho_f(\rho_p - \rho_f)g}{\mu^2} \right]^{1/3} = 0.0070 \left[ \frac{0.0012(1.4 - 0.0012)980}{(0.00018)^2} \right]^{1/3} = 2.6$$

Next, we find $u_t^*$ from the empirical diagram reported in Figure 7.7, shown again here as Figure 7.20, choosing the curve that corresponds to $\phi = 0.7$. This gives:

$$u_t^* = 0.28$$

Finally, from eqn (7.17), we obtain $u_t$:

$$u_t = u_t^* \left[ \frac{\mu(\rho_p - \rho_f)g}{\rho_f^2} \right]^{1/3} = 0.28 \left[ \frac{0.00018(1.4 - 0.0012)980}{(0.0012)^2} \right]^{1/3} = 15.5 \text{ cm s}^{-1}$$

## 7.5.3   Example E3

A *fine* spherical silica-based catalyst with the physical characteristics reported below is being investigated for the development of a new fluidised bed process. Fundamental parameters such

**Figure 7.20**  Worked example E2: empirical diagram for the calculation of the particle terminal fall velocity.

as the bed voidage and minimum fluidisation velocity, $u_{mf}$, have to be determined.

$d_p$ = particle diameter = 75 μm
$\rho_p$ = particle density = 1770 kg m$^{-3}$
$\rho_f$ = fluid density = 1.22 kg m$^{-3}$
$\mu$ = fluid viscosity = 1.8 × 10$^{-5}$ kg m$^{-1}$ s$^{-1}$

(i) Calculate the voidage of the bed, $\varepsilon$, (volume fraction occupied by the voids) when the packed bed of solid particles occupies a depth $L = 0.5$ m in a vessel of diameter $D = 0.14$ m. The mass of particle is 8.5 kg. Knowing the voidage, determine the number of particles making up the bed, assuming that they are perfectly spherical.

(ii) Determine $u_{mf}$ using the Ergun equation for pressure drop through packed beds of spherical particles. Different values of the voidage at minimum fluidisation ($\varepsilon_{mf} = 0.38, 0.42, 0.45$) are given in order to estimate the sensitivity of the calculation of $u_{mf}$ on $\varepsilon_{mf}$ when using the Ergun equation. Determine $u_{mf}$ also using the Wen–Yu equation.

*Solution:*

(i) $\varepsilon$ = volume occupied by the gas/total volume of bed

$$\varepsilon = (V_{bed} - V_{particle})/V_{bed} = 1 - V_{particle}/V_{bed}$$
$$\varepsilon = 1 - M/(\rho_p \, L \, A) = 1 - M/(\rho_p \, L\pi D^2/4)$$

Substituting the values we obtain:

$$\varepsilon = 0.38$$

Knowing the voidage, the number of particles can be determined. The total volume occupied by the particles in the bed is equal to:

$$V = AL(1 - \varepsilon)$$

which is also equal to the number of particles multiplied by the volume of a single particle, thus:

$$AL(1 - \varepsilon) = N_p \pi \frac{d_p^3}{6}$$

from which:

$$N_p = 2.2E + 10$$

(ii) The Ergun equation for pressure drop through packed beds of spherical particles (eqn (7.32)) may be applied to give $u_{mf}$ as follows:

$$\text{Ar} = \frac{150(1 - \varepsilon_{mf})}{\varepsilon_{mf}^3} \text{Re}_{mf} + \frac{1.75}{\varepsilon_{mf}^3} \text{Re}_{mf}^2 \qquad (7.32)$$

where:

$$\text{Re}_{mf} = \frac{\rho_f u_{mf} d_p}{\mu}; \quad \text{Ar} = \frac{d_p^3 \rho_f (\rho_p - \rho_f)g}{\mu^2}$$

The viscous effects dominate the fluid–particle interaction of small particles (below 100 μm), thus the inertial term of the Ergun equation can be neglected. Hence, the minimum fluidisation velocity can be obtained from:

$$u_{mf} = \frac{\varepsilon_{mf}^3 (\rho_p - \rho_f) d_p^2 g}{150(1 - \varepsilon_{mf})\mu} \qquad (7.33)$$

By substituting the given numerical values, $u_{mf}$ can be obtained for the different values of the voidage:

| $\varepsilon_{mf}$ | $u_{mf}$ $(\text{cm s}^{-1})$ |
|---|---|
| 0.38 | 0.32 |
| 0.42 | 0.46 |
| 0.45 | 0.6 |

For fine spherical particles, when using the equation proposed by Wen–Yu, $u_{mf}$ is obtained from the viscous term of eqn (7.35):

$$u_{mf} = \frac{d_p^2(\rho_p - \rho_f)g}{1650\mu} = 0.33 \text{ cm s}^{-1}$$

The results show that the values calculated using the Ergun equation are very sensitive to the value used for the voidage at minimum fluidisation. The value obtained from the Wen–Yu equation corresponds to that obtained from the Ergun eqn (7.25) with a voidage of approximately 0.38, as reported in Section 7.3.2.

## 7.5.4 Example E4

As part of the fluid-bed process development described in E3, the operative conditions and the corresponding fluidisation regime have to be determined.

Calculate the superficial gas velocity which is needed to expand the bed and obtain an average fluid-bed voidage $\varepsilon = 0.62$. Consider viscous flow regime. Use Figure 7.11 to check what fluidisation regime corresponds to the calculated superficial gas velocity.

Find the flow regime at which the fluid-bed would be operated if fluidised at 0.25 m s$^{-1}$ and 1.5 m s$^{-1}$.

*Solution:*

The Richardson–Zaki equation (eqn (7.37)) is used in fluidisation to describe the homogeneous expansion. It relates the superficial gas velocity to the fluid bed voidage and particle terminal fall velocity:

$$u = u_t\varepsilon^n \tag{7.37}$$

where for viscous flow regime, $n = 4.8$.

Knowing the particle terminal fall velocity, $u_t$, the superficial gas velocity which is needed to expand the bed and obtain an average fluid-bed voidage $\varepsilon = 0.62$ can be determined. Using a similar procedure to that adopted in E2, we can use Figure 7.11 (repeated again here as Figure 7.21) to obtain the value for $u_t$, as follows:

From the data given, we can calculate $d_p{}^*$

$d_p$ = particle diameter = 75 μm
$\rho_p$ = particle density = 1770 kg m$^{-3}$
$\rho_f$ = fluid density = 1.22 kg m$^{-3}$
$\mu$ = fluid viscosity = $1.8 \times 10^{-5}$ kg m$^{-1}$ s$^{-1}$

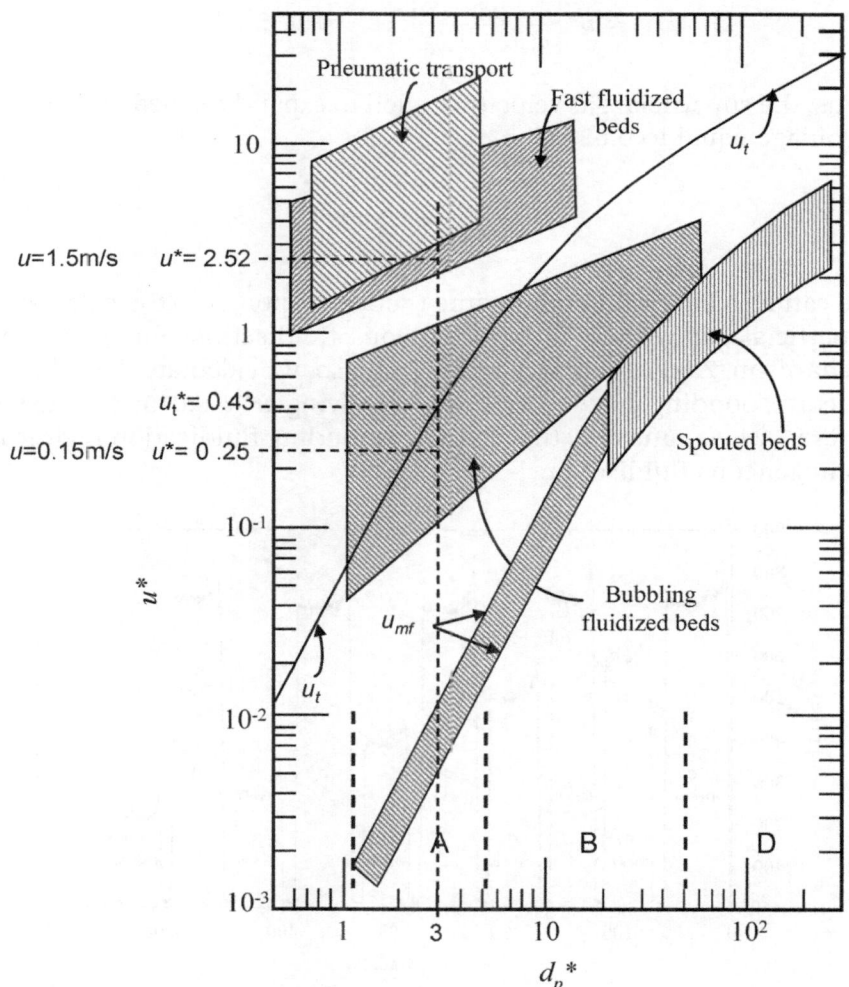

**Figure 7.21** Worked example E4: empirical diagram for the prediction of the fluidization regime.

$$d_p{}^* = \mathrm{Ar}^{1/3} = d_p \left[ \frac{\rho_f(\rho_p - \rho_f)g}{\mu^2} \right]^{1/3} = 3$$

Using the diagram in Figure 7.22, the intersect of the $u_t$ curve with the coordinate corresponding to the calculated $d_p{}^*$ gives a value for $u_t{}^* = 0.43$.

From $u_t{}^*$ we can therefore back-calculate a value for the terminal fall velocity, $u_t$:

$$u_t = u_t{}^* \left[ \frac{\rho_f^2}{\mu(\rho_p - \rho_f)g} \right]^{-1/3} = 0.26 \text{ m s}^{-1}$$

Thus, the superficial gas velocity needed to expand the bed and obtain a voidage equal to 0.62 is:

$$u = u_t \varepsilon^{4.8} = 0.026 \text{ m s}^{-1}$$

We can also check that the regime that corresponds to the calculated superficial gas velocity is homogeneous fluidisation, for which the Richardson–Zaki equation is valid. To do so, we calculate the value of $u^*$ corresponding to $u = 0.026$ m s$^{-1}$, giving $u^* = 0.015$, and using Figure 7.21 we can verify that the corresponding fluidisation regime is homogeneous fluidisation.

**Figure 7.22**  Worked example E5: experimental shear stress chart ($\sigma_c = 1106.2$ Pa).

Using Figure 7.21, the fluidisation regime corresponding to $0.30$ m s$^{-1}$ and $1.5$ m s$^{-1}$ can also be determined:

$u = 0.15$ m s$^{-1}$ $\quad u^* = 0.25$ $i e.$ Bubbling fluidisation
$u = 1.5$ m s$^{-1}$ $\quad u^* = 2.5$ $i.e.$ Fast-turbulent fluidisation

## 7.5.5 Example E5

The following example reports the experimental results obtained for the effective angle of internal friction and cohesion for three samples of powders.

Two alumina materials, A1 and A2 belonging to Geldart Group A, and ballotini belonging to Geldart group B (see Section 7.4.3) are examined. The physical characteristics of the three materials are summarised in Table 7.10. A1 is virtually free of fines (particles having size below 45 mm), while A2 contains 30% wt of fine particles. In light of this, the alumina sample A1 is expected to be more cohesive than A2. As reported in Section 7.4.3, the ballotini are instead expected to be free from any dominant effect of the interparticle forces.

For each powder, four static yield loci are obtained by applying the normal loads reported in Table 7.11, to which the corresponding consolidation stresses, $\sigma_c$, are calculated from eqn (7.41). Following the procedure described in Section 7.4.4.7, the values of the normal stresses $\sigma_i$ applied for each consolidation stress $\sigma_c$, are reported in Table 7.12 and also shown in Figure 7.22.

**Table 7.10** Worked example E5: material physical characteristics.

|  | $d_p$ ($\mu$m) | $F_{45}$ (%) | $\rho_p$ (kg m$^{-3}$) |
|---|---|---|---|
| A1 | 75 | 3.2 | 1730 |
| A2 | 49 | 30 | 1730 |
| Ballotini | 300 | — | 2500 |

**Table 7.11** Worked example E5: Consolidation stresses in internal yield loci experiments.

| Normal load ($N$:) | Normal stress $\sigma_c$ (Pa) |
|---|---|
| 300 | 1106.2 |
| 210 | 804.1 |
| 150 | 602.6 |
| 90 | 401.2 |

**Table 7.12**  Worked example E5: Shearing stresses in internal yield loci experiments.

| $\sigma_c = 1106.2$ Pa | $\sigma_c = 804.1$ Pa | $\sigma_c = 602.6$ Pa | $\sigma_c = 401.2$ Pa |
|---|---|---|---|
| $\sigma_i$ (Pa) | $\sigma_i$ (Pa) | $\sigma_i$ (Pa) | $\sigma_i$ (Pa) |
| 904.8 | 602.6 | 401.2 | 300.5 |
| 703.4 | 401.2 | 300.5 | 250.1 |
| 501.9 | 300.5 | 199.8 | 199.8 |
| 300.5 | 199.8 | 149.4 | 149.4 |
| 199.8 | 149.4 | — | — |

**Figure 7.23**  Worked example E5: experimental static yield locus – Alumina A2, $\sigma_c = 1106.2$ Pa.

Figure 7.22 shows a typical experimental shear stress chart, obtained for A2, using a compaction with normal stress equal to 1106.2 Pa. Figure 7.23 shows the corresponding yield locus from which the static angle of internal friction, $\varphi$, was worked out from the slope of the yield locus. The cohesion, $C$, is obtained from the intercept with the shear stress axis.

As outlined in Section 7.4.4.5, the static angle of internal friction and the cohesion of a granular material are a function of the consolidation stress. Therefore, they can be expressed also as a function of the major principal stress $\sigma_1$. Table 7.13 reports the values obtained for the effective angle of internal friction and the cohesion for all the powders investigated. Sample A2, which contains a large amount of fines exhibits the highest values for cohesion and effective angle of

**Table 7.13** Worked example E5: failure properties of the powders.

|  | Alumina A1 | Alumina A2 | Ballotini |
|---|---|---|---|
| C (Pa) ($\sigma_1 = 700$ Pa) | 98 | 160 | 24 |
| $\varphi_e - \varphi$ (°) ($\sigma_1 = 700$ Pa) | 22 | 25 | 6 |

internal friction. The ballotini in contrast, are less cohesive, as reflected in the lowest values for both $C$ and $\varphi_e - \varphi$.

## 7.6 Concluding Remarks

In this chapter we have introduced some of the fundamental parameters which characterise solid materials, including particle size analysis, shape, density and shear stress, and how they influence the flow properties of powders. The description of the fluid–particle interaction, for both the case of a single particle suspended in a fluid and when flow passes through a packed bed, has also been presented. We have introduced the concept of fluidisation and the principle forces acting on a particle under equilibrium conditions. We have also shown how particle characteristics influence the fluidisation behaviour and presented some industrial applications. A number of relevant worked examples have also been presented to provide the reader with the direct application of the theory.

## Nomenclature

| | | |
|---|---|---|
| $A$ | Cross-sectional area of the bed | $m^2$ |
| Ar | Archimedes number | — |
| $C$ | Cohesion | Pa |
| $C_\delta$ | Cohesion from dynamic yield locus | Pa |
| $C_D$ | Drag coefficient | — |
| $d_p$ | Mean particle diameter | m |
| $d_{pi}$ | Sieve aperture | m |
| $d_s$ | Surface diameter | m |
| $d_v$ | Volume diameter | m |
| $d_{sv}$ | Surface/volume diameter | m |
| $d_p*$ | Dimensionless particle diameter eqn (7.16) | — |
| D | Bed diameter | m |
| $F$ | Friction factor | — |
| $F_b$ | Buoyancy force | N |

| $F_d$ | Drag force on a single particle | N |
| $F_{45}$ | Fraction of fine particles below 45 μm | |
| $g$ | Acceleration due to gravity | m s$^{-2}$ |
| $K$ | Permeability | Pa s$^{-1}$ |
| $H$ | Bed height | m |
| $L$ | Length of bed | m |
| $m$ | Mass of a single particle | kg |
| $n$ | Richardson–Zaki parameter eqn (7.38) | — |
| $N_c$ | Pre-consolidation load | kg |
| $N_i$ | Normal load in shearing phase i | kg |
| $N_p$ | Number of particles in the bed | — |
| Re | Reynolds number | — |
| Re$_{mf}$ | Reynolds number at minimum fluidisation | — |
| Re$_p$ | Particle Reynolds number | — |
| Re$_t$ | Particle terminal Reynolds number | — |
| $U$ | volumetric flux | m$^3$ s$^{-1}$ m$^{-2}$ |
| $u$ | Superficial gas velocity | m s$^{-1}$ |
| $u_f$ | Fluid phase velocity | m s$^{-1}$ |
| $u_{mf}$ | Minimum fluidisation velocity | m s$^{-1}$ |
| $u_{mb}$ | Minimum bubbling velocity | m s$^{-1}$ |
| $u_t$ | Particle terminal fall velocity | m s$^{-1}$ |
| $u_t^*$ | Dimensionless particle terminal fall velocity eqn (7.17) | — |
| $V_p$ | Volume of particle | m$^3$ |
| $x_i$ | Mass fraction of particles in each size range | — |

**Greek symbols**

| $\delta$ | Dynamic angle of internal friction | Deg |
| $\varepsilon$ | Bed voidage | — |
| $\varepsilon_{mf}$ | Minimum fluidisation voidage | — |
| $\phi$ | Particle sphericity | — |
| $\phi_i$ | Angle of internal friction | Deg |
| $\varphi_e$ | Effective angle of internal friction | Deg |
| $\mu$ | Gas viscosity | Ns m$^{-3}$ |
| $\rho_{BDL}$ | Loose bulk density | kg m$^{-3}$ |
| $\rho_{BDP}$ | Packed bulk density | kg m$^{-3}$ |
| $\rho_{ABS}$ | Absolute density | kg m$^{-3}$ |
| $\rho_b$ | Bulk density | kg m$^{-3}$ |
| $\rho_g$ | Gas density | kg m$^{-3}$ |
| $\rho_f$ | Fluid density | kg m$^{-3}$ |
| $\rho_p$ | Particle solid density | kg m$^{-3}$ |
| $\Delta P$ | Bed pressure drop | Pa |

| $\sigma$ | Normal stress | $N\ m^{-2}$ |
|---|---|---|
| $\sigma_c$ | Consolidation stress | $N\ m^{-2}$ |
| $\sigma_d$ | Mean deviation | m |
| $\sigma_d/d_{50}$ | Relative diameter spread | — |
| $\sigma_1, \sigma_2$ | Major stress | $N\ m^{-2}$ |
| $\sigma_i, \sigma_{ii}$ | Shearing stress | $N\ m^{-2}$ |
| $\tau$ | Shear stress | $N\ m^{-2}$ |

# References

1. T. Allen, *Particle Size Measurements*, Chapman and Hall, London, 1996.
2. M. Rhodes, *Introduction to Particle Technology*, Wiley, London, 1998.
3. P. Lettieri, J. G. Yates and D. Newton, Homogeneous bed expansion of FCC catalysts, influence of temperature on the Richardson–Zaki index n and on the particle terminal fall velocity, *Powder Technol.*, 2002, **123**, 221–231.
4. D. Geldart, *Gas Fluidization Technology*, Wiley, London, 1986.
5. J. P. K. Seville, U. Tuzun and R. Clift, *Processing of Particulate Solids*, Chapman & Hall, London, 1997.
6. B. S. Massey, *Mechanics of Fluids*, Van Nostrand Reinhold, London, 6th edn, 1989.
7. J. M. Dallavalle, *Micromeritics*, Pitman, New York, 1948.
8. A. Haider and O. Levenspiel, Drag coefficient and terminal velocity of spherical and nonspherical particles, *Powder Technol.*, 1989, **58**, 63–70.
9. H. P. G. Darcy, *Les Fontaines Publiques de la Ville de Dijon*, Victor Dalmont, Paris, 1856.
10. L. G. Gibilaro, *Fluidization Dynamics*, Butterworth Heinemann, Oxford, 2001.
11. S. Ergun, Fluid flow through packed columns, *Chem. Eng. Prog.*, 1952, **48**(2), 89–94.
12. C. Y. Wen and Y. H. Yu, Mechanics of fluidization, *Chem. Eng. Prog., Symp. Ser.*, 1966, **62**, 100–111.
13. J. F. Richardson and W. N. Zaki, Sedimentation and Fluidization: Part I, *Trans. Inst. Chem. Eng.*, 1954, **32**, 35–53.
14. A. R. Abrahamsen and D. Geldart, Behaviour of gas-fluidized beds of fine powders Part I. Homogeneous expansion, *Powder Technol.*, 1980, **26**, 35–46.

15. J. R. Grace, Contacting modes and behaviour classification of gas–solid and other two-phase suspensions, *Can. J. Chem. Eng.*, 1986, **64**, 353–363.
16. D. Geldart, Types of fluidization, *Powder Technol.*, 1973, 7, 285.
17. J. P. K. Seville, C. D. Willett and P. C. Knight, Interparticle forces in fluidization: a review, *Powder Technol.*, 2000, **113**, 261–268.
18. J. Israelachvili, *Intermolecular and Surface Forces*, Academic Press, London, 1991.
19. R. L. Carr, Particle behaviour storage and flow, *Br. Chem. Eng.*, 1970, **15**(12), 1541–1549.
20. D. Geldart and A. C. Y. Wong, Fluidization of powders showing degrees of cohesiveness – I. Bed expansion, *Chem. Eng. Sci.*, 1984, **39**(10), 1481–1488.
21. J. Schwedes, Consolidation and flow of cohesive bulk solids, *Chem. Eng. Sci.*, 2002, **57**(2), 287–294.
22. A. E. Jenike, Gravity Flow of Bulk Solids, *Utah Eng. Exp. Stn., Bull.*, 1961, **108**.
23. A. E. Jenike, Storage and Flow of Solids, *Utah Eng. Exp. Stn., Bull.*, 1964, **123**.
24. G. Bruni, D. Barletta, M. Poletto and P. Lettieri, A rheological model for the flowability of fine aerated powders, *Chem. Eng. Sci.*, 2007, **62**, 397–407.
25. D. Kunii and O. Levenspiel, *Fluidization Engineering*, Butterworths, London, 1991.
26. J. G. Yates, Effects of temperature and pressure on gas–solid fluidization, *Chem. Eng. Sci.*, 1996, **51**(2), 167–205.

# 8   An Introduction to Process Control

Eva Sorensen

## 8.1   Introduction

All chemical plants need process control to ensure that they are operated safely and profitably, while at the same time satisfying product quality and environmental requirements. Furthermore, modern plants are becoming more difficult to operate due to the trend towards complex and highly integrated processes. For such plants, it is difficult to prevent disturbances from propagating from one unit to other, interconnected, units. The subject of process control has therefore become increasingly important in recent years and it is vital for anyone working on a chemical plant to have an understanding of both the basic theory and practice of process control.

This chapter will give a brief summary of basic control theory as applied in most chemical processing plants. It will start with a discussion of process dynamics and why an understanding of the dynamics, or *the transient behaviour*, of a process is essential in order to achieve satisfactory control. Standard feedback controllers will then be discussed and different strategies for tuning such controllers will be presented. A brief overview of more advanced control strategies, such as feed-forward control, cascade control, inferential control and adaptive control, and for which processes these may be

Concepts of Chemical Engineering for Chemists: 2nd Edition
Edited by Stefaan Simons
© The Royal Society of Chemistry 2017
Published by the Royal Society of Chemistry, www.rsc.org

suitable, will be given next. Finally, an introduction will be given to plant-wide control issues.

In most process control courses for chemical engineers, the first part of the course normally deals with the development of dynamic process models from first principles (mass and energy balances), since these are used in the analysis of process dynamics and often also for controller tuning. In this chapter, however, the focus will be on process *control* and modelling will not be considered. Neither will the chapter consider frequency response techniques for process analysis and controller tuning, as these are considered beyond the scope of this book. The emphasis will therefore be on control *concepts* and the reader is referred to standard control textbooks for more information.[1-4] In addition, although instrumentation is essential for control, it is a complete subject area in itself and is therefore not considered here.

## 8.2 Process Dynamics

The primary objective of process control is to maintain a process at the desired operating conditions, safely and efficiently, while satisfying environmental and product quality requirements. The subject of process control is concerned with how to achieve these goals. Luyben[3] gives the following process control laws:

> *First law: The simplest control system that will do the job is the best. Bigger is definitely not better in process control.*
> *Second law: You must understand the process before you can control it. Ignorance about the process fundamentals cannot be overcome by sophisticated controllers.*

The process control system should ensure that the process is maintained at its specified operating conditions at all times. To be able to do this, we must first understand the process dynamics as advised by Luyben in his Second Law.[3] The dynamics of a process tells us how the process behaves as a result of the changes. Without an understanding of the dynamics, it is not possible to design an appropriate control system for it.

In the following text, important aspects in the study of process dynamics are outlined. An example of a dynamic process is given first. *Stability* of a process is defined next, followed by a discussion of typical uncontrolled, or *open loop*, responses.

## 8.2.1 Process Dynamics Example

As an example of a dynamic process, consider the process in Figure 8.1, which is a tank into which an incompressible (constant density) liquid is pumped at a variable feed rate $F_i$ ($m^3$ $s^{-1}$). This inlet flow rate can vary with time because of changes in operations upstream of the tank. The height in the tank is $h$ (m) and the outlet flow rate is $F$ ($m^3$ $s^{-1}$). Liquid leaves the tank at the base *via* a long horizontal pipe and discharges into another tank. Both tanks are open to the atmosphere. $F_i$, $h$ and $F$ can all vary with time and are therefore *functions of time t*.

Let us now consider the dynamics of this tank, starting with the *steady state* conditions. By steady state we mean the conditions when nothing changes with time. The value of a variable at steady state is normally denoted by a subscript s. (Note that some control textbooks use a different notation, for instance, by placing a bar over the variables.) At steady state, the flow out of the tank, $F_s$, must equal the flow into the tank, $F_{i,s}$, and the height of the liquid in the tank is constant, $h_s$. At steady state we therefore have the steady-state equation:

$$F_{i,s} = F_s$$

The value of $h_s$ is that height which provides enough hydraulic pressure head at the inlet of the pipe to overcome the frictional losses of liquid flowing out and down the pipe. The higher the flow rate $F_{i,s}$, the higher the height $h_s$.

Now consider what would happen dynamically if the inlet flow rate $F_i$ changed. How will the height $h$ and the outlet flow rate $F$ vary with time? Figure 8.2 is a sketch of the problem.

The problem is to determine which curves (1 or 2) represent the actual paths that $F$ and $h$ will follow after a step change in $F_i$. The curves marked "1" show gradual increases in $h$ and $F$ to their new steady-state values. However, the paths could follow the curves

**Figure 8.1** Gravity-flow tank.

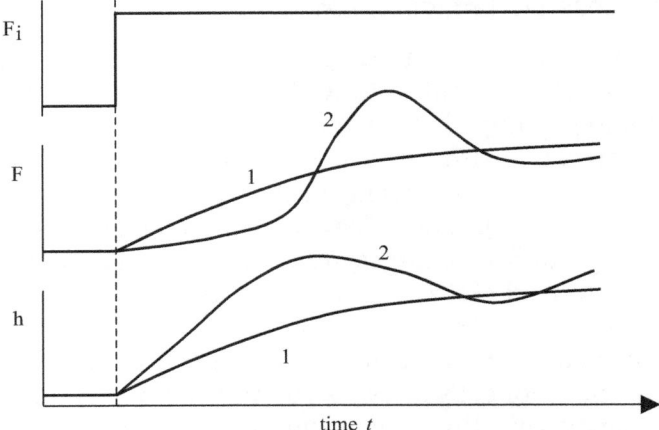

**Figure 8.2**   Dynamic responses of gravity-flow tank.

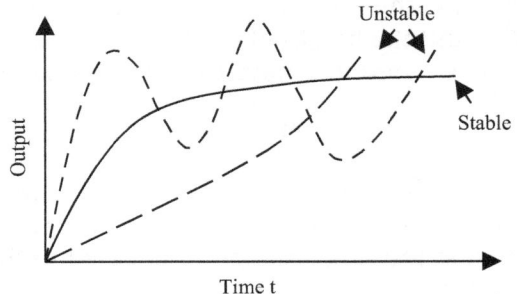

**Figure 8.3**   Stability of a process.

marked "2", where the liquid height rises above its final steady-state value. This is called *overshoot*. Clearly, if the peak of the overshoot in *h* is above the top of the tank, the tank would overflow. The steady-state calculations give no information about what the dynamic response of the system will be. Before a controller can be designed to control the tank height, the designer must determine which of the curves (1 or 2) the tank height *h* and the flow rate *F* are most likely to follow, *i.e. what are the dynamics of the process?*

## 8.2.2   Stability

A process is said to be unstable if its output becomes larger and larger (either positively or negatively) as time increases, as illustrated in Figure 8.3. Note that no real system really does this, as some constraint is usually encountered, for example, the level in a tank may overflow, a valve on a flow stream may completely shut or completely open or a safety valve may blow.

Most processes are *open loop stable*, also called *self-regulating, i.e.* stable without any controllers on the system. This means that after a change in the system caused by either a disturbance or by a deliberate change in a manipulated variable, the process will come to a new stable operating condition, *i.e.* a new steady state. This is not necessarily the desired operating condition, and in most cases it is not, but nevertheless, the new operating condition is stable. If, however, the process does not settle out to a new steady state, it is called non-self-regulating and will require control in order to ensure that the system remains safe and stable. In the tank example, as $F$ is free to vary with the liquid level in the tank, the process is self-regulating since if $F_i$ increases, the level will rise causing $F$ to increase, thereby bringing the system to a new steady state where $F$ rises to the new value of $F_i$. If, however, there is a pump on the outlet line, *i.e.* $F$ remains constant whatever the conditions within the tank, the system will be non-self-regulating as the tank would overflow if $F_i$ increased and $F$ remained constant.

What makes controller design challenging is that all real processes can be made *closed loop unstable* when a controller is implemented to steer the process to specified operating conditions. In other words, a process which is open loop stable and therefore will come to a new, although not the desired, steady state after a disturbance may become unstable when a controller is implemented to steer the process towards the desired steady state. Stability is therefore of vital concern in all control systems.

## 8.2.3   Typical Open Loop Responses

The term *process response* is used to describe the shape of the plot one would obtain if one plotted the controlled variable as a function of time after a disturbance or set point change. In the tank example, the response would typically be given by tank height $h$ as a function of time and would be referred to as "the response in $h$". Almost all chemical processes fall within a few standard response shapes, which makes analysing them easier. These standard responses are illustrated in Figure 8.4. The *order* of a process refers to the order of the differential equations describing the system, and refers loosely to the number of parts of the process in which material and energy can be contained. Simple processes can often be assumed to be of first order, while more complex processes are of second order or higher.

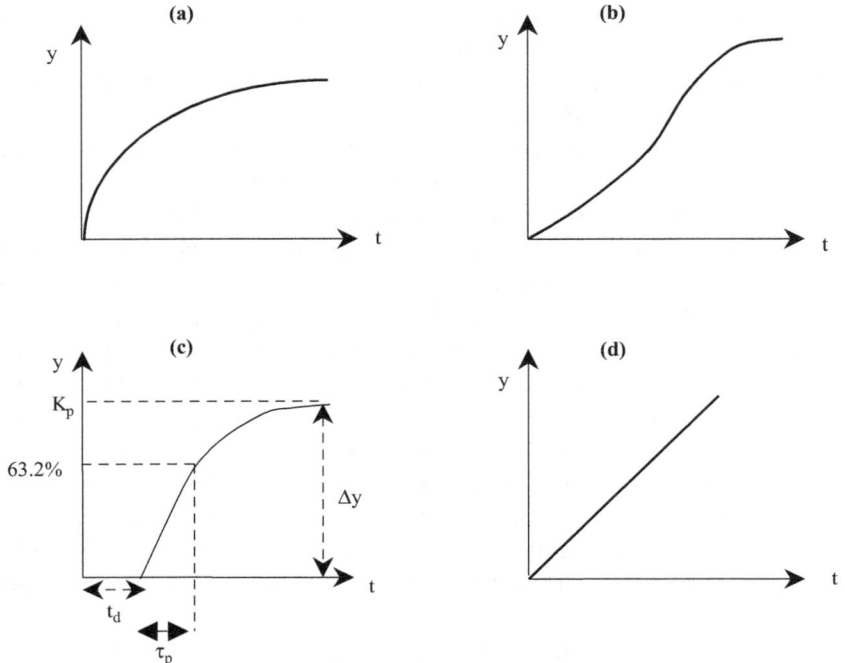

**Figure 8.4** Typical open-loop unit step responses. (a) First order, (b) Second (or higher) order, (c) first order with dead time and (d) non-self-regulating process.

In Figure 8.4, $y$ refers to the variable that is observed. For instance, in the tank example, $y$ is $h$.

*Dead time* is the length of time, $t_d$, after a disturbance occurs before any change in the output is noticed and is illustrated for a first order process in Figure 8.4(c). It occurs particularly when material or energy is being transported some distance. Clearly it means that disturbance detection is slow and control actions are made on the basis of old measurements. Dead time is the most troublesome characteristic of a system to control and can cause otherwise stable systems to become unstable when the control loop is closed.

The process gain, $K_p$, of the system is the value that the response will go to at steady state after a unit step change in the input, *i.e.* it relates the input to the output at steady state.

The *time constant*, $\tau_p$, of a process is, for a first-order system, defined as the time it takes for the response to reach 63.2% of the final response value (see Figure 8.4(c)). Theoretically, the process never reaches the new steady-state value, given by the process gain $K_p$, except at $t \to \infty$, although it reaches 99.3% of the final steady-state value when $t = 5\tau_p$. The difference between dead time and time constant must be emphasised; dead time is the time it takes before anything at all happens, while time constant is a measure of how slowly or quickly a response settles to its new steady-state value once the change has started. A slow process that has a large time constant does not necessarily have any dead time. Time constants in chemical engineering processes can range from less than a second to a couple of hours, or even days for biochemical applications.

## 8.3 Feedback Control Systems

Most plant control systems are very simple and are normally standard feedback controllers, either P-, PI- or PID controllers. These will be described in more detail in this section, together with two techniques that can be used to tune these controllers. First, however, we need to define the *control objective*, *i.e.* what do we want the controller to do, and define what we mean by *feedback control*.

### 8.3.1 Disturbance Rejection and Set Point Tracking

The operation of a process may deviate from its desired operating conditions for two different reasons:

*Disturbances*. These are changes in flow rates, compositions, temperatures, levels or pressures in the process which we can not control because they are either given by the feed conditions (assumed controlled by an upstream unit), ambient conditions, such as the weather, or utilities, such as steam or cooling water.

*Set point changes*. These are deliberate changes in the operating conditions, such as a change in polymer grade for a polymerisation reactor, change in distillate composition for a distillation column, *etc.*

The control objective is different in the two cases. In *disturbance rejection*, the control objective is to reject the disturbance as quickly as possibly, *i.e.* to bring the process back to the original steady state by counteracting the effect of the disturbance (see Figure 8.5(a)). As an example, consider a person in a shower who wants to maintain the water temperature at a constant value. To achieve this, the person may have to turn the cold water tap up to increase the flow rate of the cold water if there is a sudden increase in temperature as a result of someone else in the house using cold water, *i.e.* to fill a kettle in the kitchen. (This is an example of a disturbance caused by a shared utility, here the cold water.)

In *set point tracking*, the set point for the controller is changed and the control objective is to bring the process to the new set point as quickly as possible (see Figure 8.5(b)). For a person in the shower who wants to reduce the temperature at the end of the shower time, this means turning the hot water tap down to reduce the flow rate of hot water, thereby bringing the temperature down to the new desired level.

### 8.3.2 Feedback Control Loop

A feedback control loop is generally illustrated as shown in Figure 8.6. The *Process* refers to the chemical or physical process (the tank in the example earlier). The *Measuring Device* is used to measure the value of the variable that is to be controlled (the level indicator in the tank example or a thermometer in a shower). The *Control Element* (usually a flow control valve or the setting on a heater or cooler) changes the value of the *manipulated variable*. The manipulated variable is the variable used to change the *controlled variable* (in the tank example, the output flow rate is the manipulated variable used to change the level in the tank, which is the controlled variable).

The input to the *Feedback Controller* (P-, PI-, PID- or PD-controller, discussed later) is the difference (also called error) between the set point and the measured variable; hence the minus sign in the control

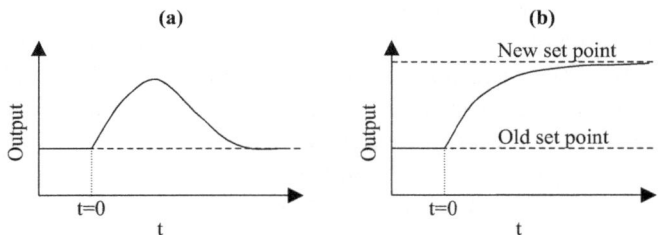

**Figure 8.5**   Process responses: (a) disturbance rejection and (b) set point tracking.

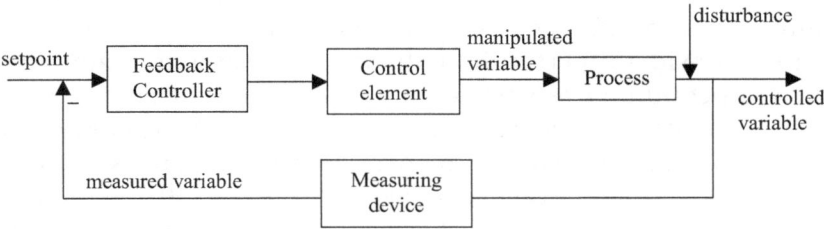

**Figure 8.6**   Feedback control loop.

loop. The controller has its name from the fact that the comparison between the controlled variable and the set point is *fed back* to the controller. The Feedback Controller determines the change required in the manipulated variable to bring the error back to zero, *i.e.* the controlled variable to its set point value, and sends a signal to the control element that executes this change. The effect of this change on the process, together with the effect of any disturbances, is then measured and compared to the set point and the loop started again.

Feedback control systems can be either *analogue*, where the controller is a mechanical device, or *digital*, where the controller is a computer or microprocessor. A digital control system will have additional elements in the control loop to convert the various signals from analogue to digital or *vice versa*.[1-11]

How often the control loop is executed is determined by the *sampling interval*, which is how often the measurement is taken. A critical pressure on an exothermic reactor may be sampled several times per second, while the level of a buffer tank may only be sampled a few times an hour. The sampling time must be chosen with care to ensure that possible changes in the process are detected early enough for the controller to take appropriate action. However, sampling too often is undesirable as it may upset the process and because a large number of data points would have to be sampled and stored.

### 8.3.3   P, PI and PID Controllers

The most commonly used controller in the process industries is the three term or PID controller. This controller is a feedback controller and adjusts the manipulated variable in proportion to the change in its output signal, $c$, from its steady state value (bias), $c_s$, on the basis of a measurement of the error in the controlled variable, $\varepsilon$, which is given by

$$\varepsilon(t) = y_{sp}(t) - y(t) \qquad (8.1)$$

where $\varepsilon$ is the error signal, $y_{SP}(t)$ is the set point and $y(t)$ is the measured value of the controlled variable, *e.g.* the height of the liquid in the tank example. The change required in the control signal $c$ is obtained from the error (calculated numerically in a digital controller or obtained by a mechanism inside an analogue controller) by the following relationship:

$$c = c_s + K_c \left( \varepsilon + \frac{1}{\tau_I} \int_0^t \varepsilon \, dt + \tau_D \frac{d\varepsilon}{dt} \right) \qquad (8.2)$$

where P is the Proportional, I is the Integral (Reset) and D is the Derivative (Preact).

The relationship involves three terms and three adjustable parameters, the *controller gain*, $K_c$, the *integral time*, $\tau_I$, and the *derivative time*, $\tau_D$ (hence the name). Finding the right values of these parameters for the best possible control action is called *tuning*. There are several techniques available for controller tuning as will be discussed later.

### 8.3.3.1 Proportional Action

Control action is proportional to the size of the error, and eqn (8.2) becomes:

$$c - c_s = K_c \varepsilon$$

That is, the change required in the manipulated variable is proportional to the error in the controlled variable. Some control literature refer to the *proportional band* which is defined as $100/K_c$. Some *offset* is associated with this control action, as will be shown later.

### 8.3.3.2 Integral Action

Control action is proportional to the sum, or integral, of all previous errors. This controller eliminates offset. Some control textbooks refer to the *reset rate*, which is defined as $1/\tau_I$.

### 8.3.3.3 Derivative Action

Control action is proportional to the rate of change of the error and so anticipates what the error will be in the immediate future.

### 8.3.3.4 Composite Control Action

Composite control actions are also possible, *i.e.* P, PI or PD, as well as PID. The most common is PI as P alone will have an offset and the derivative action in PID or PD can introduce unnecessary instability in the response.

## 8.3.4 Closed Loop Responses

When a controller is implemented on a process, the response after either a disturbance or a set point change is called the *closed loop response* as the control loop in Figure 8.6 is now closed (recall that open loop means without controller). Figure 8.7 shows typical closed loop responses of a first-order process to a unit step change in the input. Figure 8.7(a) and (b) show how Proportional control introduces some offset from the final desired steady state for both set point tracking and disturbance rejection after a change at $t = 0$. The size of the offset depends on the gain of the process $K_p$ and on the proportional gain of the controller $K_c$ and is, for a first-order process, given by:

$$\text{Offset} = \frac{1}{1 + K_p K_c}$$

The offset decreases as $K_c$ becomes larger and should theoretically go towards zero for infinitely large controller gains. However, using very large controller gains should generally be avoided as it may lead to unstable control if any deadtime, however small, is present in the system.

Figure 8.7(c) shows that Integral action removes offset but at the expense of introducing some oscillation. The level of oscillation depends on the control parameters. Systems with significant amounts of deadtime will always result in some oscillation in the closed loop response. Figure 8.7(d) shows how incorrect controller tuning may

**Figure 8.7**    Closed loop responses of first-order process. (a) P only (set point tracking), (b) P only (disturbance rejection), (c) stable PI or PID (disturbance rejection) and (d) unstable PI or PID (disturbance rejection).

lead to an unstable response whereby the changes made in the manipulated variable by the controller are too large, causing the measured variable to fluctuate more and more (for the example with a person in the shower, if the person makes a large change in cold water flow rate to compensate for the loss of cold water when the kettle is being filled in the kitchen, the water temperature may become too low. The person may then compensate with a large decrease in cold water flow rate causing the water to become too hot, *etc.*).

### 8.3.5   Controller Tuning

How do we choose the values of the controller parameters $K_c$, $\tau_1$ and $\tau_D$? They must be chosen to ensure that the response of the controlled variable remains stable and returns to its steady-state value (disturbance rejection), or moves to a new desired value (set point tracking), quickly. However, the action of the controller tends to introduce oscillations. How quickly the controller responds and with how little oscillation, depends on the application. The following are some of the available tuning methods, of which we will consider the last two in detail (see textbooks[1-11] for numerous other methods):

    (i) trial and error (not recommended);
   (ii) theoretical methods (frequency response methods, see text books);
  (iii) continuous cycling (or Ziegler–Nichols) method;
  (iv) process reaction curve (or Cohen–Coon) method.

#### 8.3.5.1   *Continuous Cycling Method (Ziegler–Nichols Tuning)*

In the *continuous cycling method*, the system (process with controller) is brought to the edge of stability under Proportional control only. Suitable values of the parameters can then be determined from the proportional gain $K_c$ found at that condition. The procedure is as follows:

    (i) close the feedback loop;
   (ii) turn on proportional action only (equivalent to setting $\tau_I = \infty$, $\tau_D = 0$ for a PID controller);
  (iii) increase the controller gain, $K_c$, until the process starts to oscillate. Continue to slowly increase the gain until the cycles continue with constant amplitude, as shown in Figure 8.8 (called *standing oscillations*);

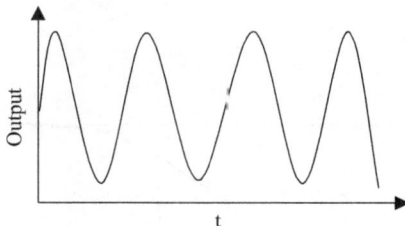

**Figure 8.8**   Standing oscillations in continuous cycling (Ziegler–Nichols) method.

**Table 8.1**   Continuous cycling (Ziegler–Nichols) tuning parameters.

|     | $K_c$ | $\tau_I$ | $\tau_D$ |
|-----|-------|----------|----------|
| P   | 0.5 $K_u$  | —          | —          |
| PI  | 0.45 $K_u$ | $P_u/1.2$  | —          |
| PID | 0.6 $K_u$  | $P_u/2$    | $P_u/8$    |

(iv) note the period of these cycles $P_u$ (distance in time between two peaks) and the value of $K_c$ at which they were obtained (called $K_u$);

(v) determine the controller settings according to the tuning (Ziegler–Nichols) rules in Table 8.1.

The settings obtained by this method are good initial estimates but are not optimal and some retuning may be necessary. Note that this method operates the system at the brink of instability and if the controller gain $K_c$ is chosen too high during the tuning procedure, the system will become unstable. The method is therefore not recommended for processes in which instability may lead to dangerous situations (*e.g.* runaway for a reactor).

### 8.3.5.2   *Process Reaction Curve Method (Cohen–Coon Tuning)*

For some processes, it may be difficult or hazardous to operate with continuous cycling, even for short periods. The *process reaction curve method* obtains settings based on the *open loop* response and thereby avoids the potential problem of closed loop instability. The procedure is as follows:

(i) disconnect the control loop between the controller and the manipulated variable (valve);

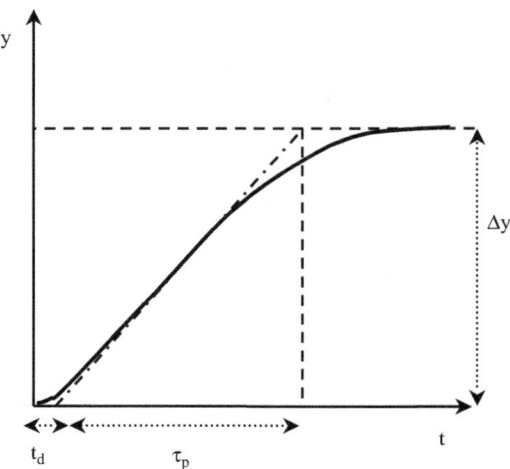

**Figure 8.9**  Process reaction curve (Cohen–Coon) method.

**Table 8.2**  Process reaction curve (Cohen–Coon) tuning parameters.

|      | $K_c$ | $\tau_I$ | $\tau_D$ |
|------|-------|----------|----------|
| P    | $\tau/Kt_d(1+t_d/3\tau)$ | — | — |
| PI   | $\tau/Kt_d(0.9+t_d/12\tau)$ | $t_d/(30+3t_d/\tau)(9+20t_d/\tau)$ | — |
| PID  | $\tau/Kt_d(1.33+t_d/4\tau)$ | $t_d/(32+6t_d/\tau),\ (13+8t_d/\tau)$ | $4t_d/(11+2t_d/\tau)$ |

   (ii) make a step change in the manipulated variable (valve opening);

  (iii) record the response of the process (should be similar to Figure 8.9);

  (iv) determine the controller settings according to the tuning (Cohen–Coon) rules in Table 8.2. The parameter $K$ is defined as

$$K = \frac{\Delta y}{\Delta c}$$

where $\Delta y$ is the final change in the output (the controlled variable) and $\Delta c$ is the initial change in the input (the manipulated variable). The tangent line is drawn as a tangent to the point of inflection. The dead time $t_d$ and process time constant $\tau_p$ are as shown in Figure 8.9.

    The settings obtained by this method are good initial estimates but are not optimal and some retuning may be necessary as for the Continuous Cycling method.

### 8.3.6 Advantages and Disadvantages of Feedback Controllers

Feedback control is the most important control technique and is widely used in the process industries. Its main advantages are as follows:

(i) Corrective action occurs as soon as the controlled variable deviates from the set point, regardless of the source and type of disturbance.

(ii) Feedback control requires minimal knowledge about the process to be controlled, in particular, a mathematical model of the process is not required, although having a model is very useful for control system design.

(iii) The PID controller is both versatile and robust. If process conditions change, re-tuning the controller usually produces satisfactory control.

However, feedback control also has certain inherent disadvantages:

(i) No corrective action is taken until a deviation in the control variable occurs. Thus, perfect control, where the controlled variable does not deviate from the set point during disturbance or set point changes, is theoretically impossible.

(ii) It does not provide predictive control action to compensate for the effects of known or measurable disturbances.

(iii) It may not be satisfactory for processes with large time constants and/or long time delays. If large and frequent disturbances occur, the process may be operated continuously in a transient state and never attain the desired steady state.

(iv) In some situations, the controlled variable cannot be measured on-line, and consequently, feedback control is not feasible.

## 8.4 Advanced Control Systems

Although widely used in the process industries, there are situations where feedback controllers are not sufficient to achieve good control. In this section, several more specialised strategies will be introduced that provide enhanced process control beyond that which can be obtained with conventional single-loop feedback controllers. As processing plants become more and more complex in order to increase efficiency or reduce costs, there are incentives for using such enhancements, which fall under the general classification of *advanced control*. This section introduces several different strategies that are widely used industrially and which in many cases utilise the principles of single-loop feedback controller design but with enhancements:

    (i) Feed-forward control
  (ii) Ratio control
 (iii) Cascade control
 (iv) Inferential control
  (v) Adaptive control

## 8.4.1  Feedforward Control

In the previous section, it was emphasised that feedback control, though widely used, has certain disadvantages. For situations in which feedback control by itself is not satisfactory, significant improvements can be achieved by adding *feedforward control*. The basic concept of feedforward control is to measure (or estimate) the disturbances on-line and take corrective action *before* they upset the process. In contrast, feedback control does not take corrective action until *after* the disturbance has upset the process and generated a non-zero error signal. A diagram of a general feedforward controller is given in Figure 8.10. A feedforward controller consists of a model of the process which is used to predict the performance of the process and this controller is therefore *not* of P-, PI- or PID-type. The model determines the value of the manipulated variable which is required to keep the controlled variable at its set point value given the value of the disturbance.

A feedforward controller has the potential of perfect control, however, in reality it has several disadvantages:

  (i) The disturbance variables must be measured on-line. In many applications, this is not feasible.

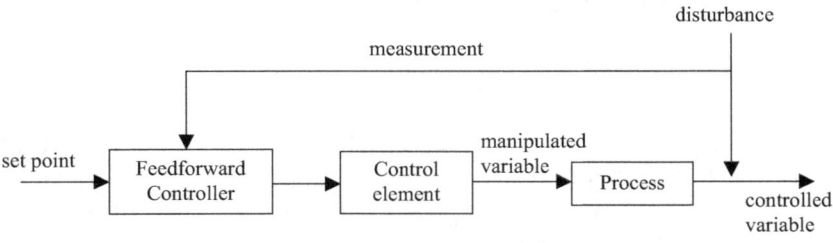

**Figure 8.10**   Feedforward control.

(ii) To make effective use of feedforward control, at least an approximate process model must be available. In particular, we need to know how the controlled variable responds to changes in both the disturbance and the manipulated variable and the quality of feedforward control depends on the accuracy of the process model.

(iii) Ideal feedforward controllers that are theoretically capable of achieving perfect control may not be physically realisable. Fortunately, practical approximations of these ideal controllers often provide effective control.

### 8.4.1.1   Feed-forward – Feedback Control

In practical applications, feedforward control is normally used in combination with feedback control. The feedforward part is used to reduce the effects of measurable disturbances, while the feedback part compensates for inaccuracies in the process model, measurement errors and unmeasured disturbances. The feedforward and feedback controllers can be combined in several different ways, as discussed in most standard control text books.

## 8.4.2   Ratio Control

*Ratio control* is a special type of feedforward control that has had widespread application in the process industries. Its objective is to maintain the ratio of two process variables at a specified value. The two variables are usually flow rates, where one is a manipulated variable and the other a disturbance. The ratio controller will change the manipulated variable as the disturbance changes to maintain the specified ratio between the two variables. Typical applications include: (1) specifying the relative amounts of components in blending operations, (2) maintaining a stoichiometric ratio of reactants into a reactor and (3) keeping a specified reflux ratio in a distillation column.

## 8.4.3   Cascade Control

A disadvantage of feedback controllers is that corrective action is not taken until *after* the controlled variable deviates from the set point. Cascade control can significantly improve the response to disturbances by employing a second measurement point and a second feedback controller. The secondary measurement is located so that it

**Figure 8.11**   Jacketed reactor. (a) single-loop control and (b) cascade control.

recognises the upset condition sooner than the controlled variable. Note that the disturbance is not necessarily measured.

As an example of when cascade control may be advantageous, consider the jacketed reactor in Figure 8.11. The reaction is exothermic and the heat generated by the reactor is removed by the cooling water in the cooling jacket. The control objective is to keep the temperature of the reacting mixture constant. Possible disturbances are the temperature of the feed and the temperature of the cooling water. The only manipulated variable available for temperature control is the cooling water flow rate.

- *Single Feedback Control.* The conventional single loop control in Figure 8.11(a) will respond much faster to changes in the feed temperature than to changes in the cooling water temperature.
- *Cascade Control.* The response of the simple feedback control to changes in the cooling water temperature can be improved by measuring the cooling water temperature and taking control action *before its effect has been felt by the reacting mixture* as shown in Figure 8.11(b). Thus if the cooling water temperature goes up, the flow rate can be increased to remove the same amount of heat, and the reaction mixture is not affected.

In cascade control, we therefore have two control loops using two different measurements but sharing a common manipulated variable. The loop that measures the controlled variable (in the example, the reacting mixture temperature) is the dominant, or *primary* control loop (also referred to as the *master* loop) and uses a set point supplied by the operator, while the loop that measures the second variable (in the example, the cooling water temperature) is called the *secondary* (or *slave*) loop and uses the output from the primary controller as its set point. Cascade control is very common in chemical processes and the major benefit to be gained is that *disturbances arising within the secondary loop are corrected by the secondary controller before they can affect the value of the primary controlled output.*

### 8.4.4   Inferential Control

In all the control systems considered so far, it has been assumed that measurements of the controlled variable were available. In some control applications, however, the process variable that is to be controlled cannot be conveniently measured on-line. For example, product composition measurements may require that a sample be sent to the plant laboratory from time to time. In this situation, measurement of the controlled variable may not be available frequently enough or quickly enough to be used for feedback control.

One solution to this problem is to employ *inferential control*, where process measurements that can be obtained more rapidly are used with a mathematical model to *infer* the value of the controlled variable, as illustrated in Figure 8.12. For example, if the overhead product stream in a distillation column cannot be analysed on-line, measurement of a selected tray temperature may be used to infer the actual composition. If necessary, the parameters in the model may be updated, if composition measurement become available, as illustrated by the second measuring device in Figure 8.12 (dashed lines).

### 8.4.5   Adaptive Control

Controllers inevitably require tuning of the controller settings to achieve a satisfactory degree of control if the process operating conditions or the environment changes significantly, *e.g.* because of heat exchanger fouling, changes in feed composition or in product specifications or because of ambient variations, such as rain storms. If these changes occur frequently, then adaptive control techniques

**Figure 8.12**   Inferential control.

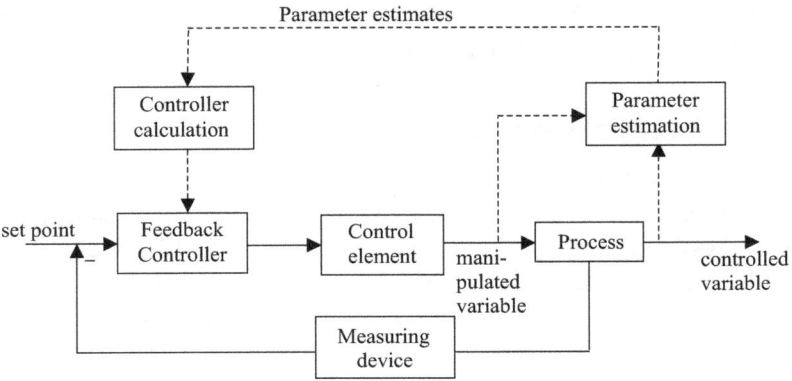

**Figure 8.13**   Adaptive (self-tuning) control.

should be considered. An *adaptive control system* is one in which the controller parameters are adjusted automatically to compensate for changing process conditions.

If the process changes can be anticipated or measured directly and the process is reasonably well understood, then *gain scheduling* may be used.[5,11] In gain scheduling, the controller gain is changed based on the measurement of a scheduling variable, usually the controlled variable or the set point. A common example of gain scheduling is pH neutralisation, where one value of the controller gain is used at low pH and another at high pH. The adaptation from one value to the other depends on the value of the pH, *i.e.* the controlled variable.

When the process changes cannot be measured or predicted, the adaptive control strategy must be implemented in a feedback manner. Many such controllers are referred to as *self-tuning controllers*, or *self-adaptive controllers*, and a typical block diagram is shown in Figure 8.13. Measurements of the controlled and the manipulated variables are used to estimate the parameters of a simple process model. This process model is then used to calculate the new control parameters based on a pre-selected tuning method.

## 8.5   Batch Control

Control systems for batch plants differ significantly from those of continuous plants discussed so far. In batch processing, there is a much greater emphasis on production scheduling of equipment to match available production equipment and raw materials with the demands for the products. Batch control systems, in contrast to continuous process control, involve *binary logic* and *discrete event analysis* applied to the sequencing of different processing steps in the same vessel, usually requiring the application of programmable logic controllers (PLCs). Feedback controllers are still used to handle set-point changes and disturbance rejection but they may require certain enhancements, such as adaptive control, to cope with the wide operating ranges because there is no steady-state operating point. A good discussion of batch control is given by Seborg *et al.*[1]

## 8.6 Plant-wide Control Issues

Plant-wide control is concerned with designing control systems for a large number of individual process units that may be highly interacting. A typical plant-wide control system will consist of many single-loop controllers as well as multi-variable controllers such as Model Predictive Control (MPC),[1,10] and may involve thousands of measurements, hundreds to thousands of manipulated variables and hundreds of disturbance variables. Fortunately, a plant with a large number of processing units can be analysed as smaller clusters of units.

Several additional issues arise from unit interactions which further distinguishes plant-wide control from the control of single units. Most modern plants will have significant heat integration between units in order to reduce energy costs. Recycling of un-reacted material is also used to improve the efficiency of the plant. Although both heat integration and material recycle can significantly reduce plant capital and operating costs, these techniques inevitably increase the amount of interaction among operating units and reduce the control degrees of freedom. Nevertheless, appropriate control strategies can deal with such undesirable consequences.[10]

It should be noted that interactions between control loops is not just limited to interacting units but will also occur within single units. A typical example is an exothermic reactor where a change in the control loop which controls the level in the reactor will have an effect on the amount of material in the reactor. This in turn will affect the heat removal requirements and, therefore, the cooling water control loop. Many strategies for reducing loop interactions, and for selecting control loops so as to minimise interactions, can be found in most standard control textbooks.[1-11]

# 8.7  Worked Example

An exothermic reaction A→B is taking place in a CSTR which has a cooling jacket with cooling water. The input stream is coming from an upstream unit. The main disturbances, controlled variables and manipulated variables are:

> *disturbances:* feed temperature $T_i$, cooling water temperature $T_{CW}$
> *controlled variables:* reactor temperature $T$, product concentration $C_B$
> *manipulated variables:* outlet flowrate $F_o$, cooling water flowrate $F_{cw}$

In order to ensure safe operation and a satisfactory quality of the reactor product, the outlet concentration $C_B$ is to be controlled. Due to possible side-reactions, the reactor temperature $T$ must also be controlled. The temperature can be measured easily but a measurement of the concentration of the product B is only available from a laboratory every hour.

Propose a control system based on feed-forward – feedback control, cascade control and inferential control to achieve these control objectives.

## 8.7.1  Solution

The control system consists of five measurements, two manipulated variables and five control loops. In the first control loop, loop 1, a temperature measurement $TT_1$ is made of the reactor temperature $T$ (where the first $T$ refers to Temperature and the second $T$ to Transmitter) which is sent to a feedback controller $TC_1$ (where the T refers to Temperature and the C to Controller) which is normally a PI-controller in order to avoid offset.

Loop 1 is cascaded with loop 2 to improve the response to disturbances in the cooling water temperature $T_{CW}$. The cooling water temperature is measured ($TT_2$) and the signal sent to a second feedback controller $TC_2$ which is normally a $P$-controller (see Figure 8.14).

The feedback control in loops 1 and 2 is combined with a feedforward controller in loop 3 which measures the inlet temperature $T_i$, calculates the change in cooling water flow rate $F_{CW}$ which is required to bring the reactor temperature $T$ back to its set point $T_S$ and sends this signal to the feedback controller (the feed-forward controller consists of a *model* of the process and is therefore *not* of P-, PI- or PID-type). The feedforward control loop will therefore theoretically eliminate any disturbances in inlet temperature $T_i$. The feedback part of

**Figure 8.14**  Control system of a CSTR.

the control system, loop 1, will compensate for any inaccuracies in the feed-forward control model as well as eliminate the effect of other, unmeasured disturbances, *e.g.* in inlet flow rate $F_i$.

The concentration of the product B, $C_B$, is not measured on-line and a measurement is only available hourly from a lab. The control of the concentration is therefore based on inferential control in loop 4 using the reactor temperature $T$. The inferential controller will then, from a model of the process, *infer* what the concentration $C_B$ is and use this inferred measurement as the signal to the controller $CC_3$ (where the first C refers to Concentration).

The model in the inferential controller is updated hourly with the actual measurement of the concentration $C_B$ from the lab in loop 5 ($CT_5$).

Note that the control system outlined above is not complete. Additional control loops to ensure safe operation would also be included, such as level control to avoid the reactor overfilling, *etc.*

Had the reactor been a batch reactor instead of a CSTR, the controllers would be combined with an adaptive control system which would have updated the controllers, *i.e.* the control parameters, as the batch progressed. The batch reactor would also have had a separate logical control system to take the reactor through the batch schedule.

# References

1. D. E. Seborg, T. F. Edgar and D. A. Mellichamp, *Process Dynamics and Control*, Wiley, Danvers, 2nd edn, 2004.
2. G. Stephanopoulos, *Chemical Process Control*, Prentice Hall, New Jersey, 1984.

3. W. L. Luyben, *Process Modeling, Simulation, and Control for Chemical Engineers*, 2nd edn, McGraw-Hill, Singapore, 1990.

4. C. A. Smith and A. B. Corripio, *Principles and Practice of Automatic Process Control*, Wiley, Danvers, 2nd edn, 1997.

5. F. G. Shinskey, *Process Control Systems*, McGraw-Hill, Singapore, 1979.

6. D. R. Coughanowr, *Process Systems Analysis and Control*, McGraw-Hill, Singapore, 2nd edn, 1991.

7. M. L. Luyben and W. L. Luyben, *Essentials of Process Control*, McGraw-Hill, New York, 1997.

8. T. E. Marlin, *Process Control: Designing Processes and Control Systems for Dynamic Performance*, McGraw-Hill, Singapore, 1995.

9. B. A. Ogunnaike and W. H. Ray, *Process Dynamics, Modeling and Control*, OUP, New York, 1994.

10. W. L. Luyben, B. D. Tyreus and M. L. Luyben, *Plantwide Process Control*, McGraw-Hill, New York, 1998.

11. B. W. Bequette, *Process Dynamics: Modeling, Simulation and Control*, Prentice Hall, New Jersey, 1998.

# 9 Economic Appraisal of Large Projects

Ken Sutherland

## 9.1 Introduction

Much of what the chemical engineer does is connected, directly or indirectly, with large plant projects – either the building of a new one or the improvement of an old one. Assuming that it is properly designed and constructed, it should work satisfactorily – and the other chapters of this book give a good understanding of how that design process is conducted.

However, the aim of the new or improved plant is normally to make a profit for the company owning it, and profitable operation is by no means assured, unless it is as carefully planned for as in the design process. This chapter will show how the planning for a profitable operation is undertaken.

The economic appraisal process is usually concerned with a *large* project, not necessarily physically large (although that is normally the case), but certainly large in respect to the sums of money required for its design and construction – sums of money that may exceed those readily available within the company for that investment. The project is also long in terms of its construction period and operating lifetime, and it is this extended lifetime that is the prime reason for the appraisal.

A small plant, which can be designed and built in a year, will normally be financed out of annual revenue; a large plant, taking several years to build and costing perhaps more than the profit

Concepts of Chemical Engineering for Chemists: 2nd Edition
Edited by Stefaan Simons
© The Royal Society of Chemistry 2017
Published by the Royal Society of Chemistry, www.rsc.org

generated in those years, has to be considered quite differently, and special techniques have to be used to determine its profitability.

## 9.2   Time Value of Money

The project to be considered, then, involves an investment that is large with respect to the available funds (costing perhaps tens or hundreds of millions of pounds), and long with respect to a company's normal yearly assessment of its financial position. It would be quite normal for such a project, say, a new oil refinery or pharmaceutical plant, to take 3 or 4 years to build, from inception to start-up, which will then operate for a long period, 10, 20 years or more. This leads to a situation where the investment in the project is made over an initial period and the revenues earned by the project do not start to appear until the investment is largely complete, but then they continue to arise over a long period after that.

It is, of course, possible to add up all of the revenues over the project's lifetime and to compare this sum with the total investments. If the revenues exceed the costs, then the project is apparently profitable, but if the reverse is true, it is not profitable and no manipulative techniques will be able to make it so. The comparison of costs and revenues, in an appropriate fashion, is the target of economic appraisal, and its first problem is to deal with the time value of money.

A pound today is not worth the same as a pound in a year's time: "worth" meaning its ability to buy an unchanging article (such as a loaf of bread). The change in value will be negative if nothing is done with the pound during the year, because inflation will inexorably decrease its real value. The change will, however, be positive if the pound is put to work, and if the return on that work (interest) is greater than the loss due to inflation.

The relation between the present value, $P$, of a sum of money, and its future value, $F$, is determined by the standard compound interest formula:

$$F = P\left(1 + \frac{i}{100}\right)^n$$

where $i$ = interest rate in percent for the period concerned (usually a year), and $n$ = the number of periods over which the calculation is being made.

In this calculation, the interest rate will be negative if inflation is working (*i.e.* $P > F$), or positive for an investment where interest is being earned (*i.e.* $F > P$).

It is this change in the value of money with time that complicates an economic appraisal – how is a pound spent at the start of construction to be compared fairly with a pound earned perhaps 15 years later? How this is done is the key to the appraisal process, and is explained later in this chapter.

## 9.3 The Company

First, some comments need to be made about the company making the investment, because the project does not exist in isolation. It will be part of a working company (even if that company is set up just to implement the project), which will be operating in accordance with the rules appropriate to the country in which it is based. Primarily, the company will exist for the purpose of making an adequate return for its owners (usually its shareholders), *i.e.* it should make an annual profit, which should increase each year in a well-managed company. Hence, it is necessary for the investment project to make its contribution to the profits.

The company, therefore, has responsibilities, not just to its stakeholders (its owners, its customers, its employees, past and present, and, increasingly these days, to its neighbours), but it also has legal responsibilities, mainly to issue an annual report, containing a set of accounts, especially a balance sheet (for a particular point in time, usually the last day of the financial year for the company) and a profit/loss account, describing the company's performance during that financial year.

From the balance sheet, it is possible to tell how "rich" the company is, and from the profit/loss account, how profitable it is (in the year in question, at least). The two documents provide very useful information as to the state of the company's capital, and hence its investment potential.

### 9.3.1 The Balance Sheet

The balance sheet, as its name implies, shows the equivalence between the assets of the company (*i.e.* what it owns), and the liabilities (*i.e.* what it owes), all calculated on the day of the balance. The total assets must equal the total liabilities.

The assets held by the company are usually grouped under two headings:

- fixed assets, which cannot easily be turned into liquid form (*i.e.* cash or its equivalent), and

- current assets, which include cash holdings and items that will yield cash during the next year.

The fixed assets include not only the property and equipment that are normally thought of as assets, but also the intangible assets, such as patent rights and any strategic investments in other companies. The current assets cover not only cash, as already stated, but also unpaid invoices for goods sold, and inventories of unused raw materials, semifinished and finished products.

The company's liabilities are listed under three headings:

- owners' funds (*i.e.* the funds in use by the company that belong to the owners),
- long-term loans (*i.e.* funds belonging to others, not due for repayment within a year of the balance sheet date), and
- current liabilities, due for payment within a year.

The owners' funds, or equity, include the issued capital (*i.e.* the face value of the company's shares), capital reserves (funds received in ways distinct from operating profits, such as premiums on the sale of shares) and revenue reserves, which are the accumulated annual profits earned. The long-term loans include not only the sums borrowed from banks and finance houses as well as bonds issued by the company, but also the provisions for future potential (but as yet unknown) costs – for example the sums a company involved in asbestos liability litigation would prudently set aside against possible penalties and costs. Current liabilities include all bills received but not yet paid (for goods and services), and any short-term loans, due within a year, such as bank overdrafts.

From the point of view of project investment planning, the items of most interest are the level of tangible investments, the holdings of cash and the balance of owners' funds and borrowed capital. These may have a controlling effect upon the way in which a new project may be funded, or even upon its practicability.

## 9.3.2 The Profit/Loss Account

The profit (or loss) account, which accompanies the balance sheet in the annual accounts, is a summary of the year's activities for the company in financial terms. Its main components are given in Table 9.1, which shows the progression from total revenue to net profit, and how each item is calculated.

**Table 9.1** Profit/loss account.

| Account item | Calculation |
|---|---|
| Total sales revenues | SR |
| Selling costs | SC |
| Factory netback | $FNB = SR - SC$ |
| Operating costs (variable and fixed) | OC |
| Overheads | OH |
| Depreciation | D |
| Production costs | $PC = OC + OH + D$ |
| Gross profit (before interest and tax) | $PBIT = FNB - PC$ |
| Interest paid | I |
| Profit before tax | $PBT = PBIT - I$ |
| Corporation tax | T |
| Net profit (profit after tax) | $PAT = PBT - T$ |
| Dividends paid to shareholders | DP |
| Retained earnings | $RE = PAT - DP$ |

A real profit/loss account will normally have more items in it than the skeleton of Table 9.1, but the format given here shows the most important items. This account is less useful in the investment planning decision, since it does not concern itself with financial resources, but the calculation of profit, as shown in Table 9.1, is a vital component of the appraisal process, as will be seen later in this chapter.

From the profit/loss account and the appropriate parts of the balance sheet, the key performance ratios for the company can be calculated, which can be used for comparison with the company's past performance, and with those of its competitors. These include return on sales (*i.e.* the profit as a percentage of the total sales revenue), the return on total assets, the return on capital employed and the return on equity. (Care must be exercised as to which profit term is used in these ratios.)

## 9.4 Sources of Money

A major part of the early stages of any investment project will involve the raising of the necessary finance for it, from internal or external sources, and at the lowest cost. A correspondingly important part of the appraisal process is therefore to examine the various sources of funds, and to try to put some relative value or cost on them.

The money required for a new project will come from one or more of the following:

- retained profits (which actually belong to the owners);
- retained depreciation funds (nominally retained to replace the depreciating assets);
- bank loan – from the company's own bankers, from an independent finance house, including venture capital sources from an international lending agency (when an overseas aid project is being considered);
- new issue of shares (ordinary/preference);
- sale of bonds or other "convertibles"; and
- government grants or long-term loans (for a project in a designated development area).

Each of these major sources, even the company's own funds, come with some kind of cost attached. The calculation of such costs is dealt with later in this chapter.

There are other ways of raising money, but they are much less likely to be of the scale necessary to finance a large project. A joint venture with another company is an increasingly important way of doing this – at the cost of losing control of the whole project. A sale of an existing asset may be a good source – especially if the company owns easily separable assets that are no longer key parts of the business.

On the smaller scale, equipment can be leased, and this concept is expanding into BOO (build, operate and own) contracts where another company – usually a plant contractor – will supply the plant, but retain ownership of it.

## 9.4.1 Owners' Funds

The funds belonging specifically to the owners of the company are increased each year by the addition of the retained earnings from the year's operations. They may also decrease, if the company makes a loss, or if some are invested in new assets, or if the company chooses to pay dividends greater than the net profit (which, given good business reasons for so doing, the company is perfectly entitled to do).

The owners' funds are all invested in the company as part of the total assets, and, as a result, are a major liability of the company. They are, of course, owned by the shareholders in direct ratio to individual shareholdings.

## 9.4.2   Shares and Bonds

Ownership in a company is conferred by issuing a share of the company in the form of a transferable certificate, and then selling that share to a person or an institution, who thus becomes an owner of that proportion of the company represented by that share. A registered company is legally bound to issue a minimum number of shares, which must be sold to its forming members. Once this minimum number is sold, a company can then issue as many shares as it likes, and sell some or all of them to whomsoever wishes to buy them.

The money raised from the sale of these shares is kept by the company (as "owners' funds") and is used to pursue the company's business. Once sold by the company, a share can be traded (on the appropriate stock market), and the company can buy back its own shares in this way if it wants to do so, at any time. The value of the share on the stock market will fluctuate daily, according to the strength of the company and the state of the world, and may fall below the face value, or rise to many times that figure, according to demand.

In order to raise new funds, a company can, at any stage in its lifetime, issue a new batch of shares and sell some or all of them. If the company is doing well and obviously expanding, then it will probably be able to sell these shares at a premium above the face value. If, on the other hand, the company is doing badly and needs the money to stave off disaster, then it will almost certainly have to offer the shares at a discount below the face value.

This is an important way of raising money, but a new share issue dilutes the existing share ownership, and may be unpopular with existing shareholders. It is normal, therefore, to offer the new issue first to existing owners.

Shares give their owners the rights to (but no guarantee of) an annual dividend, and also rights to a share of the assets if a company is wound up, but the key characteristic of a share is its ownership of a part, however small, of the company.

Companies in need of more capital can also raise it by selling bonds. A bond is a certificate, with a face value usually much larger than that of a share, which has a fixed lifetime, usually 10 years at least, at the end of which time the bond "matures" and the company pays back its face value (by then, of course, much decreased in real value). However, the company also guarantees to pay interest at a fixed percentage of the bond's face value for this lifetime, usually twice a year, and this rate is usually quite an attractive one.

Bonds can be bought and sold, like shares, on a special bond trading market, and their trading value will vary around the face value (usually without the spectacular changes that can occur with shares). The key characteristics of a bond are the guaranteed repayment of capital and annual interest payment (so long as the company survives to the maturity of the bonds), but, unlike a share, no ownership rights are conveyed with a bond.

The sale of both shares and bonds can be keyed to their intended use for a specific project, and if the project idea is a good one, whose purpose is easily conveyed to the potential investors, then the sale is correspondingly eased.

## 9.5  Valuing a Company

As part of an investment planning exercise, it is frequently necessary to put a value on the totality of an operating company. The calculation of this value can be quite difficult, but may be vital. The need arises either when the sale or purchase of the company as a whole is being considered, or when one wants to compare several companies as to size – in the present context this may be a way of determining the likely sums available from a sale for re-investment.

The simplest method of valuation starts from the obvious fact that to buy a company, one just has to buy all the outstanding shares. In principle, therefore, the value of the company is the number of shares multiplied by their individual value at any moment (the "capitalization"). Almost certainly the offered price will have to exceed the current share price to persuade people to sell their shares (unless the market is falling, in which case they may be glad to off load them). There may also have to be a premium on the capitalization to allow for intangibles, such as "goodwill".

This valuation is sound for the particular sale/purchase situation, but is very dependent upon the immediately current share price. It has no direct link to any measure of an intrinsic company value, certainly not one that can be used to compare the sizes of two or more companies.

So there must be other measures of value. The most obvious is gross sales volume – the more a company sells in any market, the bigger it must be. This is generally true, but what may be more important for comparative purposes is what lies behind these sales – are there enormous quantities of assets or capital required to achieve them, or is it a very lean company? The ratio of gross sales to asset value or to capital employed will therefore be a better measure.

Since companies exist primarily to provide income for their owners, *i.e.* to make a profit, their gross profit could be a more interesting figure than sales volume. The most profitable companies may be the best – but, again, it is the ratio of profit to assets or capital employed that may be more important.

To a large extent a company is its assets, so the purchase of a company implies the purchase of its assets. This is apparently a very real figure – fixed assets of plant and buildings, technology, people, spare cash and so on. Asset value, however, is a little nebulous – people can easily leave, for example – and some companies make very large sales off very small assets.

A key figure in valuing a company is the ratio of its current share price (as an indicator of what the world thinks of it) to the total net profit (earnings) divided by the number of shares (earnings per share). This is called the *P/E* ratio (price/earnings), and is a major tool in comparing the performance of a company within its class of similar companies.

While the individual valuations may not be all that difficult, the method to be used for determining the value of a company depends very much on the purpose of the valuation exercise.

## 9.6 The Cost of Capital

The funds that might be obtained from the various sources given in Section 9.4 come at very different costs – both real and virtual. If there are few sources available, then a simple calculation may be possible to determine an annual cost of the finance raised for the project, but it is more likely that a detailed analysis will have to be made.

The capital needed for a new major project will come from two sources: inside the company and outside it. The internal resources are the owners' funds, and this money exists in the current assets part of the balance sheet, from which it can be drawn. New capital can be raised by the sale of already issued shares held by the company, or by the sale of bonds, or by the issuance of new shares.

The normal external sources are loans from banks and other lending institutions at commercial rates of interest, and grants from governments or loans (also from overseas credit agencies) at lower than commercial rates of interest. The cost of a single source of borrowed money is easily determined – it is the annual rate of interest being charged on the loan. Where several different sources are used, a composite cost can be calculated (see below).

The cost of using the company's own funds (or, actually, the owners' funds) is less easily derived, because it contains an element of owners' future expectations and an element of risk. It is most definitely not "free", because the money can be used for a number of plans other than the one under consideration, and its cost should be set at least to equal the expected return on the most attractive of these alternatives, known as the "opportunity cost".

## 9.6.1 Cost of Company's Own Funds

The normal method of calculation for company funds is to use the *capital asset pricing model (CAPM)*. This was developed by share analysts keen to have a defence against accusations of negligence in selecting shares for clients as a means of assessing the real value of any share, in the form of risk and desirability. It essentially demonstrates one version of the direct proportionality between risk and return.

In its graphical form, risk is measured along the horizontal axis by a factor "beta", related to the volatility of the price of the share, resulting from any general disturbance of the market, relative to the average of all quoted shares on that market. This is plotted against share "yield" (dividends as a percentage of current share price). If all shares available on one market are plotted on this basis, it can be shown that they lie roughly on a straight line, the "market line", rising with increasing values of beta, as shown in Figure 9.1. Beta is lowest for long-term government bonds, and its value is set to zero for the base rate of interest. The average value for the whole market is set to 1.0, so that high risk shares have a beta value approaching 2.0.

The justification for this model is the fact that any share that sits well above the line (*i.e.* having a relatively high yield) for any reason will appear attractive to the market, investors will buy it, its price will

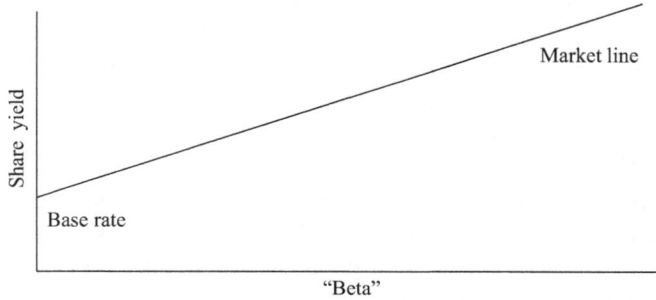

**Figure 9.1** CAPM market line.

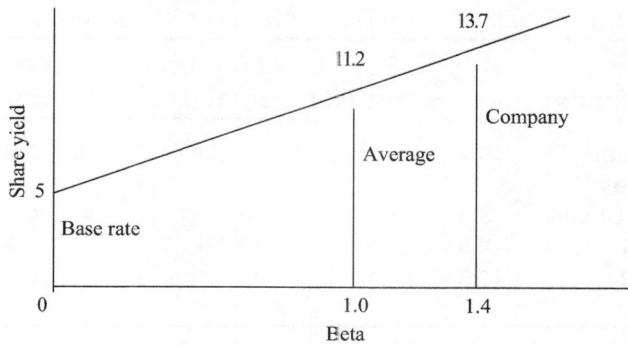

**Figure 9.2** Calculation of company's cost of capital.

rise as a result of this demand and its yield will thus fall back to the market line. Conversely, any share that is well below the line will appear unattractive, investors will sell, the price will fall and the yield will rise.

Beta values will vary from industry to industry, with "blue chip" companies near to zero, and high-risk telecom or biotech companies, for example, at the opposite end. Over the years, various types of companies have become grouped together in different regions of the market line, with different values of risk. An individual company can position itself among similar types of company (type of business being the most important characteristic), to derive its cost of capital from the resultant beta value, as exemplified in Figure 9.2. Here a company with a beta figure of 1.4 positions itself on a market line deriving from a base (beta = 0) of 5% with an average yield (beta = 1) of 11.2%. Extrapolation of the market line to the beta of 1.4 gives a yield of 13.7%, which is taken to be the "cost" of the company's own capital.

This model is so successful that large companies can use a similar method to decide which parts of the company are doing well, and which should be sold.

### 9.6.2 Average Cost of Several Sources

The cost of the company's own funds can thus be calculated, by means of the CAPM. However, it is an unusual company that does not have other sources of finance, each with its own cost (*i.e.* interest rates for different loans). The average cost of all of its sources of capital is expressed by a single figure, the *weighted annual cost of capital (WACC)*, calculated according to the relative amounts of each source of funds.

**Table 9.2**   Calculation of weighted annual cost of capital.

| Source of funds | Rate = cost (%) | Proportion of total (%) | Product = cost component |
|---|---|---|---|
| Company funds @CAPM cost | 13.7 | 25 | 3.425 |
| Government loan | 3.5 | 10 | 0.35 |
| Bank loan 1 | 8.5 | 30 | 2.55 |
| Bank loan 2 | 17.5 | 35 | 6.125 |
| WACC | | | 12.45% |

Each source of finance is tabulated both with its effective "cost" and its proportion of the total amount of money in use. The product of the percentage cost and the (decimal) fraction share of the total funds gives the component of the final weighted cost for each source, as shown by the example in Table 9.2.

Table 9.2 shows a company using its own funds (at its CAPM cost) and three other sources of capital: a cheap government loan and two commercial bank loans, one at a very high rate of interest (because the loan was made to finance what the lender considered to be a high-risk venture). In the case of bank loan 1, the entries show that the rate of interest charged on the loan is 8.5%, and that it represents 30% of the total funds. Its contribution to the weighted average is then 30% of 8.5%, or 2.55%. The total of the four contributions in the last column gives the WACC of 12.45%.

The calculated WACC is a figure that changes with time, as a loan is paid off, or another taken out, or the proportion of company funds increases. Despite this changeability, the WACC provides a good benchmark against which to measure the rate of return on a new project, and it is precisely in this way that it is used later in this chapter.

## 9.7   Cost of Construction

A major part of the investment appraisal exercise is the calculation of the total amount of capital needed:

- to design and engineer the plant;
- to buy the land (if necessary), and all the components of the plant;
- to construct and install the plant;
- to prepare the plant for operation, including staff recruitment and training;

- to provide working funds for the first few months of operation; and
- to pay interest on any money borrowed during the construction period.

The financial appraisal is normally undertaken quite early in the life of the project, well before the final designs are complete, so a great deal of the cost calculation is based on the skill of the estimator, and much of it is done by factoring on known figures (such as how much a similar plant cost in another place or at another time). As it is an estimate, the final figure will prudently include a contingency allowance – the better the estimate is believed to be, the lower this contingency needs to be.

Most of the work in the estimating process relies on the application of well-established factors to a basic figure of the cost of all of the main items of the plant, as delivered to the site entrance. This figure has thus to be calculated as accurately as possible, based upon the plant design as available at the time of the estimate. It depends upon the identification and rough sizing of the key process plant items (such as reactors, distillation columns, heat exchangers, and so on), and the determination of approximate delivered prices for these items, from charts and other information published in textbooks, journals and on the internet.

A typical set of figures is shown in Table 9.3, which includes a range of factor values for each of the installation and engineering items. Which particular figure is chosen from the range shown is based on the experience of the estimator, although the quoted mean figure is the most likely. It should be noted that some ranges are much wider than others, a situation dictated by the nature of process plant in general.

To the basic purchased equipment cost (including delivery to the site) of 100 are added each of the subsequent factors to make up the final total. The contractor's fee is necessary if the plant design and construction work is given to a contractor to undertake (which is very often the case, because the company needing the plant is unlikely to have the necessary engineers on its payroll).

The interest accrued on any necessary loan taken out during the construction period will obviously depend markedly on the loan draw down and conditions, but 10% of the fixed capital investment (FCI) is a good approximation. The working capital is more sensibly calculated in terms of production and selling costs – but until these are available, 15% of FCI is another good approximation.

The calculation embodied in Table 9.3 shows that the total amount of capital required for the project is somewhere around six times the

**Table 9.3** Total capital employed.

| Expenditure item | Range (%) | Mean |
|---|---|---|
| Purchased equipment | 100 | 100 |
| Installation of purchased equipment | 40–50 | 45 |
| Instrumentation and controls | 15–30 | 23 |
| Piping and valves | 30–75 | 52 |
| Electrical services | 15 | 15 |
| Plant buildings | 20–30 | 25 |
| Service facilities | 50–70 | 60 |
| Land | 5–10 | 7 |
| Site improvement | 10–15 | 13 |
| *Total installed plant* | | 340 |
| Engineering and supervision | 30–35 | 33 |
| Construction costs | 35–40 | 37 |
| Contractor's fee | 15–25 | 20 |
| *Total engineering* | | 90 |
| Contingency (10% of installed plant + engineering) | | 45 |
| *Fixed Capital Investment (FCI)* | | 475 |
| Interest accrued on loan during construction period (10% FCI) | | 50 |
| Start-up costs (10% FCI) | | 50 |
| *Total installed cost (TIC)* | | 575 |
| Working capital (15% FCI) | | 75 |
| *Total capital employed (TCE)* | | 650 |

cost of the purchased and delivered equipment (a ratio of 5–7 can be expected, depending upon the complexity of the plant). However, it is equally important that the way in which this capital is to be spent be determined – in order that the demands on the company's funds and borrowing capacity can be foreseen well ahead of their requirement, and also so that the way in which any loan has to be drawn upon can be seen, and hence any accrued interest be calculated.

It is normally sufficient that the capital needs be calculated on a yearly basis, and they are then assumed all to be paid out on the first day of that year (except for the start-up costs and the working capital, which are assumed to be paid on the day the plant starts to operate).

## 9.8   Revenue Calculation and Net Cash Flow

With the capital needs calculated, it is now necessary to determine what profit (if any) the plant will make. It is assumed that the project is to be implemented as a result of some sound market research that might have shown an appropriate need in the marketplace. This research might also have shown what the plant capacity should be, and should have given guidance as to the likely sales price for the product – although this will normally be a management decision based upon the market entry strategy. An established product will probably have to go in at a discount on the present market price; while a new product may be able to be priced on a cost-plus-reasonable-profit basis.

Given, then, the production rate and sales price, the total revenue for the project can be calculated (perhaps with allowance made for under-running in the first year or two as the faults are dealt with). It will cost money to get the product to market – a sales force, advertising, cost of shipment and distribution, and so on. There may also be a royalty on sales payable to the owner of the technology used in the plant. These costs, generally called selling costs, are subtracted from the revenues to derive the sum that actually gets back to the project – the "factory netback".

## 9.8.1 Operating Cost

The major annual cost for the project will probably be the cost of operating the plant, the production cost. This is normally thought of as having four components:

- variable operating costs;
- fixed operating costs;
- overheads; and
- depreciation.

The variable costs are those directly resulting from the operation of the plant and dependent upon how much product is made. They will include the costs of all raw materials and other materials consumed by the process, of all utility services (electricity, gas, water, compressed air, steam, *etc.*), plus the costs of packaging, and maybe a waste disposal cost, if that is related to production rate. These costs will vary if, for example, an underrun in the first year or two is being allowed.

The fixed operating costs are those that do not change with production level. The most important is the cost of operating, maintenance and service labour, plus its management, but also included are such items as insurances of all kinds, local taxes, waste disposal (which may be a fixed cost rather than a running cost) and energy tax (carbon levy). Apart from labour, the cost of which is based upon the nature of the plant (its complexity, its level of automation), these other fixed costs are usually obtained by factoring from the plant's capital cost.

If the project and the company are the same, then these are all the costs to be considered. However, it is usual for the project to be situated among other plants belonging to the company on a larger site, and then there are general costs for the whole site (such as a fire station, the security guard system, a canteen) to be shared among all the plants. This overhead charge (direct if it is for the same site, plus indirect for the company's other activities) is usually derived on a capital investment proportional basis.

## 9.8.2   Depreciation

This leaves just one major production cost item to deal with, that of depreciation.

Fixed assets (excluding land) used by a company usually decrease in value over time, even when properly maintained. This can be due to product obsolescence or simply wearing out (with an ultimate need to be replaced). If an asset life is less than a year, the expenditure on it is included in the profit/loss account as a cost. If, however, its life is greater than 1 year, then a way has to be found to adjust both profit/loss account and balance sheet to give recognition to the actual life of the asset.

The initial cost of the asset is shown as a fixed asset in the balance sheet. In the profit/loss account, the decrease in value is shown each year by allowing a proportion of the original value as a notional operating cost, the "depreciation allowance". A correspondingly reduced asset value is then shown in the end-of-year balance sheet, with an equivalent increase in the "cash" assets (so that the total assets are not changed).

Typically, a chemical company will depreciate its plant and equipment (on-sites) at 10–15%, and its buildings (off-sites) at 5–8% (a lower rate because they may be reusable for some other purpose when the plant's life is over). The actual values used will be what the company and its auditors think appropriate and prudent. The depreciation allowance is usually linear ("straight-line", *i.e.* same amount each year) but other methods are used, *e.g.* "declining value" (same percentage of value at start of each year), which leaves a terminal value at the end of the plant's life, which has then to be written off at one go.

Depreciation is allowed as an operating expense, and hence as part of the production cost, as far as calculation of corporation tax liability is concerned. The rates and timing of such allowance against tax can vary, but the most likely method is to count it annually against an annual tax assessment.

The things to remember about depreciation, however it is calculated, are that it is a part of the production cost, but that it does *not* represent a cash outflow, merely an internal transfer.

A complication of this picture lies in the use of the word "amortisation", which also means the progressive writing down of the value of an asset, and thus could include "depreciation" or be used instead. It is, confusingly, separated from depreciation in the currently used profit measure "EBITDA" or earnings before interest, tax, depreciation and amortisation, where amortisation refers to the writing down of a financial asset, with depreciation limited to fixed assets.

(A further complication occurs for assets that increase in value with time, such as land, or mineral resources, or even buildings. Revaluation of assets requires the efforts of skilled accountants, with matching balance sheet entries under fixed assets and capital reserves.)

### 9.8.3 Annual Profit

The total production cost (variable plus fixed operating costs, overheads and depreciation) when subtracted from the factory netback (total revenues less selling costs) leads to an operating profit, often called "gross profit", but more correctly profit before interest and tax (PBIT). From this, then, is subtracted the annual interest paid on the loans directly involved in financing the project, to give the profit before tax (PBT).

The corporation tax payment is usually a percentage of the PBT figure, and is frequently actually paid in the year following that in which the profits were earned. However, it is customary to account for it in the same year, as far as feasibility calculations are concerned.

With the tax deducted from the PBT, the net profit, net earnings or profit after tax (PAT) is determined, to be the prime source of real income for the project, but not the only one.

### 9.8.4 Net Cash Flow

The net cash flow into the project each year is made up of two items: the PAT and the depreciation allowance for that year. The depreciation is added to the PAT because, unlike all the other costs, it does not actually leave the balance sheet.

In order to carry out a proper financial appraisal, the net cash flow figures must be calculated as inputs to the project's funds for each year of its operating life. Although these sums actually arise throughout the year, they are conventionally taken as arising on the last day of the year.

## 9.9    Financial Requirement of New Project

The main object of a financial appraisal is to answer the question relating to the project, "is it worth undertaking it?". This is not the same as "will it be profitable?", but rather "will it be *adequately* profitable?".

An alternative way of deciding this is to ask, "is this the best way of spending the total required capital, or are there better ways of investing the money (*i.e.* ways that will show a greater return)?".

The company must decide if the return that is forecast to be earnable from the project is equal to or greater than some target figure of earnings that meets the company's growth goals. It is up to the finance director of the company to set this target, and then up to the project's designers to see if they can achieve it.

The financial appraisal process must be able to answer both of these basic questions: what can this project earn, and which of several alternative projects is the best, economically?

(It is recognised that greatest return may not be the only criterion, or even the most important, in making a decision on a new plant: it may be necessary to fight competitors or to take a share in a new geographical market. But greatest return is the usual wish.)

How, then, should a target figure for the return on the investment be established? The simplest is to look at current interest rates – but these are at historically low levels, and it should not be difficult for a project to exceed these rates. Next, one looks at relevant business ratios (as percentages) – say the ratio of sales to capital employed, both historically in the same company, and among competitors in the same general business group – as a guiding figure.

The most important assessment is the comparison with the company's present level of cost of capital, as measured by the CAPM for a single source, or the WACC for capital derived from several sources. As interest rates have fallen quite considerably, it is likely that these models, being historical, are higher than would be the case for capital raised now. However, to set either of these as the target rate would probably be safe, were it not for the risks involved in any new project.

In fact, the target rate, often called the hurdle rate of return (HRR – "hurdle" in the sense of needing to jump over it), is usually set at a premium above the WACC, this premium being designed to reflect the uncertainties surrounding the project. These include:

- the extra risk over the company's usual operations, arising from
    (i) factors inherent in the project,
    (ii) uncertain status of the company in the (new) business area,
    (iii) specific uncertainties relating to the business area itself, and
    (iv) exceptional risks relating to the country in which the plant is to be built (if different from the country where the company is based);
- a contingency factor to cover omissions and bias in the project proposal; and
- allowances for exceptional inflation rates or currency fluctuations in the country where the plant is to be located, if this is far from the company's main operations.

Values for this risk factor, and therefore for the HRR, vary widely with the industry and company, but, for the bulk chemicals and petrochemicals sector, it is unlikely to be less than 15% and may be over 20%.

Whatever rates are used for comparative purposes, it is absolutely vital to ensure that they are calculated on the same basis: before or after taxes have been paid, before or after interest, with or without depreciation as a cost.

## 9.10 The Appraisal Process

The project financial appraisal process is now a question of taking the calculated investments, made in the assumed way, and comparing them with the calculated cash flows, arising in the assumed pattern over the plant's operating lifetime, in such a way as to show the effective rate of return on the investment.

There are two simple factors that can be used for comparison, either with established targets or among several competing projects, which make no allowance for the ways in which money is spent or received, and hence do not allow for the time value of money:

- return on investment (RoI) = average PBIT/total capital employed, and
- payback time = point in time at which cumulative net cash flows just equals total employed capital.

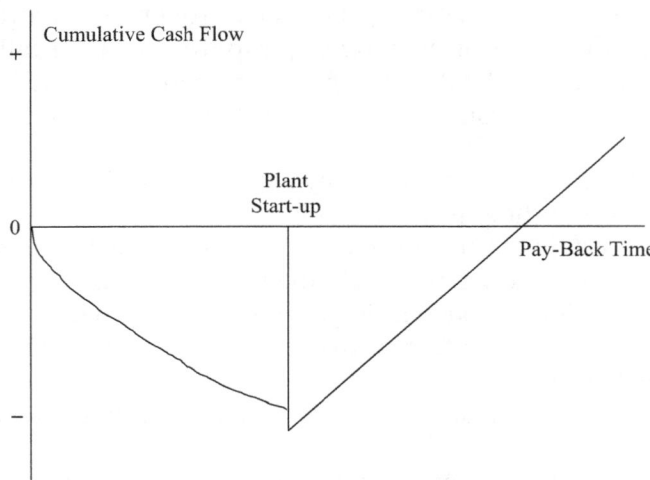

**Figure 9.3**  Payback time.

These are useful factors, despite their limitations.

A simple graph of net cash flow out of, or in to, the project's account illustrates the way in which the cumulative net cash flow moves over the lifetime of the project, and enables the payback time to be easily seen, as in Figure 9.3. The payback time may be expressed as "less than 3 years", or more precisely, such as 2.4 years or 2 years 5 months.

In general, the lower the payback time for any project, the more profitable it is likely to be, but as a complete indicator, this method fails to allow for the different patterns of payments and revenues.

## 9.10.1  The Discounted Cash Flow Process

The prime task in project appraisal is to compare the total income from the project over its productive lifetime, with the total investment needed to build it and get it running. To do this properly requires some measure that will enable the analyst to

- make a fair comparison between inputs and outputs;
- give an index by which to show the value of a new project to a company;
- enable fair comparisons among several competing projects; and/or
- enable sensitivity analyses to be made easily for different project parameters.

The discounted cash flow (DCF) method, to be described here, does enable the making of a fair comparison between inputs and outputs, and provides both an effective rate of return and an easy means of assessing the effects of different approximations.

The essence of the DCF process is that it does take into account the passage of time during the life of the project. It works by modifying all sums of money, costs or revenues, to give their nominal values at a particular point in time, so that a balance can be struck between the sum of all costs and the sum of all revenues at that point. The method works equally well at any point in time over the project's life, and the first decision to be taken is to select the most meaningful point to use.

The two balance points in time that are most commonly used are

- the commencement of work on the project, and
- the day of start-up of the plant.

There is some sense in selecting the date of the appraisal as a balance point, but this cannot easily allow for the potential delays between conception and start of construction (although it can give a good incentive to reducing delays between idea and implementation for a deserving project).

In fact, the best balance point is the date on which operation starts, because at that time, all expenditures should be complete, and no revenues have yet started to accrue. All sums of money, input or output, are then discounted, forwards or backwards, to this comparison point in time.

## 9.10.2 Present Value

The discounting process enables the conversion of any sum of money involved in the project to its "present value" (PV) at the point of comparison. (Confusingly, "present" does not necessarily mean "now", and, in fact, in the most sensible way of doing DCF analyses, it means on the day of start-up, which may be several years ahead of "now". Even the different choice of the comparison point as the day of construction commencement does not prevent confusion about the term "present", because commencement of the project may still be well ahead of the "now" of doing the appraisal.)

Choice of the day of start-up as the comparison thus requires the acceptance that the "present" is actually some time in the future, and it employs two kinds of discounting (although one is only the reciprocal of the other). The fact that at start-up all expenditures are past, and that all income is in the future, eliminates the need to use negative cash flows, and is a slightly easier concept to grasp. It is also easier to manage construction periods that are not a whole number of years in this way.

The method of calculation is based on transferring the value of money forward or backwards in time, by discounting.

The *present value* PV of a sum $S$ invested *in the past*, at an interest rate $i/100$ per period, after $n$ such periods is

$$PV = S\left(1 + \frac{i}{100}\right)^n$$

forward discounting

Correspondingly, the *present value* PV of a sum $S$ arising *in the future*, after $n$ periods, at an interest rate of $i/100$ per period is

$$PV = S\left(1 + \frac{i}{100}\right)^{-n}$$

backward discounting

Note that forward discounting *increases* the sums concerned, and that backward discounting *decreases* the appropriate amounts.

In these two discounting equations, the interest rate, $i$, measured in percentage, is called the discount factor (DF), and its choice is a complex matter, to be discussed a little later.

The calculation of "present" values requires the observance of two main conventions, that

- all cash flows are assumed to occur at one point in the year (even if they actually arise uniformly throughout it), with expenditures aggregated at the beginning of the year, and income at the end of the year; and
- all discounting is done annually, forward to the start-up date for expenditures, and backwards for all incomes.

A further convention is that working capital is paid in one sum at startup (together with those costs that are specific to the start-up process itself), and working capital (only) is repaid to the revenue account at the end of the project's life (together with any scrap value of the plant).

Tables of DFs are available in various places, but calculations are now much simplified by use of spreadsheet functions.

### 9.10.3   Net Present Value

The *net present value* (NPV) is one of the key terms in the appraisal technique – it refers to the difference between the present values of all costs and that of all revenues:

$$NPV = PV(\text{revenues}) - PV(\text{costs})$$

and may be negative or positive. Negative NPV figures usually indicate an unattractive situation.

The NPV is calculated at a particular value of the DF, but the value corresponding to $DF = 0$, *i.e.* the base case, just looking at the simple, undiscounted sums of costs and revenues, is a useful figure also.

All other things being equal, the project showing the highest value of NPV at a particular value of the DF (including zero) will be the most attractive, financially.

(If cash flow patterns are very irregular, then discounting may have to be done other than annually, but this is only needed when requiring a fairly accurate analysis, or when comparing different projects with widely varying cash flow patterns.)

The value of the DF to be used in the discounting formulae is chosen according to the purpose of the calculation. If the purpose is to demonstrate a positive NPV, or to compare NPVs for different projects, then a value will be chosen to match the company's own finances, and the value could then be

- the company's cost of equity capital (by the CAPM calculation);
- the overall cost of capital to the company (by the WACC calculation);
- the opportunity cost of capital, chosen by the company to match other development opportunities; or
- the hurdle rate of return, chosen by the company to cover all known risks.

Obviously, the last of these, $DF = HRR$, is the most sensible.

### 9.10.4   Project Lifetime

The probable life of the project will depend very much on the market place to be served by the product, but a 5 year life should be achievable, and over 10 years would normally be expected for a project of the size being considered here.

It is necessary to select a value for the effective lifetime of a project, so that the undiscounted NPV is positive (*i.e.* a lifetime longer than the payback time). A value of 10 years is conventionally chosen

- because it is a sensible real figure; and
- because the PVs of almost any practical sum and DF combination become very small for 11 years or more, and so add very little to the PV(revenues), and hence to the NPV.

An important consideration is whether or not to assign a scrap (or still-operating sale) value to the plant at the end of its economic life. Unless the plant is to be "run into the ground", there is usually some scrap value, even if only from the plant buildings, which is usually calculated as a percentage of the FCI at the start, and is then included as an additional cash flow at the end of the last year of the analysis (together with the working capital).

## 9.11   The Internal Rate of Return

There is thus an inverse relationship between NPV and DF – the higher the latter, the smaller the value of NPV. The highest value of NPV is the base figure, when the sums are not discounted (DF $= 0$), and, eventually, for increasing values of DF, NPV becomes negative.

There is then a single value of the DF for which, uniquely, the NPV is zero. This value of the DF is called the *internal rate of return* (IRR), or sometimes, the DCF rate of return. The IRR is the other key term in the DCF process

$$NPV = 0 \quad \text{for DF} = IRR$$

Because the calculation of the NPVs takes into account the ways in which the relevant sums of money have been invested or recovered, as well as their size, the IRR is characteristic of the project and its cash flows: the IRR is the rate of return achieved by investing the total capital employed in the project in the same pattern as in building the project, to produce income in the pattern of the net cash flows of the project.

In this definition, the pattern of payments or earnings is the way in which payments are made during the construction period and the way in which earnings are received once production has started. In most financial investments, especially by the private investor, the payments are of regular amounts and usually are at regular time intervals, while receipts will either be a single sum at the end of the savings period, or an annuity. In a large plant project, however, the payments will certainly be of unequal amounts, probably paid irregularly at various times during the construction period, and the receipts may well be irregular too, if only (the investors may hope) increasing every year. The IRR then determines the exact rate of return on such an irregular investment, and if the payment or receipt pattern changes, then so will the IRR – which can thus be used to compare different investment patterns, for example.

## 9.12 The DCF Appraisal Process

To determine the intrinsic value of the project (as well as an alternative means of comparing several different projects, or even several different versions of the same project), the IRR is calculated as that unique discount rate that makes the NPV equal to zero. The IRR cannot be calculated from a single equation, but must be determined iteratively, and the final stage may be graphical.

The IRR will be the ultimate goal of the DCF process, but it is usually a good idea also to determine the NPV of the project for one of the company rates of return – and the hurdle rate is the most obvious.

It is unlikely that an IRR of 15% or less will be of much interest to a company, certainly for a new development, while it is also unlikely that a rate of over 35% is feasible, unless it is for a very exotic and expensive product. The discount rates used in appraisal calculations, therefore, usually lie in the 15–35% range. In this range, the DFs for 11 years or more are so large as to make any calculation beyond 10 years rather pointless, so confirming the choice of 10 years as an economic lifetime.

The basic steps of a DCF appraisal are shown in Table 9.4, which describes a project whose total capital employment is £124 million, spent as shown over a 3 year construction period (including expenditure at year 0, *i.e.* the point of start-up). Revenues rise during the first 2 years of operation to a maximum figure from year 3 onwards.

Calculations of NPV are included in Table 9.4 for the base case (DF = 0), for the company's hurdle rate (15%), and for four other DF values, chosen to enable a graph to be plotted between the values of DF and NPV (Figure 9.4).

The NPV figure for the base case is more than twice the capital investment, so in the simplest terms the project is profitable. This

**Table 9.4** A typical appraisal calculation.[a]

| Discount rate (%) | | | | | |
|---|---|---|---|---|---|
| Period | Base | 15 | 10 | 25 | 17.5 | 20 |
| Year −3 | 25.00 | 38.02 | 33.28 | 48.83 | 40.56 | 43.20 |
| Year −2 | 45.00 | 59.51 | 54.45 | 70.31 | 62.13 | 64.80 |
| Year −1 | 30.00 | 34.50 | 33.00 | 37.50 | 35.25 | 36.00 |
| Year 0 | 24.00 | 24.00 | 24.00 | 24.00 | 24.00 | 24.00 |
| Total | 124.00 | 156.04 | 144.73 | 180.64 | 161.93 | 168.00 |
| Year 1 | 29.50 | 25.65 | 26.82 | 23.60 | 25.11 | 24.58 |
| Year 2 | 34.10 | 25.78 | 28.18 | 21.82 | 24.70 | 23.68 |
| Year 3 | 38.70 | 25.45 | 29.08 | 19.81 | 23.86 | 22.40 |
| Year 4 | 38.70 | 22.13 | 26.43 | 15.85 | 20.30 | 18.66 |
| Year 5 | 38.70 | 19.24 | 24.03 | 12.68 | 17.28 | 15.55 |
| Year 6 | 38.70 | 16.73 | 21.85 | 10.15 | 14.71 | 12.96 |
| Year 7 | 38.70 | 14.55 | 19.86 | 8.12 | 12.52 | 10.80 |
| Year 8 | 38.70 | 12.65 | 18.05 | 6.49 | 10.65 | 9.00 |
| Year 9 | 38.70 | 11.00 | 16.41 | 5.19 | 9.07 | 7.50 |
| Year 10 | 62.70 | 15.50 | 24.17 | 6.73 | 12.50 | 10.13 |
| Total | 397.20 | 188.68 | 234.88 | 130.45 | 170.68 | 155.26 |
| NPV | 273.20 | 32.64 | 90.16 | −50.19 | 8.75 | −12.74 |

[a]Notes
1. The project takes three years to build, numbered −3, −2, and −1 respectively.
2. "Year 0" applies to the moment of start-up, and is of zero length in time.
3. The column headed "Base" contains the undiscounted costs (Years −3 to 0) and revenues (Years 1 to 10), expressed in £ millions.
4. The remaining columns show these base sums discounted, forwards and backwards at the discount rates denominating the columns.
5. Working capital is paid out in Year 0, and recovered in Year 10.

means that the returns are comfortably large enough to repay any loan taken out to pay for the project.

It can be seen that the NPVs become negative between DF = 17.5% and DF = 20%, so that is where the IRR will lie, and from the graph of Figure 9.4 the value can be seen to be close to 18.5%.

## 9.12.1 Graphical Interpolation

For DCF calculations being done by hand, the final stage will most easily be a graphical interpolation – the alternative being a long-winded iteration process to find the actual value (which is not normally sought to better than one place of decimals).

The curve relating NPV to DF is not linear, as can be seen in Figure 9.4, although it can be taken as linear for very short intervals of DF. On the other hand, the NPV/DF curve is a *smooth* one, so any point not lying on a smooth curve must have been calculated or plotted incorrectly.

**Figure 9.4** Relationship between NPV and DF.

if a and b are two known values of NPV at discount factors A% and (A + 1)%,
then the IRR is found at the point (A + x)%, where:

$$a/x = b/(1 - x) \qquad \text{or} \qquad x = a/(a + b)$$

**Figure 9.5** Graphical interpolation.

The aim in selecting values of the DF to use in NPV calculations is to find two values, preferably no more than 1% apart, one of which gives a positive NPV, and the other a negative value. For this interval of DF, the NPV/DF plot can be taken as linear, and the actual value of the IRR is then calculated by triangulation (using the theory of similar triangles), as illustrated in Figure 9.5.

In the case illustrated in the previous example, the NPV figures would be calculated for DFs of 18% and 19%, and, in the example of Figure 9.5, *A* would be 18%, and *x* would be 0.5%, giving an IRR of 18.5%.

## 9.12.2 Computer Calculation

The whole repetitive process of discounting calculations is avoidable by use of a modern spreadsheet, such as Microsoft Excel, which can undertake the iteration process automatically and very rapidly. The basic annual cost and cash flow data, of course, still have to be calculated by hand (even if by a properly constructed spreadsheet), but then Excel's "GoalSeek" tool for example can produce the value of IRR directly.

However, Excel has to be treated with care when it comes to plotting points calculated by Excel onto a graph – for demonstration or interpolation purposes.

### 9.12.3   Interest During Construction

If a sum is borrowed to build any or all of the plant, then it is very likely that this loan will be called upon (or "drawn down") during the construction period. The moment any part of it is borrowed, then interest becomes due from that moment.

In all cases of such borrowing, however, the company does not expect the interest to be paid by the project as it comes due, because the project has, as yet, no revenues from which to pay the interest. The lender will be paid its interest from other funds belonging to the company, and the interest due during construction is then turned into a capital sum (capitalised), which is then added to all of the other construction costs. The loan will normally be arranged at a fixed interest rate, so the extra sum can be calculated quite easily.

### 9.12.4   Repayment of Loans

All loans have to be repaid to the lender, but it is usual for the loan terms to be such as to allow repayment over several years: at least 5, and maybe 10. In this case, the outstanding loan will decrease with each year of the plant's lifetime, and the interest payment for each year will fall accordingly.

Conventionally, the loan interest, paid in any year, is assumed due on the outstanding sum at the start of that year, while any repayment does not take effect until the end of the year.

For any project, the repayment of the loan must effectively come out of its proceeds over the years, *i.e.* the total net revenues (cash flow) must be at least as large as the amount borrowed, on top of the other costs of construction, *etc.* However, the repayment element of the loan is *not* included in a financial appraisal, because its equivalent in plant cost has already been included in the balance as part of that cost. (An alternative way of looking at this is that the company, in paying back the loan, is just replacing that part of the original cost with an equal but differently sourced sum of money.)

### 9.12.5 Effect of Plant Size

A calculated IRR is a unique index for a particular set of costs and revenues, and enables easy comparison with the corresponding index for other situations. However, it will be apparent that if all of the figures in the above example had been multiplied by two, then the same IRR would have resulted. This implies that the two plants, differing by a factor of two in size, are equally attractive financially.

This is true – and so other factors have to be brought into the decision process. Can the company afford to build a bigger plant? Can the market accept twice as much product?

It is, of course, also true that a plant of twice the output would probably cost significantly less than twice as much as the smaller one (a price factor of $2^{0.6}$, if the six-tenths rule is used), and so would be more attractive financially – but the same questions have to be asked.

## 9.13    Other Aspects

There are some other aspects of financial appraisal that need consideration.

### 9.13.1  Non-profit Projects

This chapter has been aimed at the revenue-earning project, the amounts of revenue leading to an annual cash flow, from which profitability, and hence economic feasibility, can be determined. But what if there are no revenues? A new public service investment, such as a major (non-toll) road, or an overseas aid project, such as a water pipeline system for a sub-Saharan country, generate no revenues – yet may have huge investment costs.

These kinds of project must still undergo a stringent appraisal process, to ensure that each is the best way of spending the money required. It follows that some kind of "revenue" must be calculated, and this can be done on the basis of savings on other expenditures, or benefits to the community. Such calculations are not easy, but they can be done, and "cost-benefit analysis" is a well-practised technique.

### 9.13.2   Sensitivity Analyses

A major benefit of the DCF process, especially when used in combination with Excel's calculating powers, is that it enables rapid calculation of alternative circumstances, of the "what if" variety. This allows the determination of the sensitivity of the appraisal to, say, changes in sales price (the most important variation), or raw material cost, or electricity cost, or an increase in the labour needs, or an underestimate of construction costs. If the spreadsheets are properly designed, then it may take only a few key presses on the computer to derive IRR values for different sales prices or construction costs, for example.

### 9.13.3   Inflation

It is normal, in undertaking an economic appraisal, to ignore the effects of inflation – which applies both to costs and revenues (assuming that prices rise roughly to keep pace with inflation). If inflation is included in the calculation, then a slightly higher value will normally found for the IRR, so that its exclusion presents a slightly less favourable case.

### 9.13.4   Post-construction Audit

The total expenditure on the plant, after construction and start-up, should be carefully developed and recorded for the company's files, against a need on a future project – the best possible guide to an estimate of the cost of a future plant is the detailed cost of an earlier example.

The operational costs, sales costs and resultant profitability should be recalculated once the plant is running to its full capacity, and the revenue should be checked against the forecast figure. Maintenance costs should also be carefully checked for future use.

## Appendix – Worked Example

Consider the production of acetone from $i$-propyl alcohol, at a full annual rate of 100 000 tonnes, on an existing site (*i.e.* no land purchase cost). From previous experience, the delivered equipment cost can be expected to be £14.5 million.

As this is a fluid-based process, the only solid material being the catalyst, the following equipment scale factors can be used.

## A. Total Capital Employed

|  |  | £ Millions |
|---|---|---|
| *Purchased and delivered equipment (P&DE)* |  | 14.500 |
| Land |  | 0 |
| Site improvement | 15% P&DE | 2.175 |
| P&DE installation | 40% P&DE | 5.800 |
| Pipework and valves | 40% P&DE | 5.800 |
| Instrumentation and controls | 17% P&DE | 2.465 |
| Electrical services | 12% P&DE | 1.740 |
| Buildings for plant) | 23% P&DE | 3.335 |
| Other site services) | 55% P&DE | 7.975 |
| = "Buildings"* |  |  |
| Vehicles and distribution | 7% P&DE | 1.015 |
| Installed plant |  | 44.805 |
| Process, design and detailed engineering | 30% P&DE | 4.350 |
| Construction | 36% P&DE | 5.220 |
| Contractor's fee | 17% P&DE | 2.465 |
| Royalties |  | 0 |
| Engineering |  | 12.035 |
| Contingency | 10% of IP + E | 5.684 |
| Fixed capital investment (FCI) |  | 62.524 |
| Loan interest accrued during construction (2 years) $(25 \times 2 \times 0.08) + (25 \times 0.08)$ |  | 6.000 |
| Start-up costs | 10% of FCI | 6.252 |
| Total installed cost |  | 74.776 |
| Working capital | 15% of FCI | 9.379 |
| Total capital employed |  | 84.155 |

(Funded as £50 millions loan, paid in two equal instalments, and £34.155 millions company equity.)

*Depreciation is calculated on FCI only, at different rates for "plant and vehicles" and "buildings". Of the installed plant total:

Plant and vehicles = 33.495 or 74.76%, so proportion of FCI = 46.743
Buildings        = 11.310 or 25.24%, so        = 15.781

and depreciation will then be calculated on these amounts.

## B. Production Cost

From mass balance and energy balance calculations, the raw material and energy costs are

|  |  | £ Millions |
|---|---|---|
| Raw materials and catalyst | £175/tonne product | 17.500 |
| Utilities | £21.7/tonne | 2.170 |
| Packaging |  | 0 |
| *Variable* manufacturing cost for a full year's production. |  | 19.670 |

Fixed manufacturing cost (*i.e.* expenditure on operations that is independent of amount produced).

| Process labour | 5 shifts of 5 × 52 weeks | 0.572 |  |
|---|---|---|---|
| Laboratory and maintenance labour | 25% of shift | 0.143 |  |
| Other staff | 6 × 52 weeks | 0.129 |  |
| Supervision | 20% total labour | 0.169 |  |
| Payroll costs | 27% payroll | 0.274 |  |
| Total labour |  |  | 1.287 |
| Waste disposal | Fixed fee 1% | PC* | 1% PC |
| Maintenance material costs | 4% PC |  | 4% PC |
| Communications | 0.5% PC |  | 0.5% PC |

(*PC = full year's production cost, yet to be calculated).

| Insurances | 2.5% FCI | 1.563 |
|---|---|---|
| Local taxes | 2% FCI | 1.250 |
| *Fixed* manufacturing costs |  | 4.100 + 5.5% PC |
| Overheads | 3% FCI | 1.876 |

Depreciation (straight line)

12.5% of P & V (46.743) = 5.843 (for 8 years only)
8% of buildings (15.781) = 1.262

| Total depreciation |  | 7.105 |
|---|---|---|
| Contingency | 5% PC | 5% PC |

Production cost = variable cost + fixed costs + overheads + depreciation + contingency = 32.751 + 10.5% PC

or   PC = 32.751 + 0.105 PC

0.895 PC = 32.751

or   Production cost = £36.593 millions per full year

It is prudent to allow some production shortfall in the first year of production, say 10%.

All that then changes is the variable cost, which reduces by 10%, or £1.967 millions.

So, production cost in Year 1 is 36.593 − 1.967 = £34.626 millions.

[NB In later operating years, the depreciation allowance decreases, because Plant and vehicle depreciation operates at 12.5% and therefore is not levied after Year 8. The PC thus decreases after Year 8.]

## C. Sales Price Calculation

| | |
|---|---|
| If the sales price is £SP per tonne of SP product, then gross sales revenue (in a year of full production) | 100 000 SP |
| The cost of sales = 5% sales revenue = | 5000 SP |
| Factory netback = gross revenue − selling costs = | 95 000 SP |
| Production cost in full year = | 36.593 |
| PBIT = FNB − PC = | 95 000 SP − 36.593 |
| Annual interest on loan = 8% of 50.000 | 4.000 |
| Profit before tax = PBIT − Interest | 95 000 SP − 40.593 |
| PAT = PBT less 30% tax = | 0.7 (95 000 SP − 40.593) |

Good performance will have net profit (PAT) equal to 12.5% of gross sales revenue:

| | |
|---|---|
| PAT = 12.5% gross sales = 0.125 × 100 000 SP = | 12 500 SP |

Equating these two derivations of PAT:

12 500 SP = 0.7 (95 000 SP − 40.593)
(95 000 − 12 500/0.7) SP = 40.593 millions
SP = Selling price = £526.21

A selling price of £550.00/tonne will be used.

## D. Net Cash Flow

The annual net cash flow is the annual revenue for the project that is to be balanced against the investments. It is calculated as the net profit (PAT) in each year *plus* the depreciation allowance for that year:

$$NCF = PAT + D$$

In the final year of operation, usually (and here) taken as year 10 after start-up, the NCF is increased by the recovery of the working capital (and also by any scrap value for the plant, which is ignored here).

| Component | Year 1 | Years 2–8 | Year 9 | Year 10 |
|---|---|---|---|---|
| Gross sales | 49.50 | 55.000 | 55.000 | 55.000 |
| Factory netback | 47.025 | 52.250 | 52.250 | 52.250 |
| Production cost | 34.626 | 36.593 | 30.75 | 30.75 |
| Gross profit (PBIT) | 12.399 | 15.657 | 21.50 | 21.50 |
| Loan interest | 4.00 | 4.00 | 4.00 | 4.00 |
| Profit before tax | 8.399 | 11.657 | 17.50 | 17.50 |
| Profit after tax | 5.879 | 8.160 | 12.250 | 12.250 |
| Depreciation | 7.105 | 7.105 | 1.262 | 1.262 |
| Working capital | | | | 9.379 |
| Net cash flow | 12.984 | 15.265 | 13.512 | 22.891 |

These net cash flow figures are now used as the bases for the discounted revenues.

## E. Investment Schedule

The investments are made as company equity and loan, to cover the fixed capital investment, the interest accrued on the loan during the 2 year construction period, the start-up costs (testing and commissioning, and staff recruitment), and the working capital.

These sums are conventionally considered to have been paid at the start of each of the 2 years of construction (years −2 and −1) and at the point of start-up (year 0 – a "year" of zero length). The investment schedule is thus:

| At start of | Equity | Loan | Total |
|---|---|---|---|
| Year −2 | 8.262 | 25.0 | 33.262 |
| Year −1 | 10.262 | 25.0 | 35.262 |
| Year 0[a] | 15.631 | | 15.631 |
| *Total* | 34.155 | 50.0 | 84.155 |

[a]Payment for start-up costs and working capital.

## F. Discounting Process

The investments for the construction period and the net cash flows for the 10 year operating lifetime can now be discounted to determine the net present values at various discount factors. The undiscounted payments and revenues can be seen in the left hand column:

| Disc factor | 0 | 0.05 | 0.08 | 0.09 | 0.10 | 0.15 |
|---|---|---|---|---|---|---|
| Year −2 | 33.26 | 36.67 | 38.80 | 39.52 | 40.25 | 43.99 |
| Year −1 | 35.26 | 37.03 | 38.08 | 38.44 | 38.79 | 40.55 |
| Year 0 | 15.63 | 15.63 | 15.63 | 15.63 | 15.63 | 15.63 |
| PV costs | 84.16 | 89.33 | 92.51 | 93.59 | 94.67 | 100.17 |
| Year 1 | 12.98 | 12.37 | 12.02 | 11.91 | 11.80 | 11.29 |
| Year 2 | 15.26 | 13.85 | 13.09 | 12.85 | 12.62 | 11.54 |
| Year 3 | 15.26 | 13.19 | 12.12 | 11.79 | 11.47 | 10.04 |
| Year 4 | 15.26 | 12.56 | 11.22 | 10.81 | 10.43 | 8.73 |
| Year 5 | 15.26 | 11.96 | 10.39 | 9.92 | 9.48 | 7.59 |
| Year 6 | 15.26 | 11.39 | 9.62 | 9.10 | 8.62 | 6.60 |
| Year 7 | 15.26 | 10.85 | 8.91 | 8.35 | 7.83 | 5.74 |
| Year 8 | 15.26 | 10.33 | 8.25 | 7.66 | 7.12 | 4.99 |
| Year 9 | 13.51 | 8.71 | 6.76 | 6.22 | 5.73 | 3.84 |
| Year 10 | 22.89 | 14.05 | 10.60 | 9.67 | 8.83 | 5.66 |
| PV revs | 156.24 | 119.25 | 102.97 | 98.29 | 93.92 | 76.01 |
| NPV | 72.09 | 29.92 | 10.46 | 4.70 | −0.75 | −24.16 |

From this table a number of key discoveries can be made. Firstly, the total net cash flow of £156 millions is close to twice the total investment, so there is a reasonable overall profit. However, the undiscounted payback time is about 5 years and 8 months (the total revenues for years 1–5 are still £10 millions short of covering the total costs). This is a long time before payback − 2 to 3 years would be preferable – so this project is beginning to look a little unexciting.

The NPV is negative at a discount factor of 10% (although still positive at 9%), and the NPV *versus* DF data give an IRR figure of 9.86% (by interpolation graphically between 9% and 10%).

This rate of return:

1. is higher than the interest rate being charged on the loan used to pay for the plant;
2. is higher than could probably be achieved from long-term investments on the money markets; but

3. matches the reality of bulk chemical production in western Europe; and
4. is probably below any likely figure for the hurdle rate of return for the company.

It follows that the project can be safely recommended (remember there are 5% contingencies in both plant cost and production cost calculations), but that there may well be better ways to invest £84 millions. A lot will then depend on any other reasons that there might be to want to build an acetone plant (defending a market, guaranteeing a raw material, *etc.*).

## Recommended Reading

*General coverage of process engineering economics:*
1. D. Allen, *Economic Evaluation of Projects*, IChemE, 3rd edn, 1991.
2. D. Brennan, *Process Industry Economics*, IChemE, 1998.

*Corporate finance:*
3. B. Rothenberg and J. Newman, *The Daily Telegraph Understanding Company Accounts*, Kogan Page, 3rd edn, 1993.
4. R. A. Brearly and S. C. Myers, *Principles of Corporate Finance*, McGraw-Hill, 4th edn, 1991.

*Process equipment cost data (some of which also include advice on process plant economics):*
5. M. S. Peters and K. Timmerhaus, *Plant Design and Economics for Chemical Engineers*, McGraw-Hill, 4th edn, 1991.
6. F. A. Holland and J. K. Wilkinson, *Process Economics*, McGraw-Hill, 7th edn, 1997, Section 9 of R. H. Perry, D. W. Green and J. O. Maloney, *Perry's Chemical Engineers' Handbook*.
7. *A Guide to Capital Cost Estimating*, ed. A. M. Gerrard, IChemE, 4th edn, 2000.
8. J. Sweeting, *Project Cost Estimating – Principles and Practice*, IChemE, 1st edn, 1997.
9. *(Coulson and Richardson's) Chemical Engineering*, ed. R. K. Sinnott, vol. 6, of which Chapter 6 covers "Costing and Project Evaluation", Butterworth-Heineman, 4th edn, 1999.

*Cost data on other equipment (regularly republished):*
10. *Spon's Mechanical & Electrical Services Price Book*, ed. Mott, Green and Wall, Taylor & Francis, 34th edn, 2003.

11. *Laxton's Building Price Book,* ed. V. B. Johnson, Butterworth-Heineman, 2005.

*Project management:*
12. O. P. Kharbanda and E. A. Stallworthy, *Effective Project Cost Control,* IChemE, 1985.

*Occasional articles on costs and plant economics (monthly periodicals):*
  Chemical Engineering published by Chemical Week Publishing.
  Process Engineering published by Centaur Publishing.
  The Cost Engineer published by The Association of Cost Engineers
  Oil and Gas Journal published by Penn Well Corporation.

*Regular reports on costs of chemicals and related materials (weekly periodicals):*
  European Chemical News published by Reed Business Press
  Chemical Market Reporter published by Reed Business Press.

# 10  Process and Personal Safety

D. C. Shallcross

## 10.1  Introduction

Nearly all processing facilities feature some form of hazard that has the potential for causing serious harm or damage to people, property or the environment. Good design and careful operation by trained and conscientious staff, however, make the process industries some of the safest in the world. Chemical engineers use physical, chemical and biological processes to turn one or more feed materials into more useful, valuable or less harmful materials. Sometimes these processes involve the use of toxic chemicals, elevated temperatures and pressures, dangerous machinery or work in noisy, dusty or enclosed environments. Process or equipment failure can lead to the loss of containment of material, energy or both, leading to impacts outside the site boundaries. Whilst incidents can occur during normal, routine operation, accidents can also occur during start-up or shut-down of the facility, during construction, maintenance or site close-out, or as a result of an unexpected process upset.

*Process safety* addresses the safe operation of all or part of a facility dealing with hazardous materials. *Personal safety* relates to the safety of individuals or teams of workers on site. There is obvious overlap between the two, as sound safety procedures will assist in keeping personnel safe.

A *hazard* is a source of potential harm, damage or other adverse impact to people, property and/or the environment. The hazard might

Concepts of Chemical Engineering for Chemists: 2nd Edition
Edited by Stefaan Simons
© The Royal Society of Chemistry 2017
Published by the Royal Society of Chemistry, www.rsc.org

be of a biological, chemical or physical nature, or it might be a particular situation or procedure. Hazards may be present continuously, or intermittently. The impact of a hazard on a person, property or the environment might be immediate or cumulative over an extended period of time. Hazards may result from a design flaw, a lack of knowledge and/or training, a failure of equipment, system or process, or even an extreme weather event. Hazards might be well recognized, or might only become apparent with time. An example of this last point is the once accepted use of asbestos. Decades ago, before its hazardous nature was understood, asbestos was widely used. However, now it is removed wherever it is found.

**Risk** is the potential of an event occurring that will have an adverse impact on people, property and/or the environment. Risk takes into account both the likelihood that a particular incident will occur and the magnitude of the consequences of that incident. For example, the likelihood of failure of a high-level alarm fitted in a tank might be low, but the consequences of such a failure might be high. Nonetheless, the risk might be sufficiently great to require the system to be designed to minimize the risk, possibly by the installation of a second independent alarm.

An **incident** is an event that causes harm, damage or other adverse impact on people, property or the environment. An incident might involve the loss of containment of material and/or energy, either rapidly or gradually over time. A serious incident that results in unintended injury, death, property loss or significant impact on the environment may be termed an **accident**.

The **consequence** of an event is the outcome or impact of the event on people, property and the environment. The impact of an event includes intangible consequence such as loss of reputation and loss of future earnings.

The **likelihood** of an event is a measure of the expected probability or the frequency that the event will occur. For example, based upon experience, it might be assumed that any given pump will fail on average once every two years.

**Risk analysis** is a systematic approach to understanding the nature and level of risk. The result is a quantitative estimate of the risk, based upon engineering evaluations and judgements and established numerical techniques. **Risk assessment** uses the results of risk analysis to make decisions on **risk reduction**; actions to reduce the risk of an action, process or event. **Residual risk** is the risk remaining after implementation of the risk reduction strategies. **Raw risk** is the risk that exists prior to the application of risk reduction treatments.

*Risk avoidance* is the decision not to become involved in, or to withdraw from, a risk situation.

It is impossible to make any human endeavour perfectly safe. *Acceptable risk* is the level of the potential adverse impact on people, property or the environment that individuals and/or society deem to be tolerable, given the overall benefits of the activity. Where the likelihood of an incident is very small or the consequences so slight and the benefits to society so great, then society may decide to tolerate the risk. For example, consider the risk of travelling by car. In a city of several million, every year hundreds of people die or are badly injured in car accidents. The risk that every person takes in travelling by car is accepted because of the great convenience of this form of travel. People recognize that it is never possible to make car travel perfectly safe, yet they are still prepared to risk their lives every day.

## 10.2   Safety in Process Plant Design

Safety must be actively considered in all stages of the design of a new facility, from the design overview stage (when the major process steps are being selected), to the development of operating and maintenance procedures around individual items of equipment. There are a number of strategies that design engineers can employ in designing new facilities that will help to identify potential hazards and to reduce the impact of any incident on the facility, the staff and the wider community. Design engineers also need to draw on the lessons learned from incidents and accidents around the world.

### 10.2.1   Inherently Safe Design

In the 1980s and 1990s, the inherently safe design philosophy of process design emerged in the process industries. The philosophy's four main methods to increase the safety of a process and to mitigate the impact of any incident are: minimization, substitution, moderation and simplification. Designing a facility to be inherently safe permanently eliminates or reduces hazards to either completely avoid or to reduce the consequences of incidents.

Minimization: reduce the amount of hazardous material held on site. Only sufficient amounts of the process feedstock required for efficient operation should be kept on site at any time. Equally, the finished products should be exported from the site as soon as practicable. Intermediate products should not be held in significant quantities on a regular basis. Any hazardous intermediate product should be generated and then consumed as quickly as possible, without being stored in any significant quantity. While efficient operation may require a facility to be able to store larger amounts of feedstock, product or intermediates, this should only be done as required to meet exceptional operational demands – it should not be the common practice. Reducing chemical inventories also has the advantage of reducing costs. As Kletz noted, what you don't have can't leak.[1]

Substitution: replace one material with another, safer alternative. If the desired chemical can be produced by more than one chemical pathway, then the pathway involving the production of the less hazardous intermediate should be selected. If the different intermediates are equally hazardous, then the chemical pathway involving the less extreme operating conditions might be selected. Other examples of substitution include using longer runs of welded pipe rather than pipe sections joined by flanges that can leak, using solvents that are less toxic and using water as a process heat transfer fluid rather than hot oil.

Moderation: attenuating or limiting the effects of an incident. It includes using hazardous material under less hazardous conditions, for example, by operating all or part of the process under more moderate conditions of temperature and pressure. Moderation might also call for the use of more dilute, rather than concentrated, chemicals. Moderation can also be applied in designing the layout of a facility. Locating the control room away from the operations area reduces the risk of death or injury to staff working in the control room at the time of an incident. Similarly, the road layout should be designed so that road tankers needed to enter the site to transport the finished product do not need to approach the operations area.

Simplification: make the design as simple as possible, eliminating any unneeded complexities. Simplification can occur through the reduction in process steps, the elimination of unnecessary equipment and the clearer representation of information to engineers and operators. Complexities in the design of a plant often arise to control potential hazards. If these hazards have been eliminated, then the need for the complexities may also disappear. Simpler plants are easier to understand and to operate and may be more resilient to the impact of unplanned process upsets. Simplification also means designing facilities for safer and easier maintenance, and using more reliable equipment that needs less maintenance. Computer interfaces used by operators to run the plant should also be designed to be easier to use, displaying information in an unambiguous form. Piping throughout the facility should be well labelled and easy to trace. In some cases tanks can be colour-coded to help prevent deliveries to the wrong tank. Reducing the complexity of a facility can lead to reduced construction, operating and maintenance costs.

### 10.2.2 Design Safety Studies

An important step in any design study is to identify any potential hazards that might exist in a process or operation before the process facility is built or before the procedure is implemented. A ***hazard and operability study*** (HAZOP) is a structured and systematic examination of a proposed process or operation that is undertaken to identify any potential hazards that might pose risks to people, property and the environment. This type of study can also be used to identify elements of the proposed design or process that might be inefficient and can be performed on existing facilities in an effort to identify existing hazards, or potential dangerous procedures. The HAZOP studies should be conducted by multi-disciplinary teams that include experienced process engineers. The structure of such studies has been developed to help stimulate the imagination of the participants in an effort to consider all the possible things that might go wrong in the operation of a facility. They are applied not only to normal plant operation, but also to the start-up and shut-down stages of a process. The HAZOP study calls for the engineers to apply a list of 'What if?' questions to the intended process in a logical order.

Working in a team, the engineers apply a series of guide words to a range of parameters at a series of study nodes. The guide words include:

NO or NONE
MORE OF or MORE THAN
LESS OF or LESS THAN
AS WELL AS
PART OF
INSTEAD OF
REVERSE
TOO FAST
TOO SLOW

TOO EARLY
TOO LATE
WHERE ELSE

The parameters that these guide words are applied to might include:

FLOW
TEMPERATURE
PRESSURE
LEVEL
COMPOSITION
VISCOSITY

Each guide word is applied in turn to each parameter at the study nodes.

To illustrate the use of the HAZOP study, let us consider a reaction vessel, as shown in Figure 10.1. A large stainless steel reaction vessel is used as the reactor. The reactants are continuously fed into the top of the vessel through a pipeline from another part of the process. Inside the reactor the exothermic reaction is accompanied by the production of heat. If the reactor is not cooled, then the contents of the reactor will get dangerously hot. As the rate at which the reaction occurs is a function of the temperature, increasing the temperature will cause the reaction to go faster, releasing more heat, causing the

**Figure 10.1** A simple cooled reactor.

temperature to rise even higher. If left unchecked, this spiralling increase in temperature would result in loss of control and a runaway reaction.

In order to prevent a runaway reaction, cooling water is used to control the temperature of the reactor contents. The cooling water circulates through the reactor inside a cooling coil, carrying away the excess heat. A temperature controller is used to control the flow rate of the cooling water entering the coil. Increasing the flow rate of cooling water allows more heat to be removed. The controller regulates the flow of water into the reactor in response to the temperature measured by a thermocouple located inside the reactor.

To ensure good mixing throughout the vessel, a stirrer rotates continuously within the reactor. The product stream containing some unreacted reactants is continuously withdrawn from the reactor.

We perform a HAZOP analysis on the system by placing a study node on the cooling water line into the reactor. Beginning with the parameter FLOW, we now apply each of the guide words to the flow of cooling water.

NONE: Under what circumstances might there be no flow of cooling water to the reactor? The cooling water control valve might have failed in the closed position, the controller used to control the flow rate of cooling water might have failed, the thermocouple used to measure the reactor water temperature might be faulty, the cooling water coils might have become plugged with contaminants, the entire cooling water system might have failed, the service air system used to actuate the control valve might have failed. In all cases a loss of cooling might possibly lead to a runaway reaction. Actions that can be taken to mitigate against the possibility of a runaway reaction might include the use of a control valve that only fails open, the installation of filters on the cooling water pipe line to capture contaminants, the installation of high reactor temperature alarms, and a low cooling water flow alarm. The operators might also be instructed to monitor the reliability of the cooling water system, as well as that of the controller.

MORE OF: Under what conditions might the flow of cooling water to the reactor be greater than required? The cooling water control valve might have failed in the open position, the controller used to control the flow rate of cooling water might have failed, the thermocouple used to measure the reactor water temperature might be giving readings higher than the actual temperature and so tricking the controller to send more water through the cooling water loop. In all cases, too much cooling water might cause the reaction to slow down, resulting in the reactor product stream having too little of the desired product and too great a loss of unreacted reactants from the system. Further, prolonged operation of the reactor at too low a temperature might result in an accumulation of reactants in the system that in turn may result in a runaway reaction once normal temperature is restored. Actions that might be taken in response to this condition could include the installation of a low reactor temperature alarm and a change in the operating procedure to allow proper recovery from a period of low reactor temperatures.

LESS OF: There are several reasons why too little cooling water might be flowing to the cooling loop. The cooling water line might be partially blocked due to contaminants or scale formation, the control valve might have failed partially closed due to a failure of the service air system or the cooling water supply might be compromised due to a major leak further up the line. In all cases the likelihood of a runaway reaction is increased. Similar actions to those required for the NONE case might be necessary.

AS WELL AS: Under what conditions might the cooling water be contaminated with some extra material?

PART OF: This guide word is not appropriate here. Not all guide words are relevant to each situation. PART OF is only relevant to mixtures containing two or more components. If we needed to add two components to a vessel then applying PART OF would force us to consider the case when only part of the mixture is added, *i.e.*, only one of the two required components is added.

INSTEAD OF: Is it possible that something other than cooling water is flowing through the cooling water system? How might this have occurred? What steps could be taken to ensure that it would not occur?

REVERSE: Is it possible for the cooling water to flow backwards through the loop? Could a leak in the cooling water loop allow reactants into the line, forcing the water back up the line?

TOO FAST and TOO SLOW: In this context these guide words have the same meaning as MORE OF and LESS OF.

TOO EARLY: The reactor cooling water system in this example spends most of its time operating at steady-state, under which none of the operating conditions change with time. Under steady-state operation this guide word is therefore not appropriate. However, during the start-up and shut-down phases of its operation it is appropriate to consider what might happen if any changes to the flow rate of cooling water occurs earlier than planned. For example, during the shut-down phase, if the cooling water flow rate is decreased too early while the reactants are still being introduced into the reactor, a runaway reaction might occur. Under what circumstances might the cooling water be shut off too early? What steps could be taken to ensure that this does not occur?

TOO LATE: As with the TOO EARLY guide words, TOO LATE is only really relevant during the transitional phases of the operation of start-up and shut-down. What would happen if the circulation of cooling water commenced too late during the start-up phase, or stopped too late during shut-down?

WHERE ELSE: Cooling water flows, but not through the cooling loop. Under what conditions might the cooling water flow un-intentionally into another part of the facility. What actions could be taken to prevent this from happening and to safeguard the operation of the reactor?

Having applied the HAZOP guide words to the flow rate of cooling water into the reaction vessel, we see that a few corrective actions are identified time and time again. For example, the analysis shows that low reactor temperature and high reactor temperature alarms should be installed. Similarly, a low cooling water flow alarm should be installed. The analysis also suggests that a second, independent thermocouple should be installed in the reactor in case one thermocouple fails. Other possible actions are as noted above.

Next, the analysis should be repeated for the TEMPERATURE parameter. Again, each of the guide words is applied in turn to the temperature of the incoming cooling water. Some guide words, such as NONE, have no meaning in this context. Other guide words have significant meanings.

MORE OF: Under what conditions would the temperature of the cooling water be above its design value? It might be that the incoming cooling water should be supplied at a temperature of 12 °C, but is instead supplied at 40 °C. If the incoming water is hotter than expected, then the temperature of the reaction vessel contents might rise to hazardous levels. What might cause the cooling water temperature to rise? What actions could be taken to either prevent this from happening, or to minimize the impact of such an occurrence on the safe operation of the process?

LESS OF: Having the cooling water supply temperature drop below the design temperature might also pose a hazard to the process. How might this occur? What might be the consequences of the cooling water temperature falling to perhaps 2 °C from an expected 12 °C? Processing facilities have failed when unexpectedly low temperatures have been encountered.

Once the TEMPERATURE parameter has been considered, then PRESSURE is subjected to the same series of guide words. Having completed the HAZOP analysis for the cooling water supply study node, we now consider other study nodes around the reactor, such as the reactant feed line into the reactor and the contents of the reactor.

Completing a HAZOP study after some detailed design work has been completed allows potential hazards to be identified and either eliminated, or at least reduced. In addition, elements of the process or procedure that might be inefficient may also be identified. It is always easier to change a process before it is built, rather than after. Whenever a major change is contemplated, be it a modification or expansion, then a HAZOP analysis must be employed. HAZOP studies can also be applied to existing processes or procedures as a way of improving both their safety and efficiency.

## 10.2.3 Designing for Maintenance – Designing for Safety

The need for regular maintenance of equipment, including machinery, instrumentation, cabling, piping and structures, is a requirement for both safe and efficient plant operation. Any processing facility should always be designed with ongoing maintenance in mind. If careful thought is given at the design stage, difficult and dangerous jobs can be made less troublesome and less risky.

The heavy manhole cover shown in Figure 10.2 is supported by a fixed jib. When the cover is unbolted the jib is used to swing the cover out of the way. This eliminates the need for operators to lift the heavy cover themselves. It also ensures that the cover is replaced every time in the correct orientation, with the bolt holes correctly lined up with the right bolts. The ends of the bolts are also numbered to ensure that the bolts are tightened and loosened in the correct sequence.

In a similar fashion, shell and tube heat exchangers should be located so that the tube bundle can be pulled clear of the shell for maintenance without encountering any obstruction. Some exchangers are built with their own dedicated rail-mounted block and tackle that will take the weight of the bundle as it withdrawn from the shell.

**Figure 10.2**   The fixed jib allows the heavy manhole cover to be swung out of the way to allow access to the tower. This design eliminates the need for operators to risk injury by lifting the heavy cover unassisted.

Sometimes good design for maintenance can be as simple as locating all the pumps in an area of a facility in a row at ground level. This allows maintenance to occur without the need for workers to operate at height.

### 10.2.4   Case Study – Bhopal

The world's worst industrial disaster occurred on the night of 2/3 December 1984 in the city of Bhopal, India.[2] Approximately 1 tonne of water entered a storage tank containing over 40 tonnes of methyl isocyanate (MIC), a highly reactive intermediate product generated and then used in the production of pesticide. The water began reacting with the MIC, generating modest amounts of heat, which accelerated the reaction rate. Shortly after midnight the reaction rate accelerated to such an extent that the pressure in the tank spiked, causing the hot MIC vapour to vent through the tank's automatic pressure relief system. The highly toxic MIC vapour vented out of a stack, forming a cloud that moved across the urban area, exposing over 500 000 people. Over 3700 people died from direct exposure to the gas cloud, with tens of thousands suffering temporary or permanently disabling injuries.

**Figure 10.3** The reaction scheme used by UCIL at Bhopal, showing MIC as an intermediate.[3]

The Union Carbide India Limited facility was built in the late 1960s to produce the pesticide "carbaryl" (1-naphthyl methylcarbanate). Of the two chemical pathways available to manufacture carbaryl, UCIL chose to use the one involving the manufacture of MIC, methyl isocyanate, as an intermediate (Figure 10.3). MIC is a colourless liquid at room temperature, with a boiling point of 39.5 °C. It is extremely toxic, whether inhaled, ingested or absorbed through the skin and exposure levels of 21 ppm may result in haemorrhaging and death. It reacts readily and exothermically with water. If sufficient heat is not removed fast enough, then the rate of reaction with water will increase rapidly, leading to the MIC boiling.

At the Bhopal site the MIC was stored in three large tanks. The temperature of the MIC in these two tanks was kept at a temperature below 11 °C using a refrigeration system. The need to keep the MIC below its normal boing point of 39.5 °C was understood. In January 1982, there was a catastrophic failure of a seal in the pumps used in the refrigeration system, resulting in the release of sufficient MIC to send 25 workers to hospital. No one died and there was no impact on public health outside the plant boundaries. Following the failure of the seal, the decision was made to permanently shut down the refrigeration system. It then became commonplace for the temperature inside the MIC storage tanks to range between 15 °C and 40 °C. As the MIC storage tank high temperature alarms kept on activating, the operators and management took the conscious decision to disable them. At the same time, the tank temperature indicators in the control room were rendered useless, as they were unable to indicate temperatures above 25 °C. No decision was taken to replace the tank temperature indicators with more appropriate ones with a higher temperature range. The operators then had no way of knowing the actual temperatures inside the tanks on the many occasions when the temperatures rose above 25 °C.

In the late evening of 2 December 1984, routine maintenance was taking place at the plant. This included flushing out with water some of the pipe lines that were located some distance from the MIC storage tanks. At some time before midnight a quantity of water, later estimated to be approximately 1000 litres, entered one of the MIC tanks. This initiated a set of exothermic decomposition reactions that began to pressurize the tanks, whilst generating more heat. With the high temperature alarms disabled and the temperature readout in the control room unable to display temperatures above 25 °C, the operators were unaware of the problem developing in the tank.

Just after midnight the decomposition reactions ran away, causing the pressure within the tank to spike. The elevated pressure within the tank caused the safety pressure valve to lift, allowing the toxic MIC vapour to vent to the scrubber and flare. Unfortunately, some time previously the scrubber had been decommissioned and so the MIC vapour passed through the unit without treatment. The gas flare was also out of operation, with the burner inactive. The MIC vapours therefore vented from the top of the flare stack untreated, with over 40 tonnes of the toxic compound escaping to the environment. The impact on the health of the residents was immediate and severe, with many dying before dawn.

There were many factors that led to this disaster, including poor design, poor operation and inadequate management, training and safety processes.

While the Bhopal plant had been designed before the philosophy of **inherently safe design** had been developed, the disaster is a case study of a design that was inherently unsafe. The process designers chose to utilize a process that involved the production of a highly toxic intermediate product, namely, methyl isocyanate (MIC). If they had chosen an alternative chemical pathway (Figure 10.4), the MIC hazard would never have existed at the Bhopal site. The designers also chose to include three large tanks, each capable of storing 50 tonnes of MIC. The MIC produced in the process was an intermediate product and there was no reason for its being stored in such large quantities – as soon as it was produced in one reaction it should have been consumed in the next. Indeed, the plant could have operated just as efficiently if less than 5 tonnes had been stored. Providing the capability to store up to 150 tonnes of the MIC inevitably meant that excessive quantities were stored from time to time.

**Figure 10.4** An alternative reaction scheme with a less hazardous intermediate.[3]

Beyond the poor design, the plant was operated in an unsafe manner. When the refrigeration system was decommissioned no decision was taken to replace it with an alternative cooling system. The need to maintain the stored MIC at temperatures around 15 to 20 °C remained. When the standard operating conditions inside the tank increased, rather than disconnecting the high-level alarms, the alarms should have been reset to a higher temperature and the temperature displays in the control room should have been modified to show the more elevated temperatures.

When the plant was designed and built, the scrubber and flare were important safety features, allowing the consequences of unplanned process upsets to be dealt with safely. When these two units were taken offline, the safety of the entire facility was compromised. Management should never have allowed the plant to be operated with these two units decommissioned.

The disaster of Bhopal could clearly have been avoided if the facility had been better designed, and had been operated as designed.

## 10.2.5 Case Study – Flixborough

The United Kingdom's worst industrial explosion occurred on a Saturday in June 1974 at Nypro's Flixborough facility. The explosion devastated the site, killing 28 workers and injuring 36 others. Over 2000 dwellings in the surrounding area were affected by the blast.[4-6]

The caprolactam plant located in Flixborough, England, used a process of partial oxidation of cyclohexane to produce cyclohexanol and cyclohexanone as intermediate products. The cyclohexane was recirculated through six reactors linked in series. The six reactors were arranged so that each successive reactor was at a lower height than the one before it, with reactor 1 being much higher than reactor 6.

Each reactor consisted of a mild steel outer shell surrounding and bonded to an inner stainless steel vessel. Two months before the explosion a cyclohexane leak was detected from the fifth reactor. Inspection revealed a vertical crack in the outer, mild steel shell. The leak indicated that the integrity of the inner vessel was also compromised. After the plant was shut down and the vessel depressurized and purged, the reactor was opened and a two-metre long crack was discovered in the stainless steel vessel.

The decision was taken to remove the defective reactor and to connect a replacement bypass line between the fourth and sixth reactors that spanned the gap between the missing reactor. Because the inlet and outlet points of this bypass line were at different heights, the final bypass assembly contained two 90° bends in the pipe in close proximity to one another. The bypass line also included two bellows sections. The bypass pipe assembly was supported on temporary scaffolding. Although the existing pipe lines were about 70 cm in diameter, the new bypass line was just 20 cm in diameter. The Court of Inquiry after the disaster found that "...no overall control or planning of the design, construction, testing of fitting of the assembly nor was any check made that the operations had been properly carried out".[7] The bypass line installed between the two reactors was not fit for purpose.

The site engineers did not consider it necessary to check any of the other five reactors to determine whether similar cracks existed in those other reactors.

On the days immediately prior to the accident the plant was shut down to allow a leak to be repaired in a valve associated with one of the reaction vessels. During the morning of Saturday June 1, 1974, the operators struggled to return the plant to full operation. At a little before 5 pm, the bypass line ruptured, releasing cyclohexane, which formed a vapour cloud low to the ground. When the flammable mixture encountered an ignition source, a massive explosion followed.

The control room was located within the process area of the plant and had not been designed to withstand the pressures that accompanied the explosion. As a result, the entire control room was destroyed by the explosion, killing all 18 workers present at the time. The nearby administration building was also destroyed, but because it was a Saturday the building was empty.

As the Court of Inquiry noted, the scene was set for disaster when a major modification to the design of the caprolactam plant was made with the removal of one of the six reactors. Instead of replacing the leaking reactor, the reactor was removed and a bypass line was installed. This bypass line was not designed to the same high standards that the original plant was designed. If a chain is only as strong as its weakest link, then this bypass line was the most poorly designed item in the plant.

The explosion's death toll was greatly increased by the siting of the control room within the process area. While all processing facilities are designed to avoid gas leakages and explosions, there is always the risk that a devastating explosion could occur. If such an explosion is foreseeable, then it would have been prudent to locate the control room away from the process area in a structure designed to withstand an explosion. Similarly, there was no need for the administration building to be located immediately adjacent to the process area. A safer layout of the facility would have seen this building located some distance away.

Thirty years after the Flixborough explosion, engineers at BP's refinery at Texas City located the portable office close to the process area. When an explosion occurred following a leak from the raffinate splitter tower, 15 people working in the portable offices were killed. The temporary offices were located just 37 m away from the tower. They should never have been located so close to the process area, a lesson that should have been learnt from the Flixborough disaster.[8]

At Texas City the liquid overflowed the top of the tower and ran down an overhead pipe to pressure relief valves located near the ground 45 m below. These valves opened, discharging the liquid to a blowdown drum. The liquid under pressure then vented out the top of a stack connected to the drum, forming a geyser. The liquid droplets vapourized as they rained back down to the ground, forming a dense vapour cloud. The flammable cloud found an ignition source and detonated, devastating the unit and destroying the temporary office accommodation nearby.

While the direct cause of the incident was the failure to open a valve and the subsequent failure of the instrumentation to correctly indicate that the raffinate tower was overfilling, the deaths were a direct result of the temporary office accommodation being located too close to an operating process facility. The offices were not associated with the operation of the unit and there was no need to locate them where they did.

# 10.3   Safety in Plant Operation

A comprehensive safety management system must exist in any processing facility that looks to maintain not only the personal safety of all workers, contractors and visitors, but also the safety of the process. A good safety management system will identify hazards and control risks, while also continuously conducting reviews to ensure that the all processes and procedures remain fit for purpose.

As part of the safety management system, all potential hazards must be identified, control strategies developed and implemented and the outcomes reviewed. Procedures, particularly for non-routine tasks, must be in place and all personnel must be adequately trained.

## 10.3.1   Hierarchy of Hazard Control

The hierarchy of hazard control is an approach that may be used to either entirely eliminate, or at least minimize, the exposure to hazards (Figure 10.5). Eliminating or minimizing this exposure reduces the risks to personnel of undertaking various activities, such as maintenance tasks. The six hazard controls, in the order in which they should be applied, are:

1. elimination;
2. substitution;
3. isolation;
4. engineering controls;
5. administrative controls; and
6. personal protective equipment.

**Elimination** – The best way to eliminate a risk is to completely remove the hazard. This is best done at the design stage of a process, but can also be done later, once the facility has been

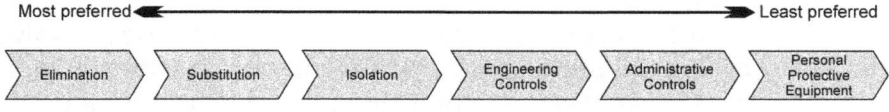

**Figure 10.5**   The hierarchy of hazard control.

built and operation has commenced. Placing equipment such as a pump on the ground rather than at an elevated position eliminates the need for operators to work at height when performing maintenance tasks. The risk of injury to workers involved in handling the heavy manhole cover shown in Figure 10.2 has been eliminated by the installation of the fixed jib. Reducing or eliminating the inventory of hazardous chemicals is another way to eliminate, or substantially reduce, risk. Two of the simplest steps that can be taken to eliminate hazards are the requirement for all personnel to remove jewellery, such as rings and chains, before entering the facility, and a total ban on matches and cigarette lighters from the facility.

**Substitution** – If a hazard cannot be completely removed, can a change be made to replace the hazard, or hazardous work practice, with a less hazardous one? At the Union Carbide plant at Bhopal, if the chemical pathway to produce carbaryl *via* MIC had been substituted with the alternative chemical pathway to chloroformate, the hazards posed by the presence of the intermediate product would have been greatly reduced. Another example of substitution would be substituting a toxic cleaning solution with one that is less toxic and is non-allergenic.

**Isolation** – This control does not eliminate the hazard, but rather physically isolates personnel from the hazard. It might involve the erection of a barrier or the use of remotely controlled equipment so that operators do not have to actually approach the hazard. The hot end of the heat exchanger, shown in Figure 10.6, is surrounded by a mesh to protect personnel from the risk of being burnt by accidental contact. Rather than send an operator into a potentially hazardous environment to manually open or close a valve, the valve might be operated remotely from the control room. Remotely controlled robots that inspect and then detonate suspected bombs are better than sending a bomb disposal expert into harm's way.

**Engineering controls** – These are modifications to equipment, such as the installations of interlocks so that equipment cannot operate until safety features are engaged. For example, equipment with rotating shafts, such as pumps or drill presses, might not operate until the safety guard is in place. Another example is the use of electrical lockouts that make it impossible to energize equipment until all the locks have been removed from the lockout hasp.

**Administrative controls** – Administrative controls address the way in which people work. These controls include the use of employee

**Figure 10.6**  The mesh guard on the end of the heat exchanger is an example of the control of a hazard by isolation.

and contractor training and site inductions, the development of specialized procedures to be used for non-routine activities and the installation of signs and warning labels identifying potential hazards to all personnel. Administrative controls do not remove the hazards, but seek to reduce exposure to the hazards. They rely on people following the set procedures and site rules. For that reason, they can be easily circumvented. Examples of administrative controls include safety systems such as the **permit to work system**, the **lockout–tagout system**, working at height permits, confined space entry permits and a regular audit of facility and activity safety.

**Personal protective equipment** – The last line of defence, and the least effective method for controlling hazards, is personal protective equipment (PPE). PPE includes safety glasses, hard hats, hearing protection, overalls, high-visibility clothing, gloves, safety footwear, dust masks and respirators. Different tasks might have different PPE requirements. It must be remembered that personnel need training in the use of certain types of PPE. In one case, a worker collapsed while wearing a respirator, as he was not getting enough air. He had never been instructed on how to properly fit the respirator, having been simply handed it without instruction. It would have taken just a minute for him to be shown how to properly fit the mask before entering the job site.

## 10.3.2   Risk Analysis and Risk Reduction

The operation of any processing facility needs to consider how the risks to the plant, personnel, the broader community and the environment can either be eliminated, or at least reduced. As noted at the beginning of this chapter, risk analysis is a systematic approach to understanding the nature and level of any risk that may be associated with the operation of a facility. The result is a quantitative estimate of the risk based upon engineering evaluations and judgements and established numerical techniques. The purpose of risk analysis is to help identify those potential hazards that pose the greatest risks, allowing corrective actions to be prioritized. Risk analysis also helps to confirm that any corrective action that is taken to minimize risk is effective.

The Consequence – Exposure – Likelihood approach to risk analysis is a three-variable approach that may be used to determine a risk rating for any potential hazard.

**Consequence** – The impact on a person if there is an interaction with the hazard can range from minor cuts and bruises that can be treated by simple first aid, through to non-permanent injury that might require immediate medical attention in a hospital by trained medical staff, to possible death. The impact on the environment of an incident occurring might range from a small loss of containment with no off-site containment, through to extensive and permanent environmental damage.

**Exposure** – This factor considers how often personnel interact with the hazard. How often must the activity be performed? How often are personnel near the hazard?

**Likelihood** – What is the likelihood that the consequences will occur if an individual is exposed to the hazard? The likelihood could range from practically impossible, through unusual, to almost certain. In estimating the likelihood that adverse consequences will occur, any safety controls in place must be considered.

Using the definitions presented in Table 10.1, values for the Consequence (C), Exposure (E) and Likelihood (L) parameters are selected. As well as the values listed, intermediate values may also be used. The Risk Score (R) is then calculated by:

$$R = C \times E \times L$$

This score is then used to determine the Risk Rating using the definitions in Table 10.2.

**Table 10.1** Risk analysis scores for the Consequence – Exposure – Likelihood approach.

| Consequence | C | Exposure | E | Likelihood | L |
|---|---|---|---|---|---|
| Catastrophe: multiple fatalities | 100 | Continuous: occurs many times each day | 10 | Almost certain: the most likely outcome if the hazard event occurs | 10 |
| Disaster: a single fatality | 50 | Frequent: occurs about once a day | 6 | Quite possible: not unusual, a 50/50 chance | 6 |
| Very serious: permanent disability or physical impairment | 25 | Occasional: occurs about once a week to once a month | 3 | Unusual but possible: would be an unusual coincidence, perhaps a 1 in 10 chance | 3 |
| Serious: non-permanent injury needing immediate offsite medical attention | 15 | Infrequent: occurs about once a month to once a year | 2 | Remotely possible: a possible coincidence, perhaps a 1 in 100 chance | 1 |
| Substantial: medical attention needed | 5 | Rare: has been known to occur | 1 | Conceivable but unlikely: has never happened after many years of exposure, but is conceivably possible | 0.5 |
| Minor: requires first aid treatment; minor cuts, bruises or bumps | 1 | Very rare: has not yet occurred | 0.5 | Practically impossible: has never happened after many years of exposure, and is virtually impossible. | 0.1 |

**Table 10.2** Risk rating.

| Risk score | Risk rating |
|---|---|
| >600 | Very high |
| 300 to 600 | High |
| 90 to 300 | Moderate |
| <90 | Low |

Suppose that in the facility a risk exists that could cause a non-permanent injury that would require immediate offsite medical attention. A person is exposed to the risk once every two weeks and the chances that the consequences will occur is around 1 in 10. From Table 10.1 we might estimate,

$C = 15$
$E = 3$
$L = 3$

Calculating the risk score we find, $R = 15 \times 3 \times 3 = 135$.

Referring to Table 10.2, we see that this results in a rating of "Moderate". The process is repeated considering all the potential hazards in the facility, with effort then concentrated on reducing the ratings.

Once the risk associated with a hazard has been identified, then steps can be taken to reduce the risk by implementing one of the safety controls discussed in the last section. A cyclic approach, as shown in Figure 10.7, is followed. After the hazards and the ways personnel might be harmed have been identified, the risks are then evaluated using the technique above. For each hazard, appropriate controls, which might include elimination, substitution, isolation or engineering, are then put in place and the risk is re-evaluated to determine whether it has actually been lowered. Any changes made to equipment, procedures, policies or training are then reviewed to

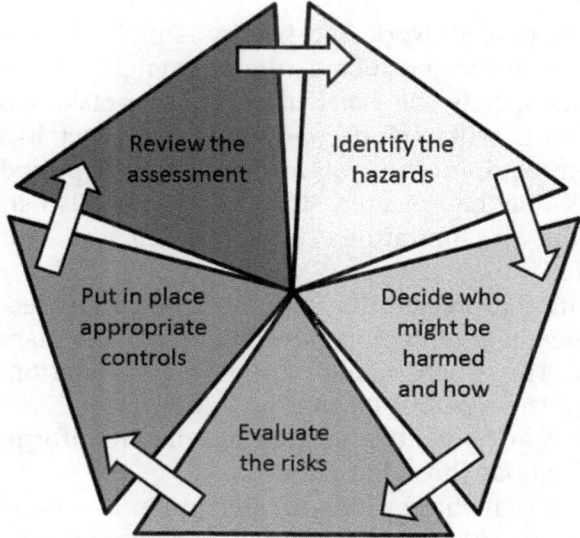

**Figure 10.7** The risk analysis and risk reduction cycle.

ensure that the change has not had any unintended consequences for the safety of the facility as a whole.

### 10.3.3   Safety Procedures – Permit to Work

The Permit to Work (PTW) system is a written or text-based system used to control and record certain types of work activities that have been identified as being potentially hazardous. It is most often used for non-routine activities, such as maintenance. It is a means of alerting stakeholders, such as managers, supervisors, operators and contractors, to work that is being carried out in the plant that might impact on their own activities. Properly used, permits to work also allow potential hazards to be identified, as well as the steps that need to be taken to manage any risks. Once the activity is complete, the permit is then returned to the issuing authority and is closed.

A permit is a written document that authorizes specific individuals to conduct carefully defined tasks with particular equipment over a specified time period and which specifies any hazards that might be encountered and the main precautions that must be taken to complete the task safely. The permit should also cross-reference any other associated tasks being undertaken by the same workers, on the same equipment, or in the same area, that might impact on the safe conduct of the task.

The main aims of the PTW system are:

- To ensure that all work in a facility is properly authorized by a designated authority, such as the duty engineer or shift leader.
- To clearly specify the exact nature of the tasks covered by the permit. This will include specifying the exact location of the work, the equipment involved and the time period over which the work may be executed. The permit may also specify certain conditions or limitations under which the work must be conducted.
- To specify the precautions that must be completed prior to the commencement of the job, or which must be in place throughout the task. This might include the physical isolation of potential risks and the type of PPE that must be worn.
- To identify any potential hazards and then to inform all parties to the work about those hazards.
- To ensure that the responsible personnel are aware of all work taking place within the unit, area or facility for which they have responsibility.

- To ensure the timing and safe continuation of the work when the work continues past the end of a shift. A permit to work handover procedure between shifts ensures that the incoming shift is aware of all active permits, *i.e.*, all active tasks currently underway.
- To ensure the overall safety when the performance of a task must be suspended for a period of time.
- To provide a formal return of the permit when the task is complete. This might mean notifying the duty engineer or shift leader that the affected equipment is safe and ready to be returned to operation.

In assessing whether a PTW should be issued, the duty engineer or shift leader should:

- Be aware of the precise nature of the task to be completed, including all associated procedures.
- Consider whether the task can be performed differently so as to be safer, or as safe but more efficient.
- Consider the impact that the task will have on surrounding areas.
- Understand how this permit might impact on other permits already issued.

Work should only ever be performed on equipment that has been depressurized and emptied of all process material, be it gas, liquid or solid. If hot, the equipment must be allowed to cool and if powered, it must be disconnected from any energy sources, such as electricity. The equipment being worked on must be physically isolated from all process lines by the use of valves and isolation plates.

The PTW should also indicate any additional PPE that is required by the task, including breathing apparatus, goggles, visors, rubber boots, rubber gloves or gas monitors.

Special procedures may also be specified if working at height or in confined spaces where ventilation may be poor.

## 10.3.4 Safety Procedures – Lockouts and Tagouts

Maintenance or other work should never be carried out on equipment that is in use or that contains stored energy, the release of which could injure people or impact the environment. In the latter case, stored energy might include material under pressure, steam or electrical power. Many accidents have occurred in the process industries when work has been conducted on pressure vessels that were still under pressure, or on electrical equipment that was still connected to the electrical supply. Other accidents have occurred when electrical equipment that had been disconnected to allow the maintenance task to be completed were then reconnected before the task was actually completed.

Lockout–tagout (LOTO) is a safety procedure that is used to ensure that the power has been fully and appropriately disconnected before a maintenance task is started and remains disconnected until the task is complete. Lockout–tagout works with at least one or more locks and a hasp that is placed in such a position that power cannot be supplied to equipment under maintenance.

For any maintenance task, a PTW is issued by the duty engineer or shift leader. As part of the permit the energy source for the equipment undergoing maintenance is identified. The power to the item is switched off and then a hasp is passed through the switch so that the power cannot be turned back on without removing the hasp. The shift leader then places their padlock through the end tags of the hasp. The maintenance worker then places their padlock that operates with a different key through the end tags, and in some cases a third padlock is placed on the hasp by the electrical supervisor, or lead contractor. Typically, the padlocks are colour-coded to indicate the lock owner.

Once the maintenance work has been conducted on the equipment, each of the key-holders must be satisfied that the work has been safely completed and that it is safe to restore power before the padlocks and the hasp may be removed.

Only the person who has placed the lock on the hasp can remove their lock. It is also important that the job of placing the lock on the hasp is not passed onto someone else. Each person involved in the process must be responsible for their actions.

"DO NOT OPERATE" tags without locks cannot be relied upon to prevent power being erroneously restored to equipment too early. Tags may become illegible if left in the open for too long, or may accidently become detached.

Modified lockout–tagout devices may also be placed on steam valves to prevent steam being admitted into a work area in which maintenance is being conducted.

## 10.3.5  Case Study – Piper Alpha

Piper Alpha was an offshore oil platform in the North Sea, 193 km north east of Aberdeen and standing in water 145 m deep. As well as producing oil and gas from its own wells, it served as a transfer station for the Claymore and Tartan oil platforms further out to sea. The oil, gas and water brought to the surface through Piper Alpha's 32 oil wells were separated on the platform. The reservoir fluids passed through the main production separators, which operated at 1.07 MPa. The oil was pumped by two booster pumps to the main oil export line and then on to the Flotta Oil Terminal in the Orkney Islands. The water produced from the production separators was further treated, removing the last traces of the oil before being dumped over the side of the platform. The gas from the separators passed through the condensate knockout drum before being compressed to 4.65 MPa in three centrifugal compressors. This gas was then further compressed to 12.0 MPa by two reciprocating compressors. From there, the gas was either exported from the platform, used to lift gas in the oil wells or sent to the flare. Condensate, which had been knocked out of the gas at various points in the process, was first compressed to 4.62 MPa by two condensate booster pumps, before being compressed to 7.58 MPa by the two condensate injection pumps (Figure 10.8). These pumps injected the condensate back into the oil for export down the pipeline to Flotta.[9]

On July 6, 1988, one of the two condensate injection pumps, pump A, was taken out of service for a full 24 month preventive maintenance. A PTW was issued in the control room to allow the pump to be electrically isolated and to allow the inlet and discharge valves to be closed. At the same time, a PTW was issued for the removal and recertification of PSV 504, the pressure safety valve on the discharge side of pump A. There were over 300 pressure safety valves aboard Piper Alpha and these had to be re-certified every 18 months. PSV 504 was the last valve requiring re-certification in the current round of work. It had been planned that the removal and recertification of the valve would coincide with the planned maintenance of the pump.[10]

PSV 504 was located above the condensate injection pump. It was not until 2 pm that the scaffolding had been erected allowing access to PSV 504 for the contractors assigned to the recertification task. The pressure relief valve was removed and blind flanges installed to seal the ends of the two pipes. PSV 504 was then removed for testing and it was not until almost 6 pm that the recertification work was complete. Because overtime had not been authorized on the job, the decision was taken not to refit the valve until the next day.

The valve recertification supervisor went to the control room to return the permit to work for the valve recertification task. As he was uncertain as to the procedure for returning a PTW when the task was

**Figure 10.8**    East elevation of the Piper Alpha platform as at 1988.
Adapted with permission from ref. 10 and 11.

incomplete, he wrote "SUSP", for suspension, on the paperwork and placed the PTW in the control room. He did not advise anyone in the control room that the job was not complete and that the valve had not been installed.[2,12]

At the same time, the shift handover was taking place in the control room. The maintenance on pump A was discussed, but nowhere on the PTW for the maintenance task was the recertification task referenced. It is uncertain whether the recertification paperwork was returned before or after the shift handover in the control room, but what is certain is that no one in the control room following handover was aware that PSV 504 was missing from the discharge side of condensate injection pump A.

At about 9:50 pm that evening, condensate injection pump B tripped out and could not be restarted. If neither of the two pumps could be re-started within 30 minutes then there was the possibility that the entire platform would have to be shut down. The maintenance PTW paperwork was consulted and it was determined that preventative maintenance work on pump A had not commenced. Operators were unaware, due to a failure in the PTW system, that the pressure relief valve was not in place and that the blind flange was not properly secured. The decision was taken to cancel the preventative maintenance task, reconnect the power to the pump, open the valves and re-start pump A. This was done before 10 pm.

Immediately the pump was started, condensate leaked from the loose flange and ignited. The room in which the pumps were located was designed to be fire-proof, but not explosion-proof. The resulting explosion generated shrapnel and severed oil lines, resulting in further fires and explosions. Two operators who had been sent to the pumps in response to the gas leak were killed instantly.

The Tartan and Claymore platforms continued to send oil and gas to the platform after the explosions on board Piper Alpha. They were unaware of any problems. Twenty minutes after the first explosion, the riser carrying natural gas from the Tartan platform failed, resulting in a massive explosion. This was followed later by the failure of a second riser, resulting in the third massive explosion within 30 minutes.

The emergency fire water system failed. Staff on board the platform were directed to wait within the accommodation block for helicopter rescues. However, due to the fires, explosions and heavy black smoke, it was impossible for helicopters to safely reach the platform. Sixty-two staff escaped from the platform but another 165 died. Two rescue personnel from a boat standing by were also killed while attempting to rescue survivors in the water. By morning, the entire platform was lost.

The inadequate PTW system was clearly a major contributor to the disaster. The permits issued for the maintenance of the condensate injection pump and for the pressure relief valve were not cross-referenced. Further, the valve recertification supervisor had not had sufficient training in PTW systems and did not properly return the permit to the shift supervisor when the decision was taken to suspend the recertification of the valve.

While the root cause of the disaster was traced to the inadequate PTW system, further issues relating to safety training and safety management were identified by an enquiry established after the event.[9]

## 10.4   Safety Vignettes

As British philosopher and statesman Edmund Burke said in the 18th century, "Those who do not learn from history are destined to repeat it". The following five case studies of incidents from around the world illustrate the importance of some of the issues presented in this chapter. They are drawn from a summary of process industry-related case studies presented by Shallcross.[13]

### 10.4.1   Risk Analyses Must Be Conducted On "Unofficial Practices"

A massive leakage of liquefied petroleum gas occurred during a non-routine operation to obtain a sample from a 1200 $m^3$ pressurized propane sphere at Feyzin, France in January 1966.[14] The sphere was one of eight on the site used to store liquefied flammable gas. Operators needed to obtain a liquid sample from the storage tank, as there was concern that the ethane content of the hydrocarbon mixture was too high. The sampling points provided on the tank were in-operable, so an unofficial, but common, practice was used, in which the samples were collected through two drainage valves located at the base of the tank.[2]

The liquid sample was to be drained from the base of the sphere through two valves connected in series, with the upper valve connected to the sphere and the lower valve open to the atmosphere. There was a short vertical length of pipe connecting the two valves. When the sampling operation was nearly complete, the upper valve was closed and then cracked open again. When no flow emerged from the cracked valve, probably because of a blockage caused by ice, the upper valve was opened fully. The blockage immediately cleared and the propane gushed out. The operator was unable to close either valve – the handle came off the upper valve and the lower valve became frozen. A massive leak of LPG ensued, with a vapour cloud 1 m deep spreading over 150 m from the tank. The operators managed to leave the immediate vicinity of the tanks and to stop traffic on a nearby road. However, the gas ignited and the storage sphere was engulfed in the fire that followed. Being heavier than air, the propane gas collected under the tank.

The fire fighters who arrived at the site were untrained in dealing with LPG fires and they failed to cool the leaking sphere, instead concentrating their efforts on cooling the other seven spheres. The sphere was fitted with a full water deluge system, but it was capable of delivering only half the amount of water that was actually required. Ten fire fighters were killed in the fire that ensued when the first tank failed. As the incident ran its course, four additional spheres failed and 18 people were killed and 81 others injured.[15,16]

The accident occurred because an operator allowed the two discharge valves to be opened simultaneously. The consequences of these actions were foreseeable.

Important lessons that can be drawn from the Feyzin disaster include:

- All procedures need to be carefully planned ahead of time and subjected to a full risk assessment. This was not done in this case as the procedure was unofficial, even though it was common practice.
- Remotely controlled isolation valves need to be fitted to all such storage tanks to prevent operators from being exposed to hazardous situations.
- The water deluge system installed should have been fit-for-purpose.
- The storage tanks should have had some form of fire-resistant insulation applied to help slow the rate of heating in the event of a fire in an adjacent tank.

## 10.4.2 Never Disable Safety Systems

The deliberate disabling of the safety systems by the engineers at the Chernobyl power station led to one of the world's worst nuclear disasters.[17] The systems were disabled so that they would not cut in when a safety experiment was being conducted on one of the operating reactions.

The Chernobyl power station comprised four water-cooled, but graphite-moderated reactors, each of which powered two turbines. A feature of this type of reactor is that they are very unstable at low power outputs and they are required to be always operated above 20% of rated output. The reactor system was designed with a number of different safety systems in place that would cause the reactor to shut down immediately should any one of a range of different conditions exist.

The power station control systems drew their power from the external power grid. In the event of an outage, back-up diesel-powered generators were available as a back-up power supply, but at that time in the Soviet Union, the only generators available would take at least a minute to start up in an emergency. Therefore, the engineers of the power station decided to test whether the turbines that were spinning down in response to an emergency shut-down might still provide sufficient power to safely control the reactors and their associated systems. The engineers and operators carefully prepared a test that would simulate an emergency shut-down. As part of the approved sequence of events that had been agreed by all participants, part of the reactor's safety systems would be disabled so that the reactor would not shut down automatically.

While the test was planned to take place in the middle of the day, it was delayed until after midnight so that the reactor could be taken offline during a period of low load. Most of the senior engineers had gone home, leaving the test to be conducted by the most junior engineers. As the test progressed, the engineers found that they had to take action, disabling more of the reactor's safety systems than were included in the original agreed test procedure. This was done in order to ensure that the reactor did not shut down prematurely.

In the early morning of April 26, 1986, the engineers operating the reactor lost control of it. Power output dropped to as low as 3% of rated output and then spiked at around 100 times the rated output. Water in the reactor quickly flashed into steam, which caused the pressure tubes to fail catastrophically. The exposed zirconium reacted with the steam to produce hydrogen, which then accumulated within the containment building. When the hydrogen exploded, debris was ejected into the air as the containment building failed. In the aftermath, at least 31 people died as a result of the release of the radiation. Moreover, it could have been so much worse.

While the senior engineers and operators had carefully planned and approved the test procedures for this systems test, they did not apply the same rigour to the decisions taken to disable more of the safety systems while the test was being conducted. The correct decision would have been to have terminated the test, restoring all safety systems and returning the reactor to normal operation, before considering a modification to the agreed test procedure. The senior management also erred by permitting the modified test to be conducted by relatively junior engineers when more experienced engineers were available.

## 10.4.3 The Importance of Safety Training

In 1987, the Dalmeny Oil Storage Terminal in Scotland consisted of ten large tanks, including seven identically sized, floating roof tanks used for the storage of crude oil. These tanks were all 78 m in diameter, 18 m in height, with a capacity of 81 000 m$^3$. In early 1987, it was found that the bottom of one of the tanks contained approximately 1000 tonnes of a thick sludge, unevenly distributed across the base of the tank. The decision was made to clean the sludge out by sending workers into the tank.[18]

A contracting company with experience in cleaning sludge from oil tanks was engaged. The tank was emptied of crude oil and then the tank covers were opened, allowing air to enter the tank. The workers were provided with breathing apparatus for use whenever they were inside the tank. The enclosed space within the tank was a confined space – a location with limited access points and often poor ventilation. Special procedures are put into place whenever such a confined space is entered. In the Dalmeny case, the workers were given face masks that were supplied with air *via* lines from external air sources. Moreover, the contractors worked in teams of four, with three working inside the tank at any one time while another stood outside the tank observing the operation. The workers used hand tools to push the sludge towards the open end of a 100 mm diameter pipeline that carried the sludge out of the tank.

Just after noon on June 11, 1987, the worker outside the tank saw a pale blue flame surrounding his three co-workers inside the tank and called for them to evacuate immediately. Two of the workers managed to escape but one did not, possibly tripping over one of the air hoses. He died in the subsequent fire that engulfed the tank. As the tank did not fail during the fire, the conflagration was contained in the tank.

After the fire, one of the men who had been working inside the tank admitted to having been smoking inside the tank and to throwing his cigarette butt onto the floor of the tank. He did not appreciate the seriousness of this action at the time, as he and others had previously been smoking within the tank during the work. The workers also admitted to frequently removing their breathing masks, as they claimed to have gotten used to the smell within the tank and found it easier to see. The workers made sure that they always wore their breathing masks when entering and leaving the tanks so as not to get into trouble for breaking safety rules. It was impossible for anyone outside the tank to see any transgressions occurring within the tank.

While it appears that the workers did not undertake formal safety training on site, they admitted to being well aware of the rules relating to the banning of smoking, the banning of carrying cigarettes, matches and lighters onto the site and the wearing of safety masks at all times within the site. However, amongst the contractor's workers, these rules were regularly flouted.

The investigators concluded that there was a poor safety culture on the site at the time of the accident. The safety training was not enforced and the workers were not fully informed of the reasons for the safety rules being in place. Management could also have reduced the risks of an incident by using a mechanical extraction system to help ventilate the tank prior to the maintenance operation.

### 10.4.4 Thermal Decomposition Leads to Explosion

Three workers were fatally injured when the cover they were unbolting from a process unit blew off the vessel, releasing hot molten plastic. The incident occurred in March 2001 at the BP Amoco Polymers plant in Augusta, Georgia, where a nylon thermoplastic polymer was manufactured.[19]

The manufacture of the thermoplastic polymer occurred in a number of stages. The raw materials were first fed to a pressurized and heated reaction vessel in which the prepolymer was formed. A high-pressure pump then sent the prepolymer to a tubular reactor, before passing the material on to the extruder and finally the pelletizer. During start-up and shut-down the effluent from the tubular reactor was sent to the polymer catch tank, a $2.8 \text{ m}^3$ horizontal cylindrical vessel.

On March 10, 2001, the entire unit was shut down for maintenance on the extruder. The next day, the polymer catch tank was opened and thoroughly cleaned of waste material. The unit was then sealed, ready for the unit to be started up the next day. The normal start-up procedure called for the extruder to be test run ahead of the system being charged with material. Unaccountably, this task was not done and the unit was started up with the raw materials flowing into the reactors. Following normal procedures, the effluent from the reactors was diverted to the polymer catch tank for the first 50 minutes of operation. When the operators attempted to start the extruder the screws would not turn. After another 20 minutes the operators abandoned the attempt to start the extruder and decided to shut down the entire process. During this period the effluent from the tubular reactors continued to fill the catch tank. As the reactors were being flushed with solvent as part of the shut-down procedure, the catch tank continued to collect the effluent: a hot, partially polymerized material. Finally, the catch tank was closed off and no additional material entered the tank.

Over its years of operation the polymer catch tank had never been filled to such an extent. The polymer near the outside of the tank solidified into a hard crust. Polymer also solidified around the drainage points and at the pressure tapping. The hot polymer within the centre of the tank began to thermally decompose, generating gases. These gases pressurized the closed-off tank. Because the pressure port and drainage points were blocked by solid polymers, there was no way for the operators to know that the tank was at an elevated pressure. Three workers were sent to open the polymer catch tank to clean it and remove the accumulated material. As usual, they began unbolting the flat end plate that was held in place by 44 bolts. When 22 had been removed, the plate blew off, venting the hot molten plastic over the three workers, who died from their injuries. The force of the release of the polymer severed an oil line, which led to an explosion and fire.

Key lessons that can be drawn from this incident include:

- The need for a more rigorous assessment of the hazards and risks associated with filling the tank to near capacity with molten plastic.
- The need for pressure sensors appropriate for the type of material stored in the tank.
- The importance of learning from past experiences when the pressure tapping and pressure relief device had become blocked with plastic.
- The operators were unaware that the polymer could thermally decompose within the tank, leading to over-pressurization.

- Operators had no direct measure of the level of the polymer in the polymer catch tank.

## 10.4.5 The Forgotten Employee

Goodyear used pressurized anhydrous ammonia as a cooling fluid in a shell and tube heat exchanger at their synthetic rubber plant in Houston, Texas. The process chemicals to be cooled were pumped through the tubes inside the exchanger while the ammonia was circulated shell-side. On June 10, 2008, operators closed an isolation valve on the shell-side so that they could replace a burst rupture disk. Closing the isolation valve also closed off a pressure relief valve that was located there to prevent pressure build-up of the ammonia on the shell side. At the conclusion of the procedure to replace the ruptured disk, the isolation valve was not re-opened as it should have been. The next day another operator closed a block valve on the ammonia shell-side loop, isolating the pressure control valve from the heat exchanger. The operator was aware that the shell-side was pressurized with ammonia, but was unaware that the isolation valve remained closed form the operations from the day before. The operator assumed that any over-pressure that might occur would be relieved by venting through the valve to the pressure relief valve.[20]

The operator connected a steam line to the tube-side of the exchanger in order to clean the exchanger. The steam quickly heated up the ammonia and with the shell-side effectively isolated, the pressure began to build up as the liquid ammonia began to evaporate. Less than an hour later the heat exchanger shell violently ruptured, spraying the area with metal shrapnel and releasing dangerous levels of ammonia to the atmosphere. Five people working nearby were exposed to the gas and required medical treatment.

A site emergency was declared and all personnel were required to assemble away from the process units, where they could be counted and checked off with a roll call. This was conducted by the company's safety team, who were also evacuating the injured workers and investigating the safety of the site.

Later that morning, the plant management declared the incident over and work began to clean up the area. It was at this point that the body of a company employee was discovered amongst the debris. She had been felled by the flying shrapnel and had died on the scene without anyone coming to her aid. Since she was a member of the plant's emergency response team, no one considered her absence during the incident unusual – the other members of the response team assumed that she was elsewhere. The company's system of accounting for staff had failed, as no one had identified that the employee was missing after the incident. Indeed, several people who had the responsibility for accounting for staff were unaware that this was their role in an emergency. Although internal company safety policies required that plant-wide emergency drills be conducted at least once every three months, no such drill had been conducted in the facility for the four years prior to the incident.

Several key lessons can be drawn from this incident:

- The correct maintenance procedures were not followed.
- The operator did not visually check that the isolation valve to the pressure relief valve was open before steaming the tubes of the exchanger.
- The head count was not correctly performed and a missing worker was not identified as being missing.
- Established safety drills were not held in the facility on a regular basis, so that the workers and management were not properly prepared to adequately respond to the incident.

## 10.4.6   No Permit to Work or Lockout–Tagout

In a pulp and paper mill, two welders were assigned to repair the tines in a repulper.[14] A repulper is an open topped vessel in which large pieces of wood pulp and recycled paper are pulped using a 400 kW impeller, fitted with blades, or tines. The resultant solution is typically a 1% to 5% aqueous suspension in a liquid. At 10 am, the operators of the repulper left for a break after finishing cleaning and draining the repulpers. A couple of minutes later, the two welders arrived. As they were required to repair the No. 3 repulper, which had just been cleaned and drained, they assumed it had been cleaned for them and they began working immediately. They had not been issued with a PTW and no locks had been installed on the electrical equipment associated with the repulper, including the impeller motor.

The operators returned at 10.15 am and saw the welding machine, but did not see any leads entering into the vessel. The operators assumed the repulper was empty and began to fill the vessel with water. Then they turned on the impellers and began to feed dry pulp into the vessel. At 2.30 pm, the welders could not be found anywhere and were notified as missing. Looking at the welders' job cards, it led the operators back to the No. 3 repulper and the vessel was immediately stopped and drained. Inside the repulper, the remaining parts of the welding equipment, rope ladder and human bones were found. The welders should have never entered the vessel without first making sure their foreman had checked that the vessel was isolated and had issued a PTW.

## 10.5   Concluding Remarks

Maintenance goes on every day in any oil refinery, petrochemical plant, or other processing facility, while the process plant continues to operate. Pumps are taken out of service while the stand-by units are brought on-line. Tanks are emptied and placed out of service for cleaning, while streams are re-routed to other tanks. Critical sensors are removed and replaced while their duty is performed by redundant units. Managing the hazards safely means that the risks are minimized and personnel remain safe. However, there comes a time when the entire facility must be shut down for an extended period for crucial maintenance to be performed. A measure of how well a processing facility is being managed is to consider the period between shut-downs. Decades ago, for most facilities, the plant would be shut down completely once a year. As better safety management practices are implemented, the period between plant-wide shut-downs has gradually extended to two, three or four years. In some cases now, petrochemical facilities are operating perfecting safely and efficiently without having been shut down for over ten years.

The message is simple: implementing comprehensive and complete safety management systems not only reduces the rate of injuries and fatalities but also leads to better, more efficient operations. Safety in the process industries should always be front of mind, and after completing every task engineers and operators should ask themselves the question, "What have I learnt from completing this task that will allow me to do it better and more safely next time?"

## References

1. T. A. Kletz, *Chem. Ind.*, 1978, **56**, 287.
2. J. Atheron and F. Gill, *Incidents That Define Process Safety*, Center for Chemical Process Safety/AIChE, New York, 2008, pp. 277–281.
3. D. W. Edwards, *Loss Prev. Bull.*, 2014, **240**, 21.
4. F. I. Khan and S. A. Abbasi, *J. Loss Prev. Process Ind.*, 1999, **12**, 361.

5. S. Mannan, *Lees' Loss Prevention in the Process Industries*, Elsevier, 3rd edn, 2005, pp. A2/1–A2/18.
6. J. E. S. Venart, *J. Loss Prev. Process Ind.*, 2007, **20**, 621.
7. R. J. Parker, J. A. Pope, J. F. Davidson and W. J. Simpson, The Flixborough Disaster: Report of the Court of Inquiry, HM Stationery Office, London, 1975.
8. U.S. Chemical Safety and Hazard Investigation Board, *Investigation Report Refinery Explosion and Fire BP*, U.S. Chemical Safety Board Report No. 2005-04-1-TX, 2007.
9. W. D. Cullen, *The Public Inquiry into the Piper Alpha Disaster*, HM Stationery Office, London, 1990.
10. D. C. Shallcross, *Educ. Chem. Eng.*, 2013, **8**, 1.
11. R. Sylvester-Evans, *IChemE Trans. B*, 1991, **69**, 3.
12. M. E. Paté-Cornell, *Risk Anal.*, 1993, **13**, 215.
13. D. C. Shallcross, *Educ. Chem. Eng.*, 2013, **8**, 12.
14. T. Kletz, *What Went Wrong? Case Histories of Process Plant Disasters*, 4th edn, Gulf Professional Publishing, Burlington, 1999.
15. S. Mannan, *Lees' Loss Prevention in the Process Industries*, Elsevier, 3rd edn, 2005, p. A1/33.
16. http://www.hse.gov.uk/comah/sragtech/casefeyzin66.htm (last accessed January, 2016).
17. S. Mannan, *Lees' Loss Prevention in the Process Industries*, Elsevier, 3rd edn, 2005, pp. A22/1–A22/10.
18. Health and Safety Executive, *The Fires and Explosion at BP Oil (Grangemouth) Refinery Ltd*, HSE Books, London, 1988.
19. U.S. Chemical Safety and Hazard Investigation Board, *Investigation Report Thermal Decomposition Incident*, U.S. Chemical Safety Board Report No. 2001-03-1-GA, 2002.
20. U.S. Chemical Safety and Hazard Investigation Board, *Case study – Heat exchanger rupture and ammonia release in Houston, Texas*, U.S. Chemical Safety Board Report No. 2008-064-1-TX, 2011.

# 11 Engineering Statistics, Process Integration, and Experimental Design

Joseph Menicucci, Elizabeth Aisenbrey and Daniella Cirillo

## 11.1 Introduction

Process design is an integral component of any chemical engineer's formal training. A traditional mechanism for learning design often involves being given an open-ended problem, designing a facility comprised of a series of processes in order to address that problem, and modelling and simulating the processes – using appropriate computer software – to predict the operating conditions, rate of production, and product yield of the designed facility. There is great utility in such endeavours, but chemical engineering practice requires not only these skills, but also the capacity to evaluate the efficacy of processes (or sub-processes) in such a facility, to optimise operational parameters of each unit in the facility and evaluate their impact and the ability to troubleshoot problems when they arise.

Chemical engineers often spend significant portions of their careers monitoring, analysing, troubleshooting, and optimising a process. Developing a complete understanding of each major chemical engineering concept is certainly important, in that it provides a foundation of knowledge that informs the daily operation as well as the optimisation of these processes. However, the impetus for such analysis exists not only in evaluation of each process in isolation, but

Concepts of Chemical Engineering for Chemists: 2nd Edition
Edited by Stefaan Simons
© The Royal Society of Chemistry 2017
Published by the Royal Society of Chemistry, www.rsc.org

also because the operation of each unit in a plant can (and likely will) have an effect on the other units.

For instance, the products from a chemical reactor may become an input to a distillation column or a reactant in a subsequent reactor. Because these processes are all interconnected, then, each individual operating parameter has the potential to affect the operating conditions of many units in a plant; processes upstream (preceding) the process in question must be able to meet the needs of that particular unit operation. Moreover, processes downstream (following) the particular process must be capable of handling the product(s) from this particular unit operation. The holistic approach to process design and optimisation that exploits the interactions between the different units to employ resources effectively and to minimise costs is known as *process integration*.

In this chapter, we will begin by looking at the large-scale production of poly(lactic acid), commonly referred to as PLA. This will allow us to explore how some of the concepts from earlier chapters of this book are interconnected and also explore the types of decisions required of engineers in pilot and production facilities. Next, we will develop the statistical tools that will allow us to monitor and evaluate the performance of engineering processes, using the production of PLA as an example. Finally, we will apply the statistical tools to not only monitor the process but also evaluate the impact of optimising engineering processes.

## 11.2   Experimental Design and Process Integration

Imagine an industrial facility designed to allow for the production of a high-purity product, Z, *via* chemical reaction of reactants X and Y.

$$X + Y \rightleftharpoons Z$$

If this reaction were carried out in a continuous stirred tank reactor (CSTR), we might envision the reactor would look like the one shown in Figure 11.1a. In this reactor, we can see the reactants entering the top of the tank and the product leaving the bottom of the tank. It is important to note that the reactor shown in Figure 11.1a must operate at ambient conditions. Therefore, both the equilibrium conversion and the rate of reaction in the CSTR will be dependent on the ambient temperature (see Chapter 2 for details). In order to achieve favourable kinetic and equilibrium conditions, it will be very important to control the temperature of the chemical reaction. The easiest mechanism

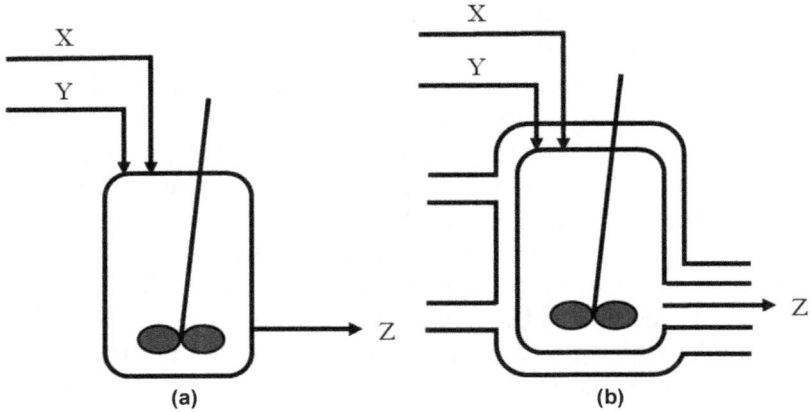

**Figure 11.1**    (a) A continuous stirred tank reactor yielding a product, Z, from reactants X and Y. (b) The same reactor, jacketed for temperature control.

by which this can be completed is using a jacketed reactor (a detailed example of jacketed temperature control in a CSTR is given in Chapter 8). Our hypothetical jacketed reactor is shown in Figure 11.1b.

Even if the chemical reactor were *as simple* as depicted in Figure 11.1b, we would still require extensive knowledge of a great number of chemical engineering principles in order to accurately evaluate the overall process. Significant knowledge of chemical re-action engineering (Chapter 2) would be required and a control scheme would need to be designed, incorporated, and tuned that would be capable of maintaining a constant reaction temperature, no matter the ambient conditions (Chapter 8). A comprehensive know-ledge of heat transfer (Chapter 4) would be needed to predict the rate of heat transfer from the jacket to the reactor (or *vice versa*), as well as heat loss from the jacket to the environment. Concepts related to fluid flow (Chapter 3) – such as the phenomena related to pumping liquid into the reactor – would also be required.

As such, no fewer than four chemical engineering principles are needed to accurately describe the chemical reactor shown in Figure 11.1b, but, in fact, there are a large number of possible overall designs for a fully operational facility built around a single chemical reaction. For example, it may be possible to purchase and use high-purity reactants (X and Y) and sell the product (Z) without further processing, but doing so may not be economically favourable. It might be that the economics favour purchasing low-quality reactants and purifying them on-site, using a carefully selected separation process(es). Moreover, it may be economically beneficial to purify the product, Z, before selling it. For instance, if purification of both products and reactants were done on-site, there could be a need for three independent separation sequences, adding significant complexity to any system model. An understanding of the principles of mass transfer (Chapter 5) required for these separations would be required for complete knowledge of this new, more highly complex, system. Each unit operation would also necessitate the design, implementation and tuning of control schemes appropriate for each process.

In this instance, we are designing an experiment to produce a certain quantity of Z in a given time frame (the rate of production of Z) at a certain, undefined, purity. The complexity of our process, therefore, depends on the factors that might prevent us from reaching the desired rate and quality of our product. Table 11.1 shows just a few factors that might limit the production rate and/or quality of our product, Z.

The level of complexity of the process is uncertain, especially because we have not clearly defined the purpose of the overall design. The importance of choosing the correct purpose for our design cannot be understated. Many resources exist that will support experimental design and problem solving, but Fogler, LeBlanc, and Rizzo have given excellent examples of this important aspect and have developed valuable heuristics in support of such decision-making.[1]

We will next illustrate aspects of decision-making related to engineering design with a comprehensive example, built around the operation of a continuous stirred tank reactor (CSTR) for the production of high-purity poly(lactic acid).

## 11.2.1   PLA Production

Poly(lactic acid) is a compostable polymer produced from renewable resources such as sugar and starch. It is only within the past decade that the large-scale production of PLA has grown in importance due to recent advances in the development of sustainable materials and renewable resources. Initially, there was little interest in the large-scale production of PLA, because its applications were limited by production techniques that could only produce high-cost, low-yield, low-molecular weight PLA. New techniques produce a high-molecular weight PLA that can be used as an environmentally friendly and renewable substitute for other petroleum-based polymers, such as polystyrene. Like many industrial-scale processes, there is a multitude

**Table 11.1** Possible limiting factors on the purity (P) and rate of production (R) of the reaction product, Z.

| Limiting factor | Chemical engineering principle(s) involved | Possible reason(s) for limitations | Possible consequences |
|---|---|---|---|
| Purity of reactants X and Y | Chemical reaction engineering, mass transfer operations | Competing reactions (*P*), reaction inhibition (*R*), catalyst poisoning (*R*) | Unwanted by-products, low purity product requiring purification |
| Bulk transport of reactants X and Y from storage tanks to the reactor | Fluid flow | Incomplete reaction (*R*), limiting reactant (*R*) | Decline in rate of production |
| Rate of reaction | Chemical reaction engineering, heat transfer, mass transfer operations | Insufficient heat supplied/removed (*R*), temperature limitations (*R*) [denaturing of an enzyme], catalyst poisoning (*R*) | Decreased yield, decreased purity, unwanted byproducts |
| Reaction equilibrium | Chemical reaction equilibrium, chemical thermodynamics, process control, heat transfer | Mass transfer limitations (*R*) (reactants, enzymatic, *etc.*), temperature effects on reaction equilibrium (*R*), insufficient heat supplied/removed, *etc.* (*R*) | Unwanted byproducts, decreased (or no) yield |
| Post-reaction purification of reaction products | Mass transfer operations | Mass transfer limitations (*P*), thermodynamic equilibria limits (*P*) | Decreased purity |

**Figure 11.2**   A black box view of the production of poly(lactic acid).

of ways to produce PLA. Here we will explore, briefly, the different design decisions, taking into consideration chemical engineering principles related to mass balances, kinetics, thermodynamics, fluid dynamics, heat transfer, process control, and economics.

### 11.2.1.1   *The Lactic Acid Feedstock*

In order to effectively look at the large-scale production of PLA, we will initially consider the process in the simplest way possible. In this case, we imagine the production of PLA as an inlet stream leading to a collection of processes that produce and purify the PLA, followed by an outlet stream taking the products away. This simple representation is shown in Figure 11.2, with the collection of processes represented by a black box. The final product in this facility will be PLA; like most production processes, however, we must make a number of design decisions that may have significant impact on both the rate of PLA production and the product purity. As we evaluate design decisions specific to the production of PLA we will, at the same time, paint a more complete picture of the processes in the black box.

---

Design Decision: Production of the Lactic Acid Feedstock

In order to produce PLA, we must first produce lactic acid, which can be manufactured *via* two very different processes: chemical synthesis or bacterial fermentation. Our first design decision then is to determine which mechanism we will use to produce the lactic acid feedstock. For the purposes of this exercise, we will assume that this production facility is not associated with any other production facilities (such as a petroleum refinery).

It turns out that the chemical synthesis of lactic acid has some major limitations, including the production limitations due to the need for a petroleum by-product, the concurrent production of the undesirable stereoisomer, D-lactic acid, and high manufacturing costs. If this production were associated with a large petroleum refinery, there would have been some incentive to choose chemical synthesis in order to use the petroleum by-products for lactic acid production, rather than considering it waste. That being said, most industrial production of lactic acid is done *via* bacterial fermentation. This method allows for high yield at standard fermentation conditions with the lowest rate of by-products and a renewable feedstock of sugar or starch. We will select bacterial fermentation as the mechanism for the production of lactic acid in this case study.

Now that we have decided on a production process for lactic acid, we can remove it from the black box and include it as its own step in an updated diagram, shown in Figure 11.3.

The fermentation production of high-quality polymerisation-grade lactic acid is not as simple as is shown in Figure 11.3, however. Significant time, resources, and engineering have gone into optimising the production of lactic acid, with specific emphasis on fermentation and purification. For instance, bacteria are required for the fermentation of dextrose and concurrent production of lactic acid. Like all living organisms, these bacteria require very specific conditions to live and, more so, for the optimum production of lactic acid. One of these factors (and a very important one) is pH.

Fermentation occurs in a (bio)reactor, where an aqueous slurry of bacteria produces lactic acid. As more lactic acid is produced, the slurry becomes very acidic. As the pH decreases, the rate of lactic acid production may slow or stop and too low a pH may even result in cell death. Therefore, the pH of the aqueous slurry must be controlled (see Chapter 8), through continual addition of a base, such as calcium hydroxide. This is only one factor to consider during fermentation, although many more can affect the productivity, yield, and purity of

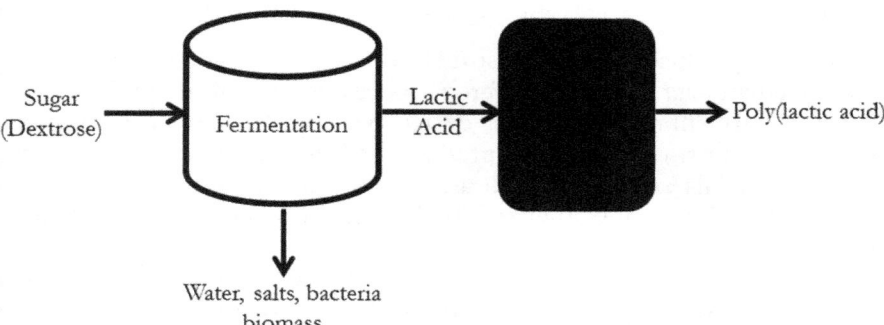

**Figure 11.3**   Production of poly(lactic acid) with an optimised fermentation process.

lactic acid. Temperature, pH and resilience to inhibitory by-products are all important when selecting the bacteria used in the CSTR; there has even been an interest in the directed evolution of more robust bacteria. Once lactic acid is produced, it must be separated from the bacteria and undesirable by-products, such as salts, in the aqueous slurry. Distillation, liquid extraction, electrodialysis and ion exchange are only a few of the separation methods that have been used for the purification of lactic acid. The desirable method is not only site-specific, but also depends on the economics, environmental impact, and life-cycle analysis.

In our CSTR, we assume that there is no accumulation of products or reactants – the system is at steady state – due to the low retention time compared to batch reactors. As such, we can perform a rate-based mole balance (see Chapter 1) on the lactic acid in the CSTR:

---

Mole Balance on Lactic Acid in a CSTR

$$C_{\text{lactic acid,in}}\dot{V}_{\text{in}} - C_{\text{lactic acid,out}}\dot{V}_{\text{out}} + r_A V_R = 0$$

where:

$C_{\text{lactic acid,in}}$ = concentration of lactic acid in the inlet [kmol m$^{-3}$]
$C_{\text{lactic acid,out}}$ = concentration of lactic acid in the outlet [kmol m$^{-3}$]
$\dot{V}_{\text{in}}$ = volumetric flow rate of the aqueous slurry in the inlet [m$^3$ h$^{-1}$]
$\dot{V}_{\text{out}}$ = volumetric flow rate of the aqueous slurry in the outlet [m$^3$ h$^{-1}$]
$r_A$ = rate of lactic acid generation [kmol m$^{-3}$ h$^{-1}$]
$V_R$ = volume of the CSTR [m$^3$]

---

The inlet concentration of lactic acid is zero (we assume no lactic acid in the feed); therefore, the mass flow rate of the lactic acid coming out of the reactor is equal to the rate lactic acid is produced *via* bacterial fermentation.

---

Simplified Mole Balance on Lactic Acid in a CSTR

$$C_{\text{lactic acid,out}}\dot{V}_{\text{out}} = r_A V_A$$

where:

$C_{\text{lactic acid,out}}$ = concentration of lactic acid in the outlet [kmol m$^{-3}$]
$\dot{V}_{\text{out}}$ = volumetric flow rate of the aqueous slurry in the outlet [m$^3$ h$^{-1}$]
$r_A$ = rate of lactic acid generation [kmol m$^{-3}$ h$^{-1}$]
$V_R$ = volume of the CSTR [m$^3$]

---

The efficiency of this process can be increased by setting multiple CSTRs in series. Specifically, the unreacted sugars from the first CSTR will be pumped to a second CSTR, where bacteria will ferment the sugars to produce lactic acid. This can be repeated in more CSTRs until all of the sugars have been reacted or until the yield of lactic acid meets production requirements.

Now that we have decided upon a mechanism for the production of lactic acid, we will take a detailed look at the design decisions related to the next step of this process: the polymerisation of lactic acid to PLA.

### 11.2.1.2 Polymerisation of Lactic Acid

---

Design Decision: Polymerisation

The polymerisation of poly(lactic acid) can be achieved by direct condensation of lactic acid, ring-opening polymerisation, or a combination of the two. The condensation reaction (removal of $H_2O$) is equilibrium-limited. As this reaction proceeds, $H_2O$ is produced; it must be removed in order to produce high-molecular weight PLA. The ring-opening polymerisation method requires an intermediate cyclisation reaction of lactic acid to lactide, followed by a catalytic ring-opening polymerisation to produce PLA from lactide.

The method chosen greatly depends on the desired end product. If a high-molecular weight PLA is not necessary, the condensation reaction may be advantageous because it eliminates an intermediate product. However, in most PLA production processes high-molecular weight polymer is desired. Cargill Dow LLC has combined the two reactions in their own economically and environmentally beneficial patented[2] process. In this process, there is an initial condensation reaction of lactic acid to produce a low molecular weight PLA prepolymer. The low-molecular weight PLA then undergoes the intermediate catalytic cyclisation reaction to produce a cyclic lactide stereoisomer of a desired molecular weight, followed by a ring-opening polymerisation reaction to produce a high-molecular weight PLA.

If we assume that the cost of licensing the combined process is not prohibitive, we will then select it for the environmental benefits it offers. Figure 11.4 now shows the individual reaction steps of the complete production process for PLA, although it is still a simplified overview of the actual process and does not include series of CSTRs for each process step, pumps, heat exchangers, separations processes, controls, *etc.* that are also required for the production of high-purity PLA.

---

The first step of the polymerisation process is to produce lactide from lactic acid. This condensation reaction is given below:

$$2\underset{\text{lactic acid}}{C_3H_6O_3} \leftrightarrow \underset{\text{lactide}}{C_6H_8O_4} + \underset{\text{water}}{2H_2O}$$

This reaction is equilibrium limited. As more water is produced, the reaction will proceed in the reverse direction, resulting in the production of lactic acid; the yield of lactide – our desired product in this

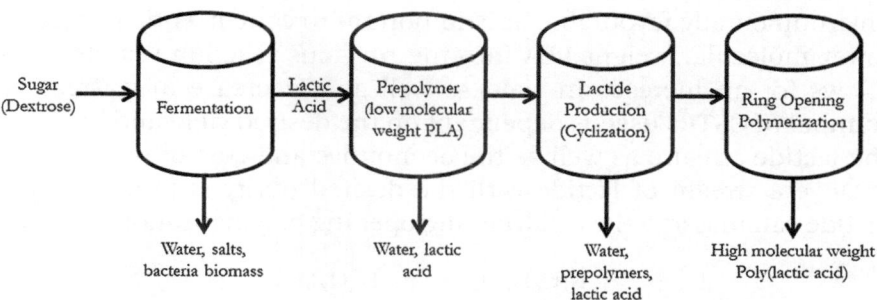

**Figure 11.4**  Diagram showing each major step in the production of (un-purified) poly(lactic acid).

step of the process – is therefore reduced. If we assume reversible elementary reactions, we can determine the final concentrations of the product and reactants leaving each CSTR.

---

Simplified Mole Balances on Lactic Acid, Lactide, and Water in a CSTR

$$C_{\text{lactic acid,out}}\dot{V}_{\text{out}} = C_{\text{lactic acid,in}}\,\dot{V}_{\text{in}} - (2k_1C_{\text{lactic acid}}{}^2 - 2k'C_{\text{lactide}}C_{\text{water}}{}^2)V_R$$

$$C_{\text{lactide,out}}\dot{V}_{\text{out}} = (k_1C_{\text{lactic acid}}{}^2 - k'C_{\text{lactide}}C_{\text{water}}{}^2)V_R$$

$$C_{\text{water,out}}\dot{V}_{\text{out}} = (2k_1C_{\text{lactic acid}}{}^2 - 2k'C_{\text{lactide}}C_{\text{water}}{}^2)V_R$$

where:

$C_{\text{lactic acid,in}}$ = concentration of lactic acid in the inlet [kmol m$^{-3}$]

$C_{\text{lactic acid,out}}$ = concentration of lactic acid in the outlet [kmol m$^{-3}$]

$C_{\text{lactic acid}}$ = concentration of lactic acid in the CSTR [kmol m$^{-3}$]

$C_{\text{lactide,out}}$ = concentration of lactide in the outlet [kmol m$^{-3}$]

$C_{\text{lactide}}$ = concentration of lactide in the CSTR [kmol m$^{-3}$]

$C_{\text{water,out}}$ = concentration of wtaer in the outlet [kmol m$^{-3}$]

$C_{\text{water}}$ = concentration of wtaer in the CSTR [kmol m$^{-3}$]

$\dot{V}_{\text{in}}$ = volumetric flow rate of the aqueous slurry in the inlet [m$^3$ h$^{-1}$]

$\dot{V}_{\text{out}}$ = volumetric flow rate of the aqueous slurry in the outlet [m$^3$ h$^{-1}$]

$k_1$ = forward reaction rate constant [m$^3$ kmol$^{-1}$ h$^{-1}$]

$k'$ = reverse reaction rate constant [m$^3$ kmol$^{-1}$ h$^{-1}$]

$V_R$ = volume of the CSTR [m$^3$]

---

The lactide balance shows us as that as the concentration of water increases, the concentration of lactide product is reduced. In order to keep the concentration of water as low as possible – and increase our yield of lactide – we can put CSTRs in series to strategically remove water. A separation step is necessary to remove water from the reactor prior to it becoming the inlet stream to the subsequent CSTR in series. In order to make this process both economically and environmentally favourable, it is important to concurrently recycle all lower molecular weight PLA into the previous reaction vessels. This allows for an increase in yield as well as a decrease in waste. The number of CSTRs used is dependent on the desired yield and purity of the lactide stream, as well as the economics and cost of production.

Once a stream of lactide with the desired purity is produced, the lactide can undergo the catalytic ring-opening polymerisation reaction.

$$N(C_6H_8O_4) \rightarrow (C_3H_4O_2)_n$$

This reaction occurs in a CSTR to ensure that the concentration and temperature inside the reactor is constant throughout. Maintaining precise controls when producing a polymer of high molecular weight is required in order to maintain a narrow molecular weight distribution (all polymers are relatively the same molecular weight and have the same number of repeat units). It is quite important to maintain a constant (and relatively high) temperature in the CSTR. This is not only because the high temperature increases the rate of reaction, but also because as the molecular weight of the polymer increases, the viscosity also increases; a high temperature is necessary to keep the viscosity low enough to allow for stirring and pumping of the product.

Again, the concentration of lactide and the desired product, poly(lactic acid), can be determined by performing a mole balance around the CSTR. Note that the reaction rate depends on the catalyst used.

---

Simplified Mole Balances on Lactide and Poly(lactic Acid) in a CSTR

$$C_{PLA,out}\dot{V}_{out} = r_A V_R$$
$$C_{lactide,out}\dot{V}_{out} = C_{lactide,in}\dot{V}_{in} - Nr_A V_R$$

where:

$C_{lactide,in}$ = concentration of lactide in the inlet [kmol m$^{-3}$]
$C_{lactide,out}$ = concentration of lactide in the outlet [kmol m$^{-3}$]
$C_{PLA,out}$ = concentration of PLA in the outlet [kmol m$^{-3}$]
$\dot{V}_{in}$ = volumetric flow rate of the aqueous slurry in the inlet [m$^3$ h$^{-1}$]
$\dot{V}_{out}$ = volumetric flow rate of the aqueous slurry in the outlet [m$^3$ h$^{-1}$]
$r_A$ = rate of PLA generation [kmol m$^{-3}$ h$^{-1}$]
$V_R$ = volume of the CSTR [m$_3$]
$N$ = stoichiometric coefficient

---

The temperature, residence time, presence of water, and pre-polymer concentrations all have an effect on the production of PLA. Temperature control is incredibly important and there are both advantages and disadvantages to operating at high temperatures. Increasing the temperature increases the reaction rate; however, high temperatures can result in hydrolytic and oxidative degradation of the PLA product and can lead to lactide reformation. Operating the CSTRs at high temperatures also results in high energy costs.

The production of a desirable molecular weight PLA depends on the residence time of the reaction. High-molecular weight PLA requires more time than a low molecular weight PLA, but longer time in the reaction vessel leads to a lower *rate of production*. This ultimately affects the economics of the plant. The residence time can be reduced by changing other factors, such as increasing temperature; however, we have already discussed the negative consequences corresponding to increasing the temperature in the CSTRs.

The prepolymer concentration can also affect the reaction. Depending on the order of the reaction, it is possible that an increase in reactant (prepolymer) concentration can result in an increase in product formation over time. However, starting with a high concentration of prepolymer can lead to a higher viscosity, resulting in issues with mixing, fluid transport, and heat transfer throughout the vessel.

Another issue specific to the production of PLA is the necessity of keeping little to no water in the reaction vessel. One attractive characteristic of PLA is that it is hydrolytically degradable, but during production it is important to limit this degradation to produce the desirable high-molecular weight PLA.

### 11.2.1.3 Parameter Optimisation, Process Integration and Process Optimisation

The production of high-molecular weight PLA involves not only identifying the optimum unit operation for each step of the process, but also evaluating the impact of, and optimising, the level of the operating parameters (such as catalyst selection, temperature, *etc.*) on the quality and yield of the final product.

Moreover, we need to optimise the *entire process*, not just a single unit (such as a reactor), if we want to maximise the purity of Z and/or the rate of production of Z. The optimisation of operating conditions throughout an entire production facility is accomplished through *integration* and *optimisation* of all processes in the facility. This would include processes not only involving chemical reactions, but also related to the separation, purification, and downstream post-processing of the PLA polymer.

The optimisation of a process facility may not only result in an increase in product purity and/or rate of production but also minimise resources use (such as water and energy) and/or (especially) cost. Comprehensive, facility-specific, models[3-5] are required for complete optimisation of a process and are therefore beyond the scope of this chapter. We will, however, explore aspects of parameter optimisation, process integration and process optimisation through worked examples later in this chapter.

The next section introduces a host of tools that will not only allow us to evaluate the performance of our process, but also help inform experimental design and allow us to better understand the impact of changing the levels of variables in our process.

## 11.3   Statistical Tools

Before we can begin to analyse a process, we must first develop a set of tools that will allow us to properly determine the implications of our results. Such tools for analysis range from very simple to very complex. For the purposes of this chapter, we will provide the tools required for a basic experimental analysis.

The first tools that we must define are the **population mean** and the **sample mean**. While both means are calculated in the same way (the sum of all measurements divided by the total number of measurements), there are implications to differentiating between these two types of means. The mean is calculated by adding each measurement and dividing by the total number of measurements. While the calculation is the same for both a **population mean** and a **sample mean**, the nomenclature is slightly different.

---

Population Mean

$$\mu = \frac{\sum_{i=1}^{N} x_i}{N}$$

where:
$\mu$ = population mean
$x_i$ = value of data point $i$
N = number of data points in the population

---

Sample Mean

$$\bar{x} = \frac{\sum_{i=1}^{N} x_i}{N}$$

where:
$\bar{x}$ = sample mean
$x_i$ = value of data point $i$
N = number of data points in the sample

---

When determining the **population mean**, *every* outcome is measured and has equal impact. A population mean is a valuable tool when a process yields very few products that have very high value. Therefore, the calculated mean will be inclusive of the entire population produced in a given time period (day, week, month, quarter, year, *etc.*).

It is important to use a **sample mean** when evaluating every data point in a population mean is not economically or practically possible. This may be the case when you are tasked with experimentally determining the mean of a very large population. The two types of samples commonly used in experimental analysis are the random sample and weighted (representative) sample.

A random sample is taken from the population in such a way that it must be irrespective of any other influencing factor. In taking a random sample, it is very important that the investigator be careful to avoid selection bias. Selection bias occurs if random sampling is, in fact, not random. The bias need not be intentional to exist; habitual tendencies or laziness may contribute to selection bias.

Imagine a company specialising in the production of PLA. This company has ten facilities in ten cities but only four on-site analytical laboratories for evaluation of the purity of PLA product. In order to prevent selection bias from occurring, a random sample must be selected for analysis not based upon proximity to the analytical laboratories, but instead by chance. If samples were only tested from the facilities with analytical laboratories on-site, selection bias is present.

A weighted (representative) sample differs from a random sample in that the number of samples tested is directly proportional to the rate of production from a particular process. If this company were to use a weighted sampling protocol for analysis, they would first determine the percentage of their overall production from each individual facility and then sample PLA from each facility in that proportion. For instance, if a particular facility accounted for 25% of the overall PLA produced, a company employing a weighted sampling protocol would collect 25% of the PLA samples for analysis from this laboratory.

Experimental analysis does not end with determination of the population or sample mean. In fact, it is not appropriate to report a statistical mean without appropriate context. If we know our desired outcome – in our current example, the production of high-purity poly(lactic acid) (85% purity) – we can, with suitable supporting statistical evidence, make the analogy that an appropriately determined statistical mean reflects the *accuracy* of our production processes. If the mean purity produced was 99.9%, the process did not *accurately* produce 85% pure PLA.

Now, imagine that not only is the mean purity 99.9%, but every sample tested was exactly 99.9% pure! The *accuracy* of the production process still remains very poor, but the *precision* of the process is, in fact, perfect. This should highlight the significant difference between accuracy and precision and especially indicate the importance of having both accurate and precise results. And while there is value in having perfectly precise results, that value is much reduced when the results are not accurate.

As you can imagine, the need for quantitative tools for these statistical measurements is acute. Two tools widely used in engineering statistical analysis are the **variance** and the **standard deviation**. These two tools are quite similar; in fact, we calculate the variance on the way to calculating the standard deviation! The calculation of variance and standard deviation is different for a population and a sample. The standard deviation, in a way, reflects the precision of experimental results in that it tells us, collectively, how our data set deviates from the mean value.

---

Population Variance

$$\sigma^2 = \frac{\sum\limits_{i=1}^{N} (x_i - \mu)^2}{N}$$

where:
$\sigma^2 =$ population variance
$x_i =$ value of data point $i$
$\mu =$ population mean
$N =$ number of data points in the sample

---

Population Standard Deviation

$$\sigma \sqrt{\frac{\sum\limits_{i=1}^{N} (x_i - \mu)^2}{N}}$$

where:
$\sigma =$ population standard deviation
$x_i =$ value of data point $i$
$\mu =$ population mean
$N =$ number of data points in the population

---

In practice, it is much more likely that a data set will be collected from a sample, rather than a population. In this case, a correction to the variance and the standard deviation is implemented. Specifically, this correction accounts for the lower level of certainty that comes from not evaluating the entire population.

---

Sample Variance

$$s^2 = \frac{\sum\limits_{i=1}^{N} (x_i - \bar{x})^2}{N - 1}$$

where:
$s^2$ = sample variance
$x_i$ = value of data point $i$
$\bar{x}$ = sample mean
N = number of data points in the sample
N − 1 = degrees of freedom in the sample

---

Sample Standard Deviation

$$s = \sqrt{\frac{\sum\limits_{i=1}^{N} (x_i - \bar{x})^2}{N - 1}}$$

where:
s = sample standard deviation
$x_i$ = value of data point $i$
$\bar{x}$ = sample mean
N = number of data points in the sample
$N − 1$ = degrees of freedom in the sample

---

When reporting evaluated data, it is incredibly important to report not only the mean, but also the standard deviation associated with that mean. Therefore, in this example, we might report that the purity of PLA from a given facility was $84.96 \pm 0.55\%$. This standard deviation still likely has little meaning to us right now, but we will continue to explore its meaning in the context of statistical probability.

The standard deviation of a population is related to statistical probability through the normal (Gaussian) distribution. Developing a better understanding of the normal distribution will also allow us to give quantitative meaning to the standard deviation that supplements (and in some way, supplants) the qualitative "precision" definition we've given it. The normal distribution can be written, mathematically, as:[6]

---

Normal Distribution

$$n(x) = \frac{1}{\sigma \cdot \sqrt{2 \cdot \pi}} \cdot e^{-\frac{(x-\mu)^2}{(2 \cdot \sigma^2)}}$$

where:
$n(x)$ = normal distribution evaluated at $x$
$\sigma^2$ = population variance
$\sigma$ = population standard deviation
$\mu$ = population mean

---

If we normalise the normal distribution function so that the total probability of all possible outcomes is equal to one (or 100%), we find that the cumulative probability up to and including any value of x depends on the (Gauss) error function, erf:

---

Cumulative Probability

$$D(x) = \frac{1}{2} \cdot \left\{ 1 + \text{erf}\left(\frac{x - \mu}{\sigma \cdot \sqrt{2}}\right) \right\}$$

where:
$D(x)$ = cumulative probability evaluated at $x$
$\sigma$ = population standard deviation
$\mu$ = population mean

---

To determine the probability of any value of $x$, we determine $D(x)$ at this value and subtract the value of $D(x)$ at the previous data point, $x - 1$. If we plot the probability at each $x$-value over the range of all possible $x$-values, we can see the probability of any given $x$-value given our mean and standard deviation. Such probabilities are either tabulated or calculated using computer software, such as Excel, so the above equations are for reference only.

When evaluating a set of data from a *sample*, there is an inherent uncertainty as to whether the calculated sample mean reflects the mean if the entire population were evaluated. As such, engineers have developed a mechanism to provide confidence in the calculated sample mean. Specifically, a confidence interval is defined such that it allows an engineer to say that the procedure used to calculate a confidence interval will be accurate a certain percentage of the time.[7]

Traditionally, engineers choose to employ a 95% confidence level in such analysis, but the level will increase or decrease depending upon the importance of the analysis. The range of the confidence interval increases – for a given sample size – as the required confidence level approaches 100%.

In order to determine a confidence interval for a single set of data, we are required to calculate the estimated standard error. The equation for the systematic standard error is:

---

Standard Error for a Single Data Set

$$\bar{s} = \frac{s}{\sqrt{N}}$$

where:

$\bar{s}$ = systematic standard error
$s$ = sample standard deviation
$N$ = number of data points in the sample

---

We will also require a t-value to calculate the confidence interval; it can be determined using Table 11.2 or by another widely available resource.[8] The t-value is defined by the number of degrees of freedom in our sample, the type of distribution (one-tailed or two-tailed), and the defined confidence level.

The number of degrees of freedom is easy to calculate for a single data set. The equation is:

---

Degrees of Freedom for a Single Data Set

$$df = N - 1$$

where:

$df$ = degrees of freedom
$N$ = number of data points in the sample

---

In a two-tailed t-test, our lower and upper limits are both bounded; assuming a 95% confidence interval, we assume uniform distribution of our bounds around our population mean. T-tables, including Table 11.2, often define $\alpha$ as 1 – confidence level, written in decimal form. Therefore, for a 95% confidence level, $\alpha = 0.05$. The equation for a two-tailed confidence interval is:

---

Two-Tailed Confidence Interval for a Single Data Set

$$(\bar{x} - t \cdot \bar{s}) \leq \mu \leq (\bar{x} + t \cdot \bar{s})$$

where:

$(\bar{x} - t \cdot \bar{s}) \leq \mu \leq (\bar{x} + t \cdot \bar{s})$ = confidence interval
$\bar{s}$ = systematic standard error
$\bar{x}$ = sample mean
$\mu$ = population mean
$t$ = t-value at a defined confidence level

---

**Table 11.2** One- and two-tailed t-table.

| Degrees of Freedom | One-Tailed Confidence Level (%) | | | | | | | | | | | |
| | 75 | 80 | 85 | 90 | 95 | 97.5 | 98.75 | 99.5 | 99.75 | 99.95 | 99.975 | 99.995 |
| | Two-Tailed Confidence Level (%) | | | | | | | | | | | |
| | 50 | 60 | 70 | 80 | 90 | 95 | 97.5 | 99 | 99.5 | 99.9 | 99.95 | 99.99 |
| 1 | 1.0000 | 1.3764 | 1.9626 | 3.0777 | 6.3138 | 12.7062 | 25.4517 | 63.6567 | 127.3213 | 636.6192 | 1273.2393 | 6366.1977 |
| 2 | 0.8165 | 1.0607 | 1.3862 | 1.8856 | 2.9200 | 4.3027 | 6.2053 | 9.9248 | 14.0890 | 31.5991 | 44.7046 | 99.9925 |
| 3 | 0.7649 | 0.9785 | 1.2498 | 1.6377 | 2.3534 | 3.1824 | 4.1765 | 5.8409 | 7.4533 | 12.9240 | 16.3263 | 28.0001 |
| 4 | 0.7407 | 0.9410 | 1.1896 | 1.5332 | 2.1318 | 2.7764 | 3.4954 | 4.6041 | 5.5976 | 8.6103 | 10.3063 | 15.5441 |
| 5 | 0.7267 | 0.9195 | 1.1558 | 1.4759 | 2.0150 | 2.5706 | 3.1634 | 4.0321 | 4.7733 | 6.8688 | 7.9757 | 11.1777 |
| 6 | 0.7176 | 0.9057 | 1.1342 | 1.4398 | 1.9432 | 2.4469 | 2.9687 | 3.7074 | 4.3168 | 5.9588 | 6.7883 | 9.0823 |
| 7 | 0.7111 | 0.8960 | 1.1192 | 1.4149 | 1.8946 | 2.3646 | 2.8412 | 3.4995 | 4.0293 | 5.4079 | 6.0818 | 7.8846 |
| 8 | 0.7064 | 0.8889 | 1.1081 | 1.3968 | 1.8595 | 2.3060 | 2.7515 | 3.3554 | 3.8325 | 5.0413 | 5.6174 | 7.1200 |
| 9 | 0.7027 | 0.8834 | 1.0997 | 1.3830 | 1.8331 | 2.2622 | 2.6850 | 3.2498 | 3.6897 | 4.7809 | 5.2907 | 6.5937 |
| 10 | 0.6998 | 0.8791 | 1.0931 | 1.3722 | 1.8125 | 2.2281 | 2.6338 | 3.1693 | 3.5814 | 4.5869 | 5.0490 | 6.2111 |
| 15 | 0.6912 | 0.8662 | 1.0735 | 1.3406 | 1.7531 | 2.1314 | 2.4899 | 2.9467 | 3.2860 | 4.0728 | 4.4166 | 5.2391 |
| 20 | 0.6870 | 0.8600 | 1.0640 | 1.3253 | 1.7247 | 2.0860 | 2.4231 | 2.8453 | 3.1534 | 3.8495 | 4.1460 | 4.8373 |
| 25 | 0.6844 | 0.8562 | 1.0584 | 1.3163 | 1.7081 | 2.0595 | 2.3846 | 2.7874 | 3.0782 | 3.7251 | 3.9964 | 4.6191 |
| 30 | 0.6828 | 0.8538 | 1.0547 | 1.3104 | 1.6973 | 2.0423 | 2.3596 | 2.7500 | 3.0298 | 3.6460 | 3.9016 | 4.4824 |

| | 0.25 | 0.2 | 0.15 | 0.1 | 0.05 | 0.025 | 0.0125 | 0.005 | 0.0025 | 0.0005 | 0.00025 | 0.00005 |
|---|---|---|---|---|---|---|---|---|---|---|---|---|
| 40 | 0.6807 | 0.8507 | 1.0500 | 1.3031 | 1.6839 | 2.0211 | 2.3289 | 2.7045 | 2.9712 | 3.5510 | 3.7884 | 4.3207 |
| 50 | 0.6794 | 0.8489 | 1.0473 | 1.2987 | 1.6759 | 2.0086 | 2.3109 | 2.6778 | 2.9370 | 3.4960 | 3.7231 | 4.2283 |
| 60 | 0.6786 | 0.8477 | 1.0455 | 1.2958 | 1.6706 | 2.0003 | 2.2990 | 2.6603 | 2.9146 | 3.4602 | 3.6807 | 4.1686 |
| 70 | 0.6780 | 0.8468 | 1.0442 | 1.2938 | 1.6669 | 1.9944 | 2.2906 | 2.6479 | 2.8987 | 3.4350 | 3.6509 | 4.1268 |
| 80 | 0.6776 | 0.8461 | 1.0432 | 1.2922 | 1.6641 | 1.9901 | 2.2844 | 2.6387 | 2.8870 | 3.4163 | 3.6288 | 4.0959 |
| 90 | 0.6772 | 0.8456 | 1.0424 | 1.2910 | 1.6620 | 1.9867 | 2.2795 | 2.6316 | 2.8779 | 3.4019 | 3.6118 | 4.0722 |
| 100 | 0.6770 | 0.8452 | 1.0418 | 1.2901 | 1.6602 | 1.9840 | 2.2757 | 2.6259 | 2.8707 | 3.3905 | 3.5983 | 4.0533 |
| 200 | 0.6757 | 0.8434 | 1.0391 | 1.2858 | 1.6525 | 1.9719 | 2.2584 | 2.6006 | 2.8385 | 3.3398 | 3.5387 | 3.9705 |
| 365 | 0.6752 | 0.8426 | 1.0379 | 1.2839 | 1.6490 | 1.9665 | 2.2507 | 2.5894 | 2.8242 | 3.3174 | 3.5123 | 3.9340 |
| 500 | 0.6750 | 0.8423 | 1.0375 | 1.2832 | 1.6479 | 1.9647 | 2.2482 | 2.5857 | 2.8195 | 3.3101 | 3.5037 | 3.9222 |
| 1000 | 0.6747 | 0.8420 | 1.0370 | 1.2824 | 1.6464 | 1.9623 | 2.2448 | 2.5808 | 2.8133 | 3.3003 | 3.4922 | 3.9063 |
| One-Tailed $\alpha$ | 0.25 | 0.2 | 0.15 | 0.1 | 0.05 | 0.025 | 0.0125 | 0.005 | 0.0025 | 0.0005 | 0.00025 | 0.00005 |
| Two-Tailed $\alpha$ | 0.5 | 0.4 | 0.3 | 0.2 | 0.1 | 0.05 | 0.025 | 0.01 | 0.005 | 0.001 | 0.0005 | 0.0001 |

In a one-tailed confidence interval, the population mean is entirely *above* or *below* a calculated limit. We define both the one-tailed confidence interval with an upper bound and the one tailed confidence interval with a lower bound.

---

One-Tailed Confidence Interval with Upper Bound for a Single Data Set

$$-\infty \leq \mu \leq (\bar{x} + t \cdot \bar{s})$$

where:
$(\bar{x} + t \cdot \bar{s}) =$ upper bound
$\bar{s} =$ systematic standard error
$\bar{x} =$ sample mean
$\mu =$ population mean
$t =$ t-value at a defined confidence level

---

One-Tailed Confidence Interval with Lower Bound for a Single Data Set

$$(\bar{x} - t \cdot \bar{s}) \leq \mu \leq \infty$$

where:
$(\bar{x} - t \cdot \bar{s}) =$ lower bound
$\bar{s} =$ systematic standard error
$\bar{x} =$ sample mean
$\mu =$ population mean
$t =$ t-value at a defined confidence level

---

There is often the need to compare data from two different units completing the same operation. In this instance, the determination of confidence intervals is not as straightforward a process. The process looks at the difference (if any) between the sample means. Assuming the sample size is different for both units, we calculate the number of degrees of freedom (used in determining the t-value) and the standard error of the difference as:

---

Degrees of Freedom for the Comparison of Two Data Sets, Assuming Unequal Sample Sizes

$$df = \frac{\left(\dfrac{s_1^2}{N_1} + \dfrac{s_2^2}{N_2}\right)}{\dfrac{\left(\dfrac{s_1^2}{N_1}\right)^2}{(N_1 - 1)} + \dfrac{\left(\dfrac{s_2^2}{N_2}\right)^2}{(N_2 - 1)}}$$

where:

$df =$ degress of freedom
$s_1 =$ standard deviation for sample one
$s_2 =$ standard deviation for sample two

$N_1$ = number of data points in sample one
$N_2$ = number of data points in sample two

Standard Error for the Difference between Two Means, Assuming Unequal Sample Sizes

$$\bar{s} = \sqrt{\frac{s_1}{N_1} + \frac{s_2}{N_2}}$$

where:
$\bar{s}$ = systematic standard error
$s_1$ = standard deviation for sample one
$s_2$ = standard deviation for sample two
$N_1$ = number of data points in sample one
$N_2$ = number of data points in sample two

We can now calculate the upper limit and lower limit for our confidence interval comparing our two means:

Confidence Interval for the Difference between Two Means

$$[(\bar{x}_2 - \bar{x}_1) - t \cdot \bar{s}] \leq \mu_2 - \mu_1 \leq [(\bar{x}_2 - \bar{x}_1) + t \cdot \bar{s}]$$

where:
$[(\bar{x}_2 - \bar{x}_1) - t \cdot \bar{s}] \leq \mu_2 - \mu_1 \leq [(\bar{x}_2 - \bar{x}_1) + t \cdot \bar{s}]$ = confidence interval
$\bar{s}$ = standard error for the difference between to means
$\bar{x}_1$ = mean of sample one
$\bar{x}_2$ = mean of sample two
$\mu_1$ = mean of population one
$\mu_2$ = mean of population two
$t$ = t-value at a defined confidence level

In this section, we have developed equations for the analysis of data from engineering processes. We will now apply these tools to evaluate the performance of our process to produce PLA, to help inform experimental design, and to allow us to better understand the impact of changing the level of variables in our process.

## 11.4   Analysis of PLA Production

In this section, we will evaluate problems in order to simultaneously demonstrate the use of the tools we have developed and further explore the impact of engineering design and optimisation on product purity and yield. Each problem in this section is designed to apply specific tools evaluated in the previous section. The formulae in Section 11.3 should be referenced when solving these problems.

**Figure 11.5** A histogram representing a population of high-purity poly(lac-tic acid) flow rates.

---

Problem 1: Production of High Molecular Weight Poly(lactic Acid) in a CSTR

**Problem:** A production facility has calculated the production rate (in kmol $h^{-1}$) of high-molecular weight PLA on a monthly basis for the past 5 years (60 months in total). The collected data, shown in Figure 11.5, is given in histogram form. Calculate the **mean, variance**, and **standard deviation** from this data.

**Solution:** The collected data can, and should, be seen as a **population**, as it contains all data in the analysed timeframe. To calculate the sample **mean**, we first take the sum of all the data.

$$\sum_{i=1}^{N} x_i = 24.1 + (2 \times 24.2) + (2 \times 24.3) + (3 \times 24.4) + (6 \times 24.5) + (3 \times 24.6)$$

$$+ (3 \times 24.7) + (5 \times 24.8) + 24.9 + (5 \times 25.0) + (3 \times 25.1) + (2 \times 25.2)$$

$$+ (6 \times 25.3) + (5 \times 25.4) + (2 \times 25.5) + (2 \times 25.6) + (3 \times 25.7)$$

$$+ 25.8 + (2 \times 25.9) + (2 \times 26.0) + 26.4 = 1502.9 \, \text{kmol} \, \text{h}^{-1}$$

The population **mean** is then:

$$\mu = \frac{1502.9 \, \text{kmol} \, \text{h}^{-1}}{60} = 25.05 \, \text{kmol} \, \text{h}^{-1}$$

The **variance** is found by taking the sum of all the squared deviations from the mean then dividing by the population.

$$\sum_{i=1}^{N}(x_i - \mu)^2 = (24.1 - 25.05)^2 + 2\times(24.2 - 25.05)^2 + 3\times(24.3 - 25.05)^2$$

$$+ 3\times(24.4 - 25.05)^2 + 6\times(24.5 - 25.05)^2 + 3\times(24.6 - 25.05)^2$$

$$+ 3\times(24.7 - 25.05)^2 + 5\times(24.8 - 25.05)^2 + (24.9 - 25.05)^2$$

$$+ 5\times(25.0 - 25.05)^2 + 3\times(25.1 - 25.05)^2 + 2\times(25.2 - 25.05)^2$$

$$+ 6\times(25.3 - 25.05)^2 + 5\times(25.4 - 25.05)^2 + 2\times(25.5 - 25.05)^2$$

$$+ 2\times(25.6 - 25.05)^2 + 3\times(25.7 - 25.05)^2 + (25.8 - 25.05)^2$$

$$+ 2\times(25.9 - 25.05)^2 + 2\times(26.0 - 25.05)^2 + (26.4 - 25.05)^2$$

$$= 16.48(\text{kmol h}^{-1})^2$$

The population **variance** is then:

$$\sigma^2 = \frac{16.48(\text{ kmol h}^{-1})^2}{60} = 0.275(\text{kmol h}^{-1})^2$$

The **standard deviation** is calculated as:

$$\sigma = \sqrt{\frac{\sum_{i=1}^{N}(x_i - \mu)^2}{N}} = \sqrt{\frac{16.48(\text{kmol h}^{-1})^2}{60}} = 0.5285 \text{ kmol h}^{-1}$$

The production rate of the above facility, over the past five years, has been $25.05 \pm 0.5285$ kmol h$^{-1}$. We can see from analysing 60 months of data that this facility has always met these minimum specifications. This *does not* mean, however, that it will always meet such specifications. In the next problem, we look at the probability that the production facility does not meet specifications in any given month.

Problem 2: Production of High-molecular Weight Poly(lactic Acid) in a CSTR, Part II

**Problem:** Assume the target mean monthly rate of (lactic acid) production in this facility described in Problem 1 should be greater than or equal to 24.0 kmol h$^{-1}$. Determine the probability that the facility does not meet the target specifications in any month.

**Solution:** From Problem 1, we know that the production rate of lactic acid over the past five years has been $25.05 \pm 0.5285$ kmol h$^{-1}$.

The **cumulative probability** calculated at the target production rate of 24.0 kmol h$^{-1}$ is then:

$$D(24.0) = D(x) = \frac{1}{2} \cdot \left\{ 1 + \text{erf}\left(\frac{24.0 - 25.05}{0.5285 \cdot \sqrt{2}}\right) \right\} = 0.023674$$

We see that, assuming normal distribution of data, there is a 2.3674% probability this facility does not meet minimum specifications in a given month. Note that this is the **cumulative probability** for all possible production rates less than or equal to 24.0 kmol h$^{-1}$. If we evaluate the cumulative probability at each possible rate of production, we can then calculate the probability of any specific outcome.

For instance, if we calculate the cumulative probability calculated at a production rate of 23.9 kmol h$^{-1}$ and subtract it from the cumulative probability at a production rate of 24.0 kmol h$^{-1}$, we find the probability of having a production rate of *exactly* 24.0 kmol h$^{-1}$ high-molecular weight PLA.

$$D(24.0) - D(23.9) = \frac{1}{2} \cdot \left\{ 1 + \text{erf}\left(\frac{24.0 - 25.05}{0.5285 \cdot \sqrt{2}}\right) \right\}$$

$$- \frac{1}{2} \cdot \left\{ 1 + \text{erf}\left(\frac{23.9 - 25.05}{0.5285 \cdot \sqrt{2}}\right) \right\}$$

$$= 0.023674 - 0.014913 = 0.00876$$

If we repeat this calculation for each possible production rate, we can generate a probability distribution curve for the data in Figure 11.5. This probability distribution curve is shown in Figure 11.6.

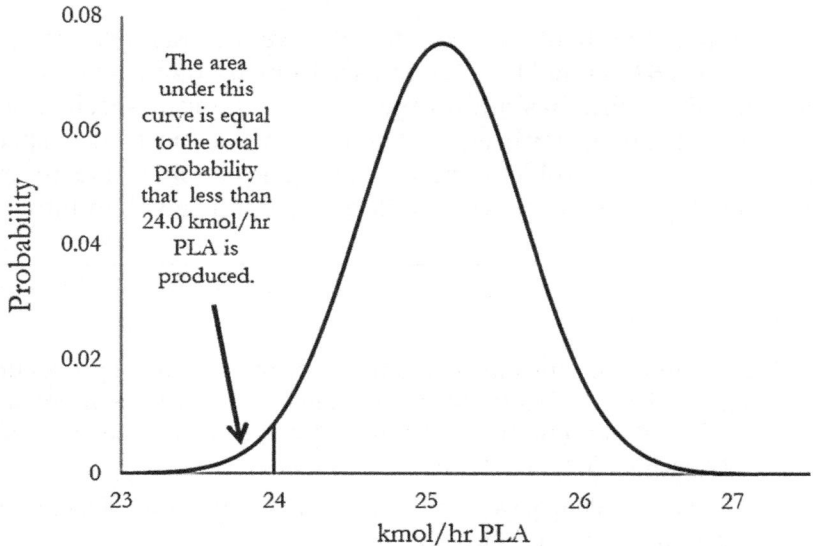

**Figure 11.6**  Probability distribution of the production of high-purity poly(lactic acid).

**Figure 11.7** Probability distribution for the production of high-purity poly(lactic acid) over a range of standard deviations.

Imagine that the mean monthly rate of PLA production in this facility remained at 25.05 kmol h$^{-1}$, but had a significantly different data distribution (and thus standard deviation). The probability distributions at standard deviations of 0.26425 kmol h$^{-1}$, 0.5285 kmol h$^{-1}$, and 1.0570 kmol h$^{-1}$ are shown in Figure 11.7. Note that the tall, narrow distribution represents a standard deviation of 0.26425 kmol h$^{-1}$ and reflects data that is generally very close to the mean. The short, wide distribution represents a standard deviation of 1.0570 kmol h$^{-1}$; it reflects data that is much more regularly distributed both close to and (relatively) far from the mean.

In the next problem, we will again look at the production of poly(lactic) acid but, on this occasion, we will assume that we have developed a new process that allows for the simultaneous conversion of lactic acid to lactide with a separations process facilitating the removal of the water by-product. This technique – coupling two or more processes into one process or technique – is known as process intensification.

---

Process Intensification

In process intensification, the overall process or technique is completed in a more optimised fashion. A paper by Stankiewicz and Moulijin[9] highlights a number of different types of process intensification methods and examples thereof. For instance, multifunctional reactors are an example of a type of a process-intensifying method; many of these processes – as in our example – couple a chemical reaction and a separation technique in a single process, such as reactive distillation or reactive crystallization.[9]

*"Process intensification consists of the development of novel apparatuses and techniques that, compared to those commonly used today, are expected to bring dramatic improvements in manufacturing and processing, substantially decreasing equipment-size/production-capacity ratio, energy consumption, or waste production, and ultimately resulting in cheaper, sustainable technologies.[9]"*

---

### Problem 3: Evaluating a Prototype Unit

**Problem:** A production facility is testing a prototype CSTR/separations unit for the polymerisation of lactic acid to lactide. Over one month, they randomly measured the yield of lactide in this new process 60 times. The mean yield of lactide was $61.120 \pm 5.191\%$. What is the highest confidence level our sample data supports if we estimate that the yield in this new unit is greater than or equal to 60%?

**Solution:** Because the data collected was only representative (not inclusive) of the population, the given mean and standard deviation are a **sample mean** and a **standard deviation of a sample**.

Moreover, the problem statement implies that we should be looking for the **one-tailed confidence interval with a lower bound**.

$$(\bar{x} - t \cdot \bar{s}) \leq \mu \leq \infty$$

We can re-write this, substituting known values:

$$(61.120 - t \cdot \bar{s}) \leq 60\% \leq \infty$$

Next, we must calculate:

$$\bar{s} = \frac{s}{\sqrt{N}} = \frac{5.191}{\sqrt{60}} = 0.67$$

The highest confidence level will coincide with when the lower bound is equal to, but not less than, our assumed mean:

$$(61.120 - t \cdot 0.67) = 60$$

Solving for t in this equation, we find that this corresponds to a t-value of 1.6711. Before we can find the corresponding confidence interval, we must first calculate the degrees of freedom of this sample:

$$df = 60 - 1 = 59$$

We interpolate between the t-values at 50 and 60 degrees of freedom (using Table 11.2) in order to find the t-value associated with 59 degrees of freedom.

$$t_{59} = \left[\frac{t_{60} - t_{50}}{60 - 50}\right] \cdot [59 - 50] + t_{50}$$

We see that a **one-tailed** confidence level of 95% will return the required t-value

$$t_{95} = \left[\frac{1.6706 - 1.6759}{60 - 50}\right] \cdot [59 - 50] + 1.6759 = 1.6712$$

This analysis indicates that we have identified the t-value to be valid at the "standard" 95% confidence level that engineers generally apply. In the next problem, we will again look at a **one-tailed confidence interval** to see if the concentration of lactic acid in a CSTR is maintained at an acceptable level.

---

**Problem 4: Revisiting a Prototype Unit**

**Problem:** A production facility is monitoring the lactic acid production (*via* fermentation) in a CSTR. In order to avoid pH inhibition, the concentration of lactic acid should not be higher than 2.00 g L$^{-1}$. In 60 random samples taken over one month, the mean lactic acid concentration in the CSTR was found to be $1.96 \pm 0.22$ g L$^{-1}$. Can we indicate, at a 95% confidence level, that the mean (population) lactic acid concentration in the CSTR is less than or equal to 2.00 g L$^{-1}$?

**Solution:** In this instance, we must see if the **one-tailed confidence interval with an upper bound** is true, given our data.

$$-\infty \leq \mu \leq (\bar{x} + t \cdot \bar{s})$$

Again, we find that a **one-tailed** confidence level of 95% will return a t-value of 1.6712:

$$t_{95} = \left[\frac{1.6706 - 1.6759}{60 - 50}\right] \cdot [59 - 50] + 1.6759 = 1.6712$$

We must next calculate the standard error before we can write our confidence interval:

$$\bar{s} = \frac{s}{\sqrt{N}} = \frac{0.22}{\sqrt{60}} = 0.028402$$

For our claim to be true (that the mean lactic acid concentration in the CSTR is less than or equal to 2.00 g L$^{-1}$), the upper bound must be less than or equal to 2.00:

$$-\infty \leq \mu \leq (1.96 + 1.6712 \cdot 0.028402) \text{ g L}^{-1}$$

When calculated, we see that the upper limit on our **one-tailed confidence interval** bounding our true population is greater than 2.00:

$$-\infty \leq \mu \leq 2.0075 \text{ g L}^{-1}$$

Therefore, we cannot indicate, at a 95% confidence level, that the mean (population) lactic acid concentration in the CSTR is less than or equal to 2.00 g L$^{-1}$. In the next example, we will use statistical tools to compare the performance of two separate PLA production facilities.

---

### Problem 5: Comparing Two Reactors

**Problem:** Imagine you are the process engineer in charge of quality control of two industrial facilities designed for the production and purification of poly(lactic acid). Table 11.3 shows the average impurity level (%) of D-lactic acid determined from 365 random samples from each facility. Assume that the targeted impurity level is 3.5% and that too high of an impurity level will greatly decrease the value of the product and too low an impurity level has too high a production cost. The data is presented in Table 11.3 with respect to each quarter as well as the year as a whole. Develop a confidence interval – at a 95% confidence level – that bounds the difference between the two population means in both facilities.

**Solution:** First, calculate the number of degrees of freedom:

$$df = \frac{\left(\dfrac{s_1^2}{N_1} + \dfrac{s_2^2}{N_2}\right)^2}{\dfrac{\left(\dfrac{s_1^2}{N_1}\right)^2}{(N_1 - 1)} + \dfrac{\left(\dfrac{s_2^2}{N_2}\right)^2}{(N_2 - 1)}}$$

Even though the formulae we developed earlier were for samples of different sizes, we can use them when the sample size is the same:

$$df = \frac{\left(\dfrac{0.175^2}{365} + \dfrac{0.143^2}{365}\right)^2}{\dfrac{\left(\dfrac{0.175^2}{365}\right)^2}{364} + \dfrac{\left(\dfrac{0.143^2}{365}\right)^2}{364}} = 700$$

**Table 11.3** Average D-lactic acid impurity level (%) in two reactors in parallel.

| Report | Reactor 1 | Reactor 2 |
|---|---|---|
| 1st Quarter | $3.351 \pm 0.138$ | $3.513 \pm 0.140$ |
| 2nd Quarter | $3.455 \pm 0.149$ | $3.519 \pm 0.141$ |
| 3rd Quarter | $3.552 \pm 0.140$ | $3.495 \pm 0.151$ |
| 4th Quarter | $3.623 \pm 0.143$ | $3.490 \pm 0.140$ |
| Year in review | $3.496 \pm 0.175$ | $3.504 \pm 0.143$ |

Next, calculate the systemic standard error:

$$\bar{s} = \sqrt{\frac{s_1}{N_1} + \frac{s_2}{N_2}}$$

$$\bar{s} = \sqrt{\frac{0.175}{365} + \frac{0.143}{365}} = 0.0295$$

After this, we interpolate to find the t-value (at a 95% confidence level) using Table 11.2 at 700 degrees of freedom:

$$t_{95} = \left[\frac{1.9623 - 1.9647}{1000 - 500}\right] \cdot [700 - 500] + 1.9647 = 1.9638$$

Finally, we find our confidence interval:

$$[(\bar{x}_2 - \bar{x}_1) - t \cdot \bar{s}] \leq \mu_2 - \mu_1 \leq [(\bar{x}_2 - \bar{x}_1) + t \cdot \bar{s}]$$
$$[(3.504 - 3.496) - 0.058] \leq \mu_2 - \mu_1 \leq [(3.504 - 3.496) + 0.058]$$
$$-0.0500 \leq \mu_2 - \mu_1 \leq 0.066$$

The confidence interval offers us a mechanism by which we can compare the actual percentage of D-lactic acid impurity in each reactor. Had we hypothesised that there was a trivial (or no difference) between these means, our data would not have supported that hypothesis.

The annual data for each reactor described in Problem 5 is plotted in Figure 11.8. If you look carefully at Table 11.3, you should note that the *annual* standard deviation in Reactor 1 is larger than that in Reactor 2. Looking closely at Figure 11.8, it should be obvious to you why

**Figure 11.8**　Impurities in two independent CSTRs. The percentage of impurity in Reactor 1 is increasing with respect to passing time.

this is the case! The percentage of the impurity in Reactor 1 is increasing very slowly, but consistently, over the course of a year.

In Section 11.2 of this chapter we introduced the concepts of parameter optimisation, process integration and process optimisation. In the next two problems, we will evaluate the impact of optimising a single operating parameter (Problem 6) and explore the implications of this change on a downstream process (Problem 7).

---

### Problem 6: Parameter Optimisation

**Problem:** Imagine it is your responsibility to optimise the fermentation production of lactic acid in an industrial facility. After careful evaluation of this process, you have decided to optimise the process by installing state-of-the-art pH controllers in the bioreactor. Prior to this installation, you have determined that the rate of production of lactic acid leaving the reactor is $25.0 \pm 0.50$ kmol h$^{-1}$, based upon a sample size of 365. After installation and tuning of the new controllers, you find the rate of production to be $26.0 \pm 1.0$ kmol h$^{-1}$, based upon a sample size of 31. Can we indicate, at a 95% confidence level, that the mean product yield has increased because of this change?

**Solution:** First, calculate the number of degrees of freedom:

$$df = \frac{\left(\dfrac{0.50^2}{365} + \dfrac{1.0^2}{31}\right)^2}{\dfrac{\left(\dfrac{0.5^2}{365}\right)^2}{364} + \dfrac{\left(\dfrac{1.0^2}{31}\right)^2}{30}} = 377.56$$

Next, calculate the systemic standard error:

$$\bar{s} = \sqrt{\frac{s_1}{N_1} + \frac{s_2}{N_2}}$$

$$\bar{s} = \sqrt{\frac{0.5}{365} + \frac{1.0}{31}} = 0.1815$$

After this, we interpolate to find the t-value (at a 95% confidence level) using Table 11.2 at 377.55 degrees of freedom:

$$t_{95} = \left[\frac{1.9647 - 1.9665}{500 - 365}\right] \cdot [377.56 - 365] + 1.9665 = 1.9663$$

Finally, we find our confidence interval:

$$[(\bar{x}_2 - \bar{x}_1) - t \cdot \bar{s}] \text{ kmol h}^{-1} \leq \mu_2 - \mu_1 \leq [(\bar{x}_2 - \bar{x}_1) - t \cdot \bar{s}] \text{ kmol h}^{-1}$$

$$[(26.0 - 25.0) - 0.357] \text{ kmol h}^{-1} \leq \mu_2 - \mu_1 \leq [(26.0 - 25.0) + 0.357] \text{ kmol h}^{-1}$$

$$0.643 \text{ kmol h}^{-1} \leq \mu_2 - \mu_1 \leq 1.357 \text{ kmol h}^{-1}$$

Because both the lower and upper bounds are positive, we have evidence to support our hypothesis that the optimisation of the process controllers has resulted in a higher yield of lactic acid. Note that while we have *improved* the rate of lactic acid production, this does not mean that we have *optimised* the rate of lactic acid production. It may be possible, for instance, to further improve the yield by changing another operating parameter in this CSTR.

In this problem, we have seen an increase in the production rate of lactic acid by installing new pH controllers in the bioreactor. In order to *optimise* the use of the new controllers, we would need to fine-tune the controllers, collect data after this change, and see if this change (using the same statistical tools used earlier in this problem) has statistically improved the overall rate of lactic acid production. Note that parameter optimisation is an *iterative process*; in order to optimise pH control in the bioreactor, many tuning parameters must be tested and evaluated to find the *best* parameters. Note that there is no inherent promise of optimisation when making process or level changes; it is quite possible that the original operating equipment or level set points were, in fact, already optimised. Generally speaking, attempts at parameter optimisation (and subsequent process integration and process optimisation) should only be initiated when supported by sound theoretical reasoning.

Moreover, just because you have optimised pH control does not mean that you have optimised the production of lactic acid in the bioreactor. For instance, a similar iterative optimisation process would need to be implemented for *temperature control* in the bioreactor. A comprehensive experimental design protocol, such as 2k factorial experimental design, is required to evaluate multiple parameter changes simultaneously.[10,11]

Finally, when we introduced the concept of process integration, we mentioned that complicated models are required to optimise an entire industrial process. In this next worked example, we will see how changing an operating parameter in one process can have an impact on another process in the production facility. When making a change in operating parameters in a unit, we must first attempt to predict the impact of such a change both upstream and downstream from that unit but also, after implementation, monitor – through data collection – the impact on all other units in the facility to ensure that there are no unexpected negative consequences. These consequences may be related to the production rate and/or quality of the final product, but may also be related to the safety and/or environmental impact of the process. Moreover, there is often a delicate balance between product purity and rate of production that must be addressed when optimising a process.

Problem 7: Process Integration

**Problem:** In Problem 6, we optimised the production of lactic acid in a bioreactor. Calculate the additional water (kmol h$^{-1}$) that must be removed in the first downstream polymerisation reaction to maintain a 60% rate of conversion of the lactic acid to lactide in the polymerisation reaction.

**Solution:** Even though the solution to this problem is straightforward, the implications are significant. From the stoichiometry of the reaction, we see that 2 kmol water are produced for every 2 kmol of lactic acid converted:

$$\underset{\text{lactic acid}}{2C_3H_6O_3} \leftrightarrow \underset{\text{lactide}}{C_6H_8O_4} + \underset{\text{water}}{2H_2O}$$

If we assume that the optimisation of the process controls described in Problem 6 results in an hourly increase in the lactic acid production bounded by our confidence interval, we can calculate the additional water that must be removed each hour as:

$$0.6 \times 0.643 \text{ kmol h}^{-1} \leq \text{additional water removed} \leq 0.6 \times 1.357 \text{ kmol h}^{-1}$$

$$0.386 \text{ kmol h}^{-1} \leq \text{additional water removed} \leq 0.814 \text{ kmol h}^{-1}$$

As is quite often the case in industrial facilities, a change in operating parameters upstream from one unit necessitates a change (or changes) downstream from that unit. In this example, excess water must be removed in order to maintain a specified conversion in the polymerisation unit. If this water is not removed, the percentage yield of lactide, and ultimately PLA, would decrease due to inhibitory effects. This would work counter to the optimisation described in Problem 6 and, depending on the magnitude of inhibition, might actually decrease the overall hourly yield of PLA!

Note that the removal of this additional water necessitates a change in the operating parameters of another unit in this facility. This parameter change might also have an impact on the units upstream and downstream from this unit and would also likely need to be optimised!

Finally, process optimisation is not likely to be as simple as making a change and evaluating the overall improvement in your process. As mentioned in Section 11.2 of this chapter, there is often a financial incentive to decrease the production rate in order to improve the product purity. This may be accomplished by optimising operating parameters for a *specific outcome*, such as to obtain PLA that is 99.9% pure. There are likely numerous processes (such as production costs, production rates, *etc.*) that must operate *sub-optimally* in order to optimise the economics of the overall process, while also meeting environmental and safety regulations.

## 11.5    Concluding Remarks

In this chapter, we first investigated design decisions related to the large-scale production of poly(lactic acid). This allowed us to not only further explore some of the concepts developed in earlier chapters of this book but also evaluate the type of problem-solving required of engineers in pilot and production facilities. Finally, we developed and applied the basic statistical tools required to evaluate the operation and integration of processes.

Building upon the tools you've developed in this chapter, you should feel confident not only further analysing and comparing data – such as the data found in Table 11.3 (it is interesting to compare the performance of Reactor 1 in the first quarter of the year to the performance in the fourth quarter) – but also using the analysed data to brainstorm possible reasons related to the fundamental concepts learned earlier in this text that might cause such a marked change. Of course, we hope these tools become the foundation on which you will begin to evaluate the performance, integration and optimisation of the pilot-scale and/or industrial-scale facilities where you work.

## References

1. S. Fogler, S. LeBlanc, B. Rizzo, *Strategies for Creative Problem Solving*, Pearson, 3rd edn, 2014.
2. I. D. Fridman, J. Kwok, R. J. Downey, S. P. Nemphos, inventors; Lactide polymerization, United States of America patent US5357034, 1994 October 18.
3. H. Tokos and Z. N. Pintarič, Synthesis of batch water network for a brewery plant, *J. Cleaner Prod.*, 2009, **17**(16), 1465–1479.
4. H. Tokos, Z. Novak-Pintarič, Y. Yang and Z. Kravanja, Multilevel strategies for the retrofit of large-cale industrial water system: a brewery case study, *AIChE J.*, 2012, **58**(3), 884–898.
5. H. Tokos, Z. N. Pintarič and P. Glavič, Energy Saving Opportunities in Heat Integrated Beverage Plant Retrofit, *Appl. Therm. Eng.*, 2010, **30**(1), 36–44.
6. W. Feller, *An Introduction to Probability Theory and Its Applications*, New York, Wiley, 3rd edn, 1971.
7. J. Neyman, Outline of a Theory of Statistical Estimation Based on the Classical Theory of Probability, *Philosophical Transactions of the Royal Society of London. Series A, Mathematical and Physical Sciences*, 1937 August, **236**, 767.

8. D. S. Soper Student t-Value Calculator. [Online], 2015. Available from: http://www.danielsoper.com/statcalc.
9. A. Stankiewicz and J. Moulijn, Process Intensification: Transforming Chemical Engineering, *Chem. Eng. Prog.*, 2000, 22–34.
10. Cooper M. Loose Change and Dishwasher Optimization: Creative Applications of Engineering Statistics. In Paper presented at 2013 ASEE Annual Conference; 2013; Atlanta.
11. D. C. Montgomery, *Design and Analysis of Experiments*, Wiley Global Education, 8th edn, 2012.

# Subject Index